the ways
and
means of
statistics

the ways and means of statistics

Leonard J. Tashman
University of Vermont

Kathleen R. Lamborn
The Upjohn Company

HARCOURT BRACE JOVANOVICH, INC.
New York · San Diego · Chicago · San Francisco · Atlanta

to Peter and Jacqueline

L.J.T.

to my parents

K.R.L.

Library of Congress Catalog Card Number: 78–61383

ISBN: 0–15–595132–7

Printed in the United States of America

Illustration credits appear on page 517.

preface

The Ways and Means of Statistics is concerned with communicating information through statistics. From the first page to the last, the text places a premium on helping the student to articulate the meaning of statistical information in a manner that is both technically precise and intelligible to the layman.

Surprisingly little mathematical "baggage" is found in the book. *Ways and Means* has been written in English, not in Algebra. The student whose mathematical background fulfills only the bare prerequisite should not feel anxious: Recall of only the simplest concepts of elementary algebra is required—not only in the early chapters, but throughout the text.

Despite its nonmathematical presentation, *Ways and Means* prepares the student to deal not only with textbook-case exercises but with the complexities of statistical problems as well. Every chapter includes illustrative examples that serve as models of statistical investigations. In working through these examples, the student becomes our partner in research, joining us both in the development of the evidence and in the presentation of the findings. This partnership experience is intended to better equip the student to analyze new problems appropriately and creatively.

In statistics especially, a little bit of knowledge can be a dangerous thing. *Ways and Means* stresses the limitations as well as the capabilities of statistical procedures. Every chapter concludes with a Caveats section, which warns the student to recognize the danger of drawing conclusions from incomplete models, inappropriate designs, and/or insufficient evidence. "Beware of *simple* regression results," explains a Caveat in Chapter 7, and a Caveat in Chapter 13 emphasizes the critical nature of the experimental design in two-sample tests.

We know the student will find *Ways and Means* clear and understandable. We hope it will be enjoyable as well. We have tried never to be long-winded or unnecessarily sophisticated. Numerous diagrams as well as cartoons and photographs lend variety to the narrative and make it more vivid. The examples are lively, varied in content, and, as much as possible, pertinent to current issues.

exercises

Exercises follow chapter sections for immediate reinforcement of fundamental statistical concepts. Numerical solutions to representative exercises follow the Appendix Tables.

An asterisk (*) appears before certain exercises to indicate that the student should save solutions for subsequent use. These exercises permit students to refer to their previous work on a set of data as they learn to apply new methods to the same data set, thereby preserving the continuity of the statistical presentation.

Issues for Analysis close each chapter. Each Issue is designed to provide food for thought, in that it requires a synthesis of the tools and knowledge the student has acquired to that point and, in many cases, the application of personal intuition and common sense. Some Issues anticipate procedures that are formally presented in later chapters but that nevertheless can be approached with techniques the student has already acquired.

Finally, the *Instructor's Manual* contains five data sets with accompanying problem sets and analytical exhibits, which are designed, in part, for use as comprehensive review problems. (See the *Instructor's Manual* for details.)

topical
organization

For many courses, Chapters 1–10 will form the core and absorb 70–80 percent of a semester. An average of two or three additional chapters can be covered in the remainder of the semester; virtually the entire book can be completed in a two-quarter course. The prerequisite structure is

CHAPTER	PREREQUISITE
11	1–10
12	1–4; 8–10
13	1–4; 8–10
14	1–4; 8–10; 13.1–13.3
15	1–4; 8–10; 13.1–13.3

Several features of the topical organization should be noted here:

1. Correlation and regression (including multiple linear regression) are given substantial emphasis. The topic is presented in Chapters 5–7 without the encumbrance of the probability model. We return to the subject in Chapter 11 to show applications of the principles of estimation and hypothesis testing to multiple regression investigations.

The early introduction of correlation and regression has worked well for us. We've found it pedagogically efficient and highly motivating for our students. They feel the sense of doing "substantial" data exploration even prior to midsemester. Alternatively, Chapters 5–7 can be deferred without loss of continuity.

2. Inferential procedures have been organized by *purpose* rather than *form*. For example, Chapter 13, "Two-Group Comparisons," includes both nonparametric and parametric tests and treats paired as well as unpaired data. This should enable instructors to present a systematic overview of this subject area without chapter splicing.

3. The essential probability theorems have been reduced to a brief section in Chapter 8, "Sampling," to avoid a lengthy transition between descriptive and inferential statistics. However, the probabilistic foundation of classical statistical inference is developed thoroughly throughout Parts 3 and 4. In particular, we stress the concept of the sampling distribution and the laws that govern the chance variation of sample statistics.

Leonard J. Tashman
Kathleen R. Lamborn

acknowledgments

The Ways and Means of Statistics was five years in the making, during which time we received valuable assistance and encouragement from many people.

Special thanks go to Dorothy Lamborn for the many hours she spent reviewing the manuscript for clarity and checking the numerical accuracy of the examples; to David Ketcham for his careful work on the data sets and graphics for the *Instructors' Manual*; and to them both for their patient and diligent efforts in preparing exercise solutions.

We benefited substantially from the suggestions of professional colleagues, including Professors Alan Agresti, University of Florida; Louis Bush, San Diego City College; Kenneth R. Eberhard, Chabot College; Jeff Hooper, Mesa Community College; Ron Koot, Pennsylvania State University; and Chester Piascik, Bryant College; and from the editorial help provided by Suzanne Moyer. Our thanks also go to Caroline Harnly for her work on the index.

We are grateful to many of our former students for their assistance in the development of materials, including Anne Bickford, Anne Searle Chan, Kathleen Dell, Diane Dwyer, Anne Harris, and Karen Bednarski. Cindy Shattuck, Lynn Wells, Catherine Anderson, and Kerra Desseau turned our handwritten manuscript into presentable typed copy. Barbara Boczany and Lana Turner helped with the proofreading.

Finally, we would like to express our appreciation to Gary Burke, Edwin Greif, and Jacqueline Tashman for their constant encouragement and wise counsel during the stages of textbook development.

contents

the ways
and
means of
statistics

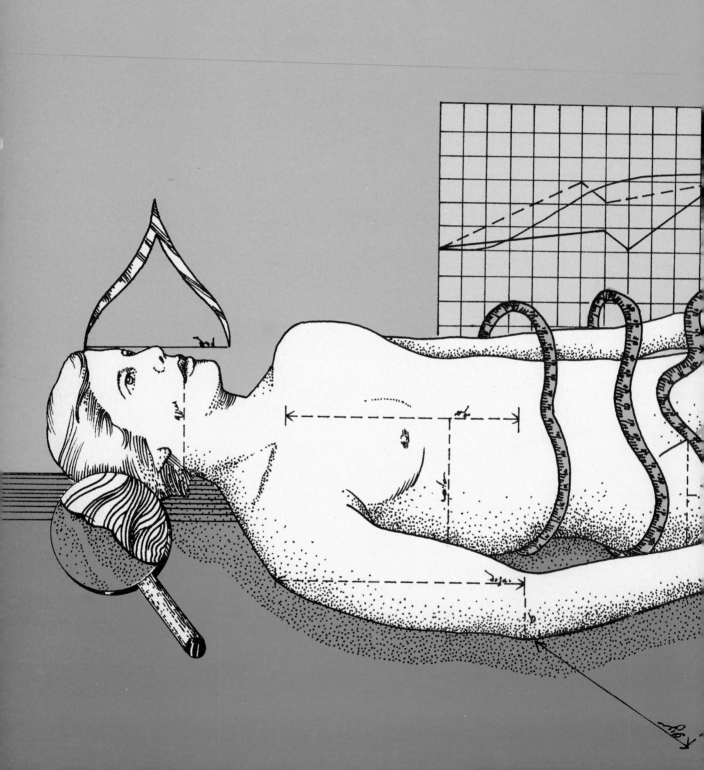

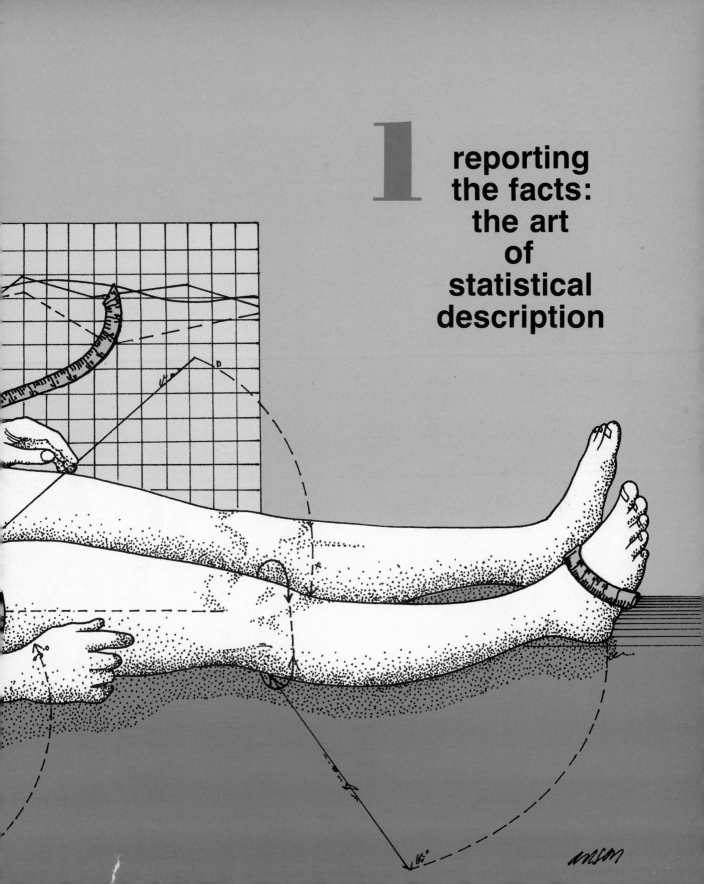

1
reporting
the facts:
the art
of
statistical
description

Behind the Headline

As you begin this book, you have some impressions of what statistics is all about. You know that you are going to work with *data* (numeric information) and that you are going to study methods for making sense out of the numbers. You've used statistical language when you've spoken about an *average* level, a *range* of values, or a *correlation* between events. You probably know that *random sampling* is a good way to collect data and have mused about the *law of averages* when the game hasn't gone your way. You may have read articles reporting *statistically significant* results without completely understanding this expression.

So the language of statistics is not completely new to you. The purpose of *The Ways and Means of Statistics* is to teach you more about this language—to expand your vocabulary and make it more precise. This book won't give you the expertise of a professional statistician, but it will provide you with statistical tools to apply in your discipline. It also will help you to follow the logic and perceive the strengths and weaknesses of the various types of statistical studies you're likely to read.

The purpose of the subject of statistics is to *make the most of data*. The statistical system allows us to clarify what is inherently confusing. It tells us how to collect data, how to make the data presentable, how to report the facts hidden in the data, and how to draw scientific conclusions about the issues under examination.

So let's bite into the food of the statistical system: *data*.

data

"It is an error to argue in front of your data. You find yourself insensibly twisting them around to fit your theories."

Sherlock Holmes to Doctor Watson
in *The Adventure of Wisteria Lodge*

1.1   data: a few definitions

Data are bits of information (numbers or facts). A *piece of data,* also called a score or an observation, conveys a piece of information about some topic or event. The topic or event itself is called a *variable.* For example, *your age* is a piece of data about the variable *age of people.* A collection of pieces of data pertaining to one or more variables is called a *data set.* The data set represented in Chart 1-1 contains four variables:

> *Variable 1: Period of Revolution.* The time it takes each planet in our solar system to complete one full revolution around the sun.
>
> *Variable 2: Number of Satellites.* The number of "moons" that orbit each planet.
>
> *Variable 3: Relative Size.* 1 = smallest size (diameter); 9 = largest size (diameter).
>
> *Variable 4: Distance from Sun Index.* Earth = 1.000) The mid-orbit distance of a planet from the sun divided by the mid-orbit distance of Earth from the sun.

5

chart 1-1

PLANET	PERIOD OF REVOLUTION	NUMBER OF SATELLITES	RELATIVE SIZE	DISTANCE FROM SUN INDEX
Mercury	87.97 days	0	1	0.387
Venus	224.70 days	0	4	0.723
Earth	365.25 days (1 year)	1	5	1.000
Mars	1.88 years	2	3	1.524
Jupiter	11.86 years	14	9	5.203
Saturn	29.45 years	10	8	9.539
Uranus	84.01 years	5	7	19.180
Neptune	164.79 years	2	6	30.070
Pluto	247.70 years	1	2	39.440

For each of these variables, we see nine pieces of data, corresponding to the nine planets in our solar system. We say, therefore, that the planetary data set in Chart 1-1 contains nine observations on each of four variables.

Variables can be classified according to the type of process by which their observations are generated. The processes associated with the four variables in Chart 1-1 are:

1. *A measurement process:* Each observation is recorded from a measuring device, such as a ruler, scale, clock, or thermometer. Example: *period of revolution.* We have *measured* the length of time necessary for each planet to complete one revolution around the sun. Observations that result from a measurement process carry *units of measurement. Period of revolution* in Chart 1-1 is reported in units of days—if the period is shorter than 1 year—or in units of years otherwise.

2. *A counting process:* Each observation results from counting the number of objects or events. Example: *number of satellites.* We have *counted* the number of moons that orbit each planet. The unit of each count is implied by the name of the variable; therefore, Uranus has five *satellites.*

3. *A ranking process:* Each observation results from assigning a rank. Example: *relative size.* We have ranked the planets from smallest to largest based on their diameters. Observations that result from a ranking process do not have explicit units of measurement. Each observation simply represents a position.

4. *An indexing process:* Each observation results from re- cording the value of a measurement relative to a base- line measurement. Example: *distance from sun index.* The mid-orbit distance from Earth to the sun (roughly 93 million miles) is the baseline measurement. Saturn's score (9.539) reveals that Saturn is about 9.5 times far- ther from the sun than Earth is. Venus' score (0.723) in- dicates that the distance between Venus and the sun is 72.3 percent of the distance between Earth and the sun. Earth's score is 1.000, since the score is found by divid- ing Earth's distance from the sun by itself.

(The planetary data set is a *small* data set, containing only nine observations on each variable. Because it is small in size, we can extract a planetary fact of interest merely by viewing Chart 1-1. The chart by itself is adequate for reporting the plan- etary facts.)

But in dealing with a large set of data, something more must be done to make the important facts stand out.

EXERCISES

1-1   For each state in the United States (including the District of Columbia), we are shown the dollar amount of the state government's grant to insti- tutions of higher education during 1965 and 1975.
(a) How many *variables* are there in the data set? Name them.
(b) How many *observations* are there on each variable?
(c) What is the distinction between a *variable* and an *observation* on a variable?

1-2   Classify each of the following variables according to whether the observa- tions on the variable are *measurements, counts, ranks,* or *index numbers.*
Variable 1: Blood pressure of each newly admitted hospital patient.
Variable 2: Number of tourists last year, by state.
Variable 3: Ratio of the annual population of a country to its popula- tion in 1900, by year.
Variable 4: Each nation's position in order of the number of Olympic medals won in 1976.
Variable 5: Number of *A*'s on a student's record, by semester.
Variable 6: High temperature on April 7, 1978, by city.
Variable 7: Weight of each individual divided by the "normal" weight for the individual's height, age, and frame.

1-3   Selected data from the Consumer Price Index (CPI) are provided below. The index number for the baseline year (1967) is 100, or 100 percent.

YEAR	CPI NUMBER
1965	95
1970	116
1975	161

(a) For every $100 spent in 1967 to purchase the commodities whose prices are included in the CPI, how many dollars had to be spent in 1965? In 1970? In 1975?

(b) If we divide the index number for 1975 (161) by the index number for 1970 (116), we obtain the quotient 1.388, or 138.8 percent. What does this result tell us?

1.2   organizing a large set of data

Members of the United States Supreme Court are appointed to serve until death or disablement. Of course, a justice may choose to resign or retire from the bench. It would be interesting to know how long terms on the U.S. Supreme Court tend to last and how frequently each cause of termination of service occurs.

The historical record of the 92 U.S. Supreme Court justices whose terms ended prior to January 1, 1976, is presented in Chart 1-2. We can see that this data set contains 92 observations on each of 3 variables: *year oath taken, length of service,* and *cause of termination.* The observations on *length of service* are measured in units of years and represent the number of full years of service.

(The observations on *cause of termination* result from still another process of generating data—the *classification* process. Each observation denotes a category rather than a numeric value. Stated another way, each observation defines a *quality* rather than a *quantity.* Disablement is distinguished from resignation in that the latter is voluntary; the former is imposed by a justice's peers on the bench.)

The observations in Chart 1-2 are ordered according to the year in which each justice took the oath of office. If we wish to report facts about the variables *length of service* and *cause of termination,* this method of presentation has notable deficiencies. Consider the following questions:

chart 1-2

NAME	YEAR OATH TAKEN	LENGTH OF SERVICE (Years Completed)	CAUSE OF TERMINATION
1. William Cushing	1789	20	Death
2. James Wilson	1789	8	Death
3. John Blair, Jr.	1789	6	Resignation
4. John Jay	1789	5	Resignation
5. John Rutledge	1789	2	Resignation
6. James Iredell	1790	8	Death
7. Thomas Johnson	1791	1	Resignation
8. William Paterson	1793	13	Death
9. Samuel Chase	1796	15	Death
10. Oliver Ellsworth	1796	4	Resignation
11. Bushrod Washington	1798	31	Death
12. Alfred Moore	1799	4	Resignation
13. John Marshall	1801	34	Death
14. William Johnson	1804	30	Death
15. Brockholst Livingston	1806	16	Death
16. Thomas Todd	1807	18	Death
17. Joseph Story	1811	33	Death
18. Gabriel Duvall	1812	22	Resignation
19. Smith Thompson	1823	20	Death
20. Robert Trimble	1826	2	Death
21. John McLean	1829	32	Death
22. Henry Baldwin	1830	14	Death
23. James M. Wayne	1835	32	Death
24. Roger B. Taney	1836	28	Death
25. Philip P. Barbour	1836	5	Death
26. John Catron	1837	28	Death
27. John McKinley	1837	15	Death
28. Peter V. Daniel	1841	19	Death
29. Samuel Nelson	1845	27	Resignation
30. Levi Woodbury	1845	5	Death
31. Robert C. Grier	1846	23	Resignation
32. Benjamin Curtis	1851	6	Resignation
33. John A. Campbell	1853	8	Resignation
34. Nathan Clifford	1858	23	Death
35. Noah H. Swayne	1862	18	Retirement
36. Samuel F. Miller	1862	28	Death
37. David Davis	1862	14	Resignation
38. Stephen J. Field	1863	34	Retirement
39. Salmon P. Chase	1864	8	Death
40. William Strong	1870	10	Retirement
41. Joseph P. Bradley	1870	21	Death
42. Ward Hunt	1873	9	Disabled
43. Morrison R. Waite	1874	14	Death
44. John Marshall Harlan	1877	33	Death
45. William B. Woods	1881	6	Death
46. Stanley Matthews	1881	7	Death

(continued)

chart 1-2
(continued)

NAME	YEAR OATH TAKEN	LENGTH OF SERVICE (Years Completed)	CAUSE OF TERMINATION
47. Horace Gray	1882	20	Death
48. Samuel Blatchford	1882	11	Death
49. Lucius Quintus C. Lamar	1888	5	Death
50. Melville W. Fuller	1888	21	Death
51. David J. Brewer	1890	20	Death
52. Henry B. Brown	1891	15	Retirement
53. George Shiras, Jr.	1892	10	Retirement
54. Howell E. Jackson	1893	2	Death
55. Edward D. White	1894	26	Death
56. Rufus W. Peckham	1896	13	Death
57. Joseph McKenna	1898	26	Retirement
58. Oliver Wendell Holmes	1902	29	Retirement
59. William R. Day	1903	19	Retirement
60. William H. Moody	1906	3	Disabled
61. Horace H. Lurton	1910	4	Death
62. Charles Evans Hughes	1910	27	Retirement
63. Willis Van Devanter	1911	26	Retirement
64. Joseph R. Lamar	1911	4	Death
65. Mahlon Pitney	1912	10	Disabled
66. James C. McReynolds	1914	26	Retirement
67. Louis D. Brandeis	1916	22	Retirement
68. John H. Clarke	1916	5	Resignation
69. William Howard Taft	1921	8	Retirement
70. George Sutherland	1922	15	Retirement
71. Pierce Butler	1923	16	Death
72. Edward T. Sanford	1923	7	Death
73. Harlan Fiske Stone	1925	20	Death
74. Owen J. Roberts	1930	15	Resignation
75. Benjamin N. Cardozo	1932	6	Death
76. Hugo L. Black	1937	34	Retirement
77. Stanley F. Reed	1938	19	Retirement
78. Felix Frankfurter	1939	23	Retirement
79. William O. Douglas	1939	36	Retirement
80. Frank Murphy	1940	9	Death
81. James F. Byrnes	1941	1	Resignation
82. Robert H. Jackson	1941	13	Death
83. Wiley B. Rutledge	1943	6	Death
84. Harold H. Burton	1945	13	Retirement
85. Frederick M. Vinson	1946	7	Death
86. Thomas C. Clark	1949	17	Retirement
87. Sherman Minton	1949	7	Retirement
88. Earl Warren	1953	16	Retirement
89. John M. Harlan	1955	16	Retirement
90. Charles E. Whittaker	1957	5	Retirement
91. Arthur J. Goldberg	1962	3	Resignation
92. Abe Fortas	1965	4	Resignation

1. How long was the maximum term of service?
2. How many justices served less than 30 years?
3. What percentage of the justices served between 5 and 10 years?
4. What proportion of the terminations were due to death?

The answers to these questions can be found in Chart 1-2, but you have to search for them. This takes effort and increases the chance for error.

Chart 1-2 is also too long. Readers who wish to determine the key facts about a variable's behavior are discouraged by seemingly interminable lists of observations.

How then can we reorganize the data in Chart 1-2 to highlight important facts about *length of service* and *cause of termination*?

(A useful way to present the data) on *length of service* (is in the form of an *array*.)

the array

(An *array* is an arrangement of the observations on a variable from the lowest to highest values (an array in *ascending* order) or from the highest to the lowest values (an array in *descending* order). Chart 1-3 presents the observations on *length of service* in ascending order.

Now let's see how this array helps us to answer the questions we just asked about *length of service*.

1. *How long was the maximum term of service?* We can immediately see from the last observation in the array that the maximum term of service spanned 36 years (William O. Douglas, 1939–1975). It would take perhaps a half minute to search through Chart 1-2 to identify the value 36 as the largest observation.
2. *How many justices served less than 30 years?* The first 82 observations in the array have values of less than 30, so the answer is 82 justices.
3. *What percentage of the justices served from 5 to 10 years?*[1] There are 92 justices listed in the array. Justices 13 through 34 (22 of them) served terms of 5 to 10 years. 22/92 = 23.9 percent, or approximately 24 percent.

[1] When the word "to" connects the first and last values denoting an interval, the last value is not included in the interval. For example, the interval "5 to 10 years" includes the years 5 *up to* 10, or years 5, 6, 7, 8, and 9.

chart 1-3

JUSTICE	LENGTH OF SERVICE (Years Completed)	JUSTICE	LENGTH OF SERVICE (Years Completed)	JUSTICE	LENGTH OF SERVICE (Years Completed)
1. Byrnes	1	32. Wilson	8	63. Gray	20
2. Thomas Johnson	1	33. Murphy	9	64. Thompson	20
3. Howell Jackson	2	34. Hunt	9	65. Cushing	20
4. Trimble	2	35. Pitney	10	66. Fuller	21
5. John Rutledge	2	36. Shiras	10	67. Bradley	21
6. Goldberg	3	37. Strong	10	68. Brandeis	22
7. Moody	3	38. Blatchford	11	69. Duvall	22
8. Fortas	4	39. Burton	13	70. Frankfurter	23
9. Joseph Lamar	4	40. Robert Jackson	13	71. Clifford	23
10. Lurton	4	41. Peckham	13	72. Grier	23
11. Moore	4	42. Paterson	13	73. McReynolds	26
12. Ellsworth	4	43. Waite	14	74. Van Devanter	26
13. Whittaker	5	44. Davis	14	75. McKenna	26
14. Clarke	5	45. Baldwin	14	76. White	26
15. Lucius Lamar	5	46. Roberts	15	77. Hughes	27
16. Woodbury	5	47. Sutherland	15	78. Nelson	27
17. Barbour	5	48. Brown	15	79. Miller	28
18. Jay	5	49. McKinley	15	80. Catron	28
19. Wiley Rutledge	6	50. Samuel Chase	15	81. Taney	28
20. Cardozo	6	51. John M. Harlan	16	82. Holmes	29
21. Woods	6	52. Warren	16	83. William Johnson	30
22. Curtis	6	53. Butler	16	84. Washington	31
23. Blair	6	54. Livingston	16	85. Wayne	32
24. Minton	7	55. Clark	17	86. McLean	32
25. Vinson	7	56. Swayne	18	87. John Marshall Harlan	33
26. Sanford	7	57. Todd	18	88. Story	33
27. Matthews	7	58. Reed	19	89. Black	34
28. Taft	8	59. Day	19	90. Field	34
29. Salmon Chase	8	60. Daniel	19	91. John Marshall	34
30. Campbell	8	61. Stone	20	92. William O. Douglas	36
31. Iredell	8	62. Brewer	20		

So the array has helped us to find answers, but it still contains a long list of observations. If our principal concern is to highlight the pattern of lengths of service, we might want to condense this array and display the data graphically.

frequency
distributions

(In an array, individual attention is given to each observation, because each observation is placed on a separate row. To condense the array—to reduce the number of rows—we can group observations with closely related values into *class inter-*

vals (intervals of similar values). We can then list each class interval on a row and report the number of observations whose values lie within that interval. This number is the *frequency* with which observations in that class occur, and the result is called a *frequency distribution*. The frequency distribution for the variable *length of service* is presented in Chart 1-4.)

By giving each class interval a width of 5 years, we have condensed our length of service data from an array of 92 rows to a frequency distribution of 8 rows.

We could engineer a further condensation by widening the class intervals to spans of say 10 years (0 to 10, 10 to 20, and so on).(There is no set rule governing the width of each class interval; but, for visual impact, it's probably wise to choose a width such that the frequency distribution contains no more than a dozen rows.)

(In addition to (and sometimes instead of) the actual number of scores within each class interval, charts often report the percentage of the total number of scores in each interval. When the total number of observations is large, percentages are more meaningful to the reader than actual numbers. These percentages are called *relative frequencies*.)

(At times it also may be useful to present data as a *cumulative frequency distribution*. The cumulative frequency entry for a class interval (number or percent) represents the frequency of observations located within this or any class interval of smaller values.)

(Computer printouts of frequency distributions usually include the absolute frequency (number of observations) and the

chart 1-4
Frequency Distribution of *Length of Service* Observations

GENERAL LABEL (Specific Label)	CLASS INTERVAL (Length of Service)	FREQUENCY OF OBSERVATIONS (Number of Justices)
	0 to 5 years*	12
	5 to 10 years	22
	10 to 15 years	11
	15 to 20 years	15
	20 to 25 years	12
	25 to 30 years	10
	30 to 35 years	9
	35 to 40 years	1
	Total	92

* When the word "to" connects the first and last values denoting an interval, the last value is not included in the interval. For example, the interval "5 to 10 years" includes the years 5 *up to* 10, or years 5, 6, 7, 8, and 9.

chart 1-5

CLASS INTERVAL (Length of Service)	FREQUENCY (No. of Justices)	RELATIVE FREQUENCY (Percent of Justices)	CUMULATIVE FREQUENCY (Cumulative No. of Justices)	CUMULATIVE RELATIVE FREQUENCY (Cumulative Percent of Justices)
0 to 5 years	12	13.0	12	13.0
5 to 10 years	22	23.9	34	36.9
10 to 15 years	11	12.0	45	48.9
15 to 20 years	15	16.3	60	65.2
20 to 25 years	12	13.0	72	78.2
25 to 30 years	10	10.9	82	89.1
30 to 35 years	9	9.8	91	98.9
35 and over	1	1.1	92	100.0
	92	100.0		

relative frequency (percentage of observations), as well as the corresponding cumulative frequency, as illustrated by Chart 1-5. We have added headings in parentheses to clarify the meaning of the computer headings.

From the information in the fourth row, for example, we can report that:

1. 15 justices served terms lasting 15 to 20 years.
2. 16.3 percent of the justices served terms lasting 15 to 20 years.
3. 60 justices served terms of up to 20 years.
4. 65.2 percent of the justices served terms of up to 20 years.

We have seen how frequency distributions have helped to clarify the pattern of *length of service.* Frequency distributions also can be used with the variable *cause of termination.* Because *cause of termination* is a categorical variable, we re-

chart 1-6

CATEGORY (Cause of Termination)	FREQUENCY (No. of Justices)	RELATIVE FREQUENCY (Percent of Justices)
Death	48	52.2
Retirement	24	26.1
Resignation	17	18.5
Disablement	3	3.3
	92	100.1*

* The discrepancy from 100 percent is due to the rounding of these percentages to one decimal place.

place the class intervals with the names of the categories and report the frequency of observations within each category, as shown in Chart 1-6.*)*

Now we can see how often each cause of termination occurred. For example, 48 (52.2 percent) of the justices died while in office.

EXERCISES

1-4
Referring to Chart 1-1 (page 6):
(a) For which variables could this chart be considered an array?
(b) Reorder the planets into an array in ascending order of *relative size.*
(c) Array the planets in descending order of their *number of satellites.*

1-5
The number of farms (to the nearest 1,000) in each New England state in 1970 and 1976 is listed below:

| | NUMBER OF FARMS | |
STATE	1970	1976
Connecticut	27,000	4,000
Maine	59,000	8,000
Massachusetts	38,000	6,000
New Hampshire	29,000	3,000
Rhode Island	5,000	1,000
Vermont	33,000	7,000

(a) Array these data so that the states are listed in descending order of the number of farms in 1976.
(b) Indicate the changes in state ranking (based on *number of farms*) that occurred between 1970 and 1976.

1-6
A linguistic analysis of a recent novel produced the following relative frequency distributions of the length of words (number of syllables):

NUMBER OF SYLLABLES	RELATIVE FREQUENCY OF WORDS	CUMULATIVE RELATIVE FREQUENCY OF WORDS
1	0.34	0.34
2	0.29	0.63
3	0.24	0.87
4	0.12	0.99
5	0.00	0.99
6	0.01	1.00
	1.00	

What proportion of the words contained:
(a) 2 syllables?
(b) 2 or 3 syllables?
(c) At least 2 syllables?
(d) At most 3 syllables?
(e) 4 or more syllables?

1-7   The frequency distribution of the size of industrial firms in a certain country is

SIZE (No. of Employees)	FREQUENCY ( )	RELATIVE FREQUENCY ( )	CUMULATIVE FREQUENCY ( )	CUMULATIVE RELATIVE FREQUENCY ( )
Less than 10	4	___	___	___
10 through 49	10	___	___	___
50 through 99	15	___	___	___
100 through 999	12	___	___	___
1,000 or more	9	___	___	___
	50			

(a) Give specific labels to the columns whose general labels are Frequency, Relative Frequency, Cumulative Frequency, and Cumulative Relative Frequency, respectively.
(b) Calculate the relative frequencies, cumulative frequencies, and cumulative relative frequencies for each class interval.
(c) What percentage of the industrial firms employed between 100 and 999 workers?
(d) How many industrial firms employed less than 100 workers?
(e) What percentage of the industrial firms employed less than 1,000 workers?

*1-8   The following table lists the annual unemployment rate in the United States (number unemployed as a percent of the civilian labor force), 1948–1974:

YEAR	UNEMPLOYMENT RATE (%)	YEAR	UNEMPLOYMENT RATE (%)	YEAR	UNEMPLOYMENT RATE (%)
1948	3.8	1957	4.3	1966	3.8
1949	5.9	1958	6.8	1967	3.8
1950	5.3	1959	5.5	1968	3.6
1951	3.3	1960	5.5	1969	3.5
1952	3.0	1961	6.7	1970	4.9
1953	2.9	1962	5.5	1971	5.9
1954	5.5	1963	5.7	1972	5.6
1955	4.4	1964	5.2	1973	4.9
1956	4.1	1965	4.5	1974	5.6

(a) Prepare an array listing the annual unemployment rates in ascending order.

* An exercise with an asterisk indicates that students should save their results for use in future exercises.

(b) Prepare a frequency distribution and a relative frequency distribution of the unemployment rate data. Use the class intervals: 0.0 to 1.0; 1.0 to 2.0; 2.0 to 3.0; 3.0 to 4.0; 4.0 to 5.0; 5.0 to 6.0; 6.0 to 7.0. Give specific labels to the Frequency and Relative Frequency columns. Note that scores on a common boundary are placed in the higher class interval.

1-9   The following chart presents the frequency distributions for the length of service of Supreme Court justices whose terms began prior to the conclusion of the Civil War (1865). The relevant observations pertain to the first 39 justices listed in Chart 1-2.

CLASS INTERVAL	FREQUENCY	RELATIVE FREQUENCY	CUMULATIVE FREQUENCY	CUMULATIVE RELATIVE FREQUENCY
0 to 5	5	12.8	5	12.8
5 to 10	9	23.1	14	35.9
10 to 15	3	7.7	17	43.6
15 to 20	6	15.4	23	60.0
20 to 25	5	12.8	28	71.8
25 to 30	4	10.3	32	82.1
30 to 35	7	17.9	39	100.0
35 to 40	0	0	39	100.0
	39	100.0		

(a) Consolidate this chart into four class intervals: 0 to 10; 10 to 20; 20 to 30; and 30 to 40.
(b) Using these four class intervals, prepare a similar frequency chart for justices 40 through 92, all of whom began service following the end of the Civil War.
(c) Can you identify any notable distinctions between the pre- and post-Civil War distributions?

1.3 frequency graphs

It often has been said that a picture is worth a thousand words, and we find several types of frequency graphs — pictures of the data — in common use, including the *histogram*, the *frequency curve*, and the *line graph*. All three have the same fundamental structure.

1. A *horizontal axis* labeled with the name of the variable. The possible values for the variable are listed along this axis.

2. A *vertical axis* labeled with the number (or percentage) of observations. Movement up the vertical axis denotes increasing frequency.*)*

(3. A *graphic device* (bars, lines, a curve) to depict the frequency with which different values occur. *)*

Charts 1-7 and 1-8 present two types of frequency graphs for the variable *length of service*. Accordingly, the horizontal axes are labeled "Length of Service" and the vertical axes are labeled "Number of Justices."

(The *histogram* in Chart 1-7 uses the device of rectangular bars to represent frequency. The width of each rectangle represents the width of a class interval, and the height of each rectangular bar represents the frequency with which observations fall into that class interval. So the taller rectangles depict the more usual terms of service. The values that are placed along the horizontal axis indicate the *beginning* point of each class interval, so that an observation positioned right on the border between two intervals (for example, a term of service of 5 years) falls into the *higher* class interval.)

(The *frequency curve* in Chart 1-8 is drawn by locating the center (or midpoint) of each class interval and placing a dot above each of these points at the height representing the frequency for that class interval. Finally, the dots are connected by a smooth curve, which is anchored to the horizontal axis, as shown in Chart 1-8.)

(The *area under a frequency curve* within an interval provides us with an impression of the *percentage* of the scores that lie within that interval. Think of the total area under the curve in Chart 1-8—the light blue area—as depicting 100 percent of the scores in the distribution. The portion of the total area within the 5-year class interval centered on 17.5 years—the dark blue area—then presents the percentage of scores within this class interval of 15 to 20 years. At times throughout this textbook, we will need to measure the relativity frequency of scores within an interval on the basis of the area under the curve within that interval.)

The two graphs present the same basic picture: Terms of 5 to 10 years occur more frequently than either shorter or longer terms; after approximately 20 years of service, increasingly long terms occur with decreasing frequency; and a term of 35 years or longer is extremely unusual.

chart 1-7
Histogram

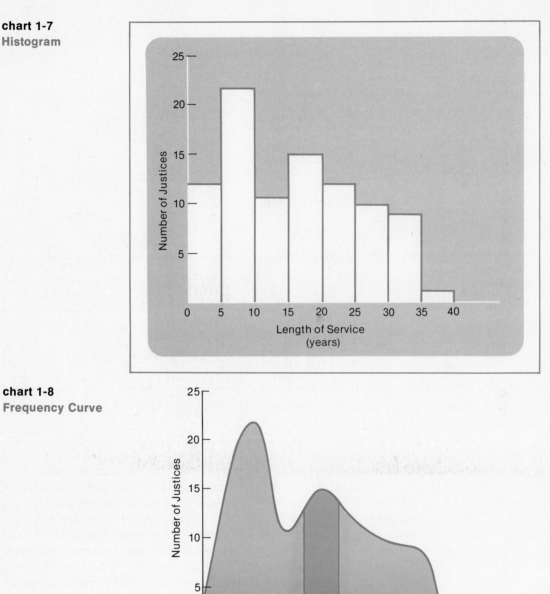

chart 1-8
Frequency Curve

chart 1-9
Minimum Age to
Marry, 1976

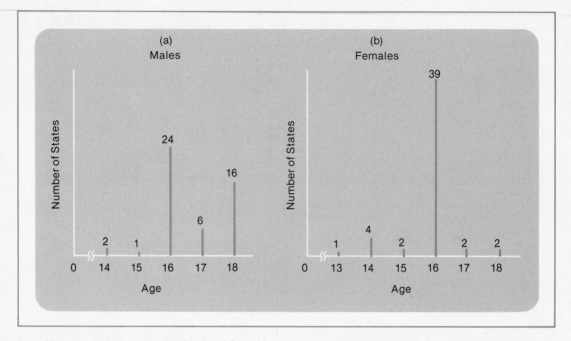

Note: Of the 51 states (including the District of Columbia), two set no minimum age for males and one sets no minimum age for females.
Source: Women's Bureau, U.S. Department of Labor.

To construct either a histogram or a frequency curve, the observations on a variable must first be grouped into class intervals. When the observations take on a limited number of values, grouping is inappropriate. For example, state governments set the minimum age for entering marriage at one of six values: 13, 14, 15, 16, 17, or 18 years. In cases like this, a useful picture of the data is the *line graph,* as illustrated in Chart 1-9, where each line indicates the frequency of observations associated with a particular value (rather than within a class interval). At a quick glance, we can see that in 1976, the marriage laws of most states required minimum ages of 16 or 18 years for males, and 16 years for females.

(Line graphs also are used with categorical data. In Chart 1-10, this device is applied to the *cause of termination* vari-

chart 1-10
Cause of
Termination

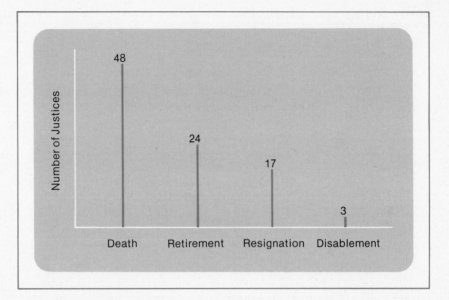

able. The entries on the horizontal axis are categories rather than numerical values and are conventionally indicated at equally spaced intervals.)

EXERCISES

1-10   Prepare a line graph of the observations on *number of satellites* in Chart 1-1 (page 6). Label the axes.

1-11   Prepare a line graph of the relative frequency distribution of word lengths (number of syllables) in Exercise 1-6 (page 15). Label the axes.

1-12   The U.S. Bureau of Census reported the 1975 marital status of persons 14 years and older to be

MALES		FEMALES
TOTAL NUMBER: 76 million		TOTAL NUMBER: 84 million
29.5%	Single	22.8%
64.8%	Married	60.4%
2.4%	Widowed	12.1%
3.3%	Divorced	4.8%
100.0%		100.1% (rounding error)

(a) Using a two-color scheme, prepare a line graph showing the percentage of males and females within each marital status category.

(b) Were there more male or female *singles* in the United States in 1975?

*1-13   Referring to the results of Exercise 1-8 (page 16), prepare a histogram of the relative frequency distribution of annual U.S. unemployment rates. Label the axes.

1-14   The following data from the U.S. Bureau of Prisons reflect the annual number of executions in the United States, 1932–1967. Using the class intervals 0 to 25, 25 to 50, and so on, prepare a histogram from these data.

YEAR	NUMBER OF EXECUTIONS	YEAR	NUMBER OF EXECUTIONS
1932	140	1950	82
1933	160	1951	105
1934	168	1952	83
1935	199	1953	62
1936	195	1954	81
1937	147	1955	76
1938	190	1956	65
1939	160	1957	65
1940	124	1958	49
1941	123	1959	49
1942	147	1960	56
1943	131	1961	42
1944	120	1962	47
1945	117	1963	21
1946	131	1964	15
1947	153	1965	7
1948	119	1966	1
1949	119	1967	2

Caveat: Facts versus Inferences

A caveat is a warning to beware of some danger. The Latin *caveat emptor* ("let the buyer beware") warns the consumer that personal knowledge and care offers the surest protection against deceitful sales practices. Throughout this book, we will present caveats to warn you against the unwarranted use or interpretation of statistical concepts.

(Our first caveat is

> Do not present an *inference* as a *fact*. Do not imply that the observed scores on a variable necessarily reflect the unobserved scores on that variable.)

EXAMPLE 1

We have observed (Chart 1-2) the length of service for each Supreme Court justice *whose term ended prior to January 1, 1976*. No such observations were recorded for justices still active at the start of 1976 or for justices appointed subsequently.

FACT: Approximately 24 percent of the justices (see Chart 1-5) whose terms expired prior to January 1, 1976 served terms of 5 to 10 years.

INFERENCE: We may infer that about one-quarter of the Supreme Court justices who enter service in the next 30 years will serve terms of between 5 and 10 years.

(The inference will be correct only if the unobserved (future) frequency distribution of this variable does not change from the observed (past) frequency distribution; in other words, if the past proves to be a reliable guide to the future.)

EXAMPLE 2

The voting preferences of 100 registered voters were solicited by telephone. Of the voters polled, 54 favored Candidate A, 40 favored Candidate B, and 6 were undecided.

FACT: Candidate A is currently preferred by a majority (54 out of 100) of the voters surveyed.

INFERENCE: If the election were held at the time of the survey, Candidate A would win.

The inference assumes that the unobserved preferences of the voters (the preferences of those who were not surveyed plus the preferences of the undecided voters) would divide in virtually the same proportion that the preference of the surveyed voters did.

These examples illustrate the distinction between a fact and an inference. (A *fact* is a statement supported by actual scores on a variable. An *inference* combines a fact with an assumption or judgment about the unobserved scores on the variable.)

"Reporting the Facts" is the title we have given to the first part of this book, because an ability to report the facts is required before inferences can be formulated.

EXERCISE

Dissect each of the following inferential statements into its two components: (1) the fact based on actual behavior, and (2) the assumption about unobserved behavior.
(a) We can expect approximately 20 percent of the country's retail clothing establishments to declare bankruptcy within three years after beginning business operations, according to evidence developed in a national study of all U.S. retail outlets opened in 1973.
(b) There is a 60 percent chance that a rape victim will refrain from carrying her case to court, according to police records.
(c) The album "Swinging 70s" became the 5th best seller in the country last month, based on a survey of the 100 largest record outlets.

COMING
TO
TERMS

Array   A list of the observations on a variable, either in ascending order (from lowest to highest values) or in descending order (from highest to lowest values).

Distribution   The arrangement of observations on a variable that emphasizes how frequently certain values occur. An abbreviation for the *distribution of observations on a variable.*

Frequency Expressions
Frequency: The number of observations in a category or class interval.
Relative Frequency: The proportion or percentage of observations in a category or class interval.
Cumulative Frequency: The number of observations less than or equal to a given value.
Cumulative Relative Frequency: The proportion or percentage of observations less than or equal to a given value.

Frequency Graph   The graphic presentation of a frequency distribution. A variable's possible values are listed on the horizontal axis, and the frequencies of occurrence of these values are listed on the vertical axis. The graphic devices used to depict frequency include:
Histogram: Rectangular bars depict the frequency of scores in designated class intervals. Used for grouped data.
Frequency Curve: A curve connecting the midpoints of class intervals at a height representing the frequency of scores in each class interval. Used for grouped data.

Line Graph: Vertical lines depict the frequency with which each value occurs. Used for categorical data or when it is inappropriate to group data.

Observations   Pieces of data (information) on a variable. Also called *scores*. Types of observations include:

Measurements: Data recorded from a measuring device.

Counts: Data resulting from a counting process.

Ranks: Data specifying relative positions in a distribution.

Index Numbers: Ratios of measurements to a baseline measurement, usually set at 1.0 or 100.

Categories: Data resulting from a classification process.

Scores   Synonym for *observations*.

Variable   The topic about which data are collected.

ISSUES
FOR
ANALYSIS

1. To determine how student attendance rates vary with class size, the mathematics department in a large university asked its instructors in Math 101 (Analytical Geometry and Calculus I) to record attendance in each class during the semester. The basic results follow:

SECTION:	A (JENKINS)	B (CALDWELL)	C (DEVONSHIRE)	D (KRASNOW)
Class Size	24	35	52	100
Average Number of Absentees per Class	4	7	13	28

(a) Determine the absentee *rate* for each class (Average Number of Absentees per Class divided by Class Size), and add these rates to the table.

(b) Briefly describe the observed relationship between class size and absentee rates.

(c) What assumptions must be made to permit the *inference* that class absenteeism generally worsens as class size increases?

2. *(Continuation of Exercise 1-8, page 16)* Assume the time is early in 1975 and you are asked to make a rough forecast of the nation's unemployment rates during 1975 and 1976 based on historical records from 1948 to the present. You sense that unusually high unemployment rates will be recorded in the next two years. Would the historical record permit you to forecast two successive years in which the unemployment rate climbed above 6 percent? Explain your answer. Then check a recent *Economic Report of the President* for the actual unemployment rates (All Workers category) during 1975 and 1976.

Behind the Headline

In Chapter 1, you learned about some useful devices for organizing and displaying a set of data. Arrays, frequency tables, and graphs do lend perspective to the data at hand, but these organizational tools often are little more than the first step in reporting statistical facts. Particularly when you examine a large mass of data, you must summarize the data numerically to highlight important characteristics. In two words, you must *condense* and *focus* the data.

What can you use to describe a large number of observations in a few words? You could start with an *average*, a number that represents the usual or "ordinary" observations. "The average American feels" is a phrase we often hear from politicians and public relations experts. If you use an aver-

age, obviously you do not tell everything there is to know about a set of data, but you do give some idea about where the center of the data lies and of what values represent the "middle of the road." Moreover, an average allows us to compare different distributions. Then you can say that the scores in one distribution are higher than the scores in another distribution *on the average*.

So an average is our first recourse in an effort to condense and focus the data. But when we quote an average, we must define it. People often refer to different averages from the same set of data, depending on what point they wish to prove. We will now investigate several types of averages and see what each of them really describes.

averages

> *"A fool takes in all the lumber of every sort that he comes across, so that the knowledge which might be useful to him gets crowded out, or at best is jumbled up with a lot of other things so that he has difficulty in laying his hands upon it. It is of the utmost importance, therefore, not to have useless facts elbowing out the useful ones."*
>
> Holmes to Watson
> in *A Study in Scarlet*

2.1 types of averages

Two rival business firms are trying to hire a rising young executive. Firm A points out to him that its average executive earns $20,000 a year, which is $2,000 a year more than the average executive earns at Firm B. Firm B tells him that its executive salaries average $26,000 compared to $24,000 in Firm A. Neither firm is distorting the facts, but the two firms are using different kinds of averages to represent their salary distributions. Firm A is using a *median;* Firm B is using an *arithmetic mean.* Two other measures of the average are the *mode* and the *midrange.* We will discuss each measure in turn.

the
midrange

When home heating oil became scarce, homeowners began calculating how much fuel they would need for the winter. They paid increasing attention to the daily weather report on degree-day units. Degree-day units represent the extent to which the temperature has fallen below 65°F. For each day, the daily high temperature (H°) and the daily low temperature (L°) are added together and the total is divided by 2 to obtain

27

the day's average. The degree-day units for that day are then determined by subtracting the day's average from 65°.

(In this case, the average used to represent daily temperature levels is the *midrange*.)

(MIDRANGE   The value *halfway* between the smallest and the largest observations of the distribution.)

Suppose that the hourly temperatures shown in Chart 2-1 were recorded during a particular day in January. We can identify the midrange by noting that the day's lowest temperature was 10°, reached at 3 A.M. and the day's highest temperature was 36°, reached at 1 P.M. The value halfway between 10° and 36° is 23°, the midrange. The easiest way to find this value is to add the smallest and the largest observations together and divide by 2: $(10° + 36°)/2 = 23°$. The midrange of 23° is both 13° above the low and 13° below the high. On this day in January, the weather bureau would record $65° − 23° = 42$ degree-day units. It is estimated that a gallon of fuel oil is burned for every 12 degree-day units, so homeowners could expect to consume $42/12 = 3.5$ gallons of fuel heating their homes on this day.

Chart 2-2 displays the hourly temperature readings in a frequency distribution and shows the location of the midrange.

(As a representative of a distribution, the midrange has the virtue of being simple to compute: Only two observations, the

chart 2-1

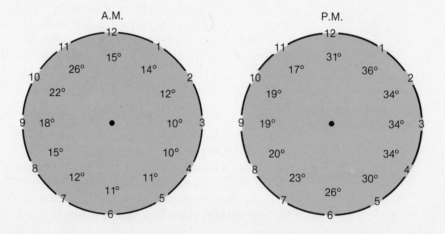

chart 2-2
Frequency Distribution (Histogram) of Hourly Temperatures

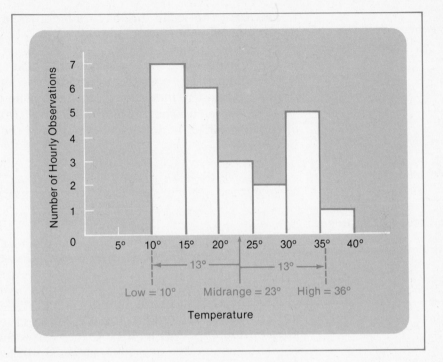

high and the low, must be considered. However, this apparent virtue is frequently a handicap. Because the midrange is based only on the two most extreme observations and ignores the great mass of data between the two extreme scores, it is often positioned far to one side of a distribution.)

the mode

Two duffers spent most of the day watching professional golfers play the 15th hole—a difficult dog-leg of 432 yards, rated as a par 4. As the last pro began to tee off, the duffers wagered a beer on how many strokes the pro would take to play the hole. Rather matter of factly, one duffer chose 4 strokes, because par is supposed to be the most likely score for a professional golfer. But the other had been keeping track of the actual results for the day and knew that the most frequent score on the 15th hole *that day* was 5 strokes. So the second duffer bet on a bogie (one over par). (Both choices are based on a type of average called the *mode*.)

(MODE   The value that occurs *more frequently* than any other value in a distribution.)

(In a game of chance, you have the best chance of winning if you choose the most frequently occurring outcome. In statistical terms, the *modal* outcome is the one on which to bet.)

The scores for the day on the 15th hole (excluding the score of the last golfer, which was the object of the wager) are given in Chart 2-3.

chart 2-3
15th Hole
432 Yards Par 4
Number of Strokes

	ACE 1	EAGLE 2	BIRDIE 3	PAR 4	BOGIE 5	DOUBLE BOGIE 6	TRIPLE BOGIE 7
Number of Golfers	0	1	8	29	63	21	3

As the second bettor had observed, more pros had scored a bogie 5 than any other score, including par. The mode of this distribution of scores is, therefore, 5 strokes.

The last golfer scored a birdie 3 on the 15th hole. The friendly bettors went Dutch treat for the beer.

The frequency distribution of the golf scores is given in the line graph in Chart 2-4. (The mode is the value associated with the tallest line on the graph. Also note that the midrange is 4.5 strokes—the value midway between the lowest score (2 strokes) and highest score (7 strokes) on the hole.)

It may seem strange to say that the average number of strokes is 4.5 when we know that no golfer can ever score between 4 and 5 strokes on a hole. This phenomenon happens frequently when averages are used: The average family size is 3.7 children, washing machines are repaired on the average of 0.5 times per year, and remember, the average statistician has 1.75 wives. (In actuality, the mode is the only type of average that will always be an *observable* value.)

(When the data are displayed through a histogram or a frequency curve, the exact location of the mode (or modal value) is lost because the data have been grouped into class intervals.) Charts 2-1 and 2-2 present the same hourly temperature data. In Chart 2-1, we can identify the mode as 34°, because this temperature value occurred more than any other value during that day. But we cannot determine that 34° is the modal value

Is there an average family?

chart 2-4
Frequency Distribution of Golf Scores

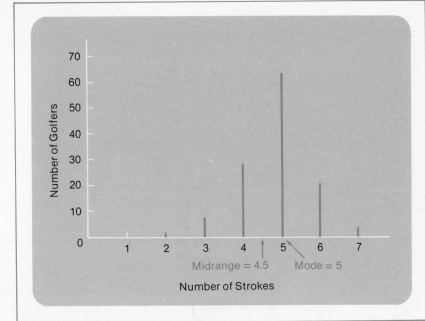

from the histogram in Chart 2-2.(Rather, the tallest rectangular bar in the histogram depicts the *modal interval*—the class interval that includes the largest number of scores.)The modal interval of hourly temperatures on this day is 10–15°.

(When it is desirable to display the data through a histogram or frequency curve, rather than a line graph, the modal interval is probably more informative than the modal value.)

The mode of the temperature distribution (34°) is only 2° less than the day's high and is a higher temperature than the temperatures reached in 20 of the 24 hours. On the other hand, the modal interval indicates that the temperature hovered between 10° and 15° over a significant portion of the day (7 hours).

the median

After receiving a flood of mail from voters who were irate over the high cost of electricity, a state legislator commissioned a study of utility costs in the district. Of the 21 households surveyed, 50 percent had received electricity bills of $28.19 or

chart 2-5

HOUSEHOLD	BILL	HOUSEHOLD	BILL
1	$ 6.75	12	$29.37
2	9.29	13	29.89
3	10.18	14	30.01
4	14.77	15	30.45
5	16.53	16	33.18
6	18.91	17	34.70
7	22.14	18	34.95
8	23.93	19	36.11
9	26.61	20	41.04
10	26.64	21	79.25
11	28.19		

more for the previous month. (In this study, the *median* was chosen to represent the average of the distribution.)

(MEDIAN   The value of the middle score in the distribution.)

The electricity bills received by the 21 households surveyed are arrayed from lowest to highest in Chart 2-5.

From this array, we can see that the middle score is the bill of the 11th household. Therefore, the median is $28.19. Ten of the households paid less than $28.19 and ten paid more.

(When the distribution has an even number of observations, there are two middle scores. In such cases, the median is defined as the value midway between the two middle scores.)

If the household that received a bill of $79.25 had not been included in the survey, then the median electricity bill would have a value midway between the 10th and 11th largest bills; that is, the median would be ($26.64 + $28.19)/2 = $27.42. Again, there are the same number of households with bills smaller than the median as there are households whose bills exceed the median.

Chart 2-6 presents the frequency distribution of the electricity bills of all 21 households. (The median is the value that divides the area of the histogram into two equal parts. This means that 50 percent of the area within the rectangular bars lies on either side of the median. (When data must be grouped to draw a histogram, a degree of inexactness is introduced. For this reason, the median calculated from the data may not be *precisely* the 50 percent point on the graph, but the two values should be close.))

Chart 2-6 also shows the midrange and the modal interval of the distribution of the electricity bills of all 21 households. The midrange is ($6.75 + $79.25)/2 = $43.00, a larger value than any bill except that of the 21st household.

Here the handicap of the midrange is evident. The electricity bill of the 21st household ($79.25) is so much greater than any other household's bill that it pulls the midrange far above the center of the distribution and makes it unrepresentative of most of the scores. When the extreme score of the 21st household is excluded from the distribution, the midrange becomes $23.90, a value that lies much closer to the center of the distribution.

The distribution of electricity bills has no mode because each value occurs only once. However, when the observations are grouped into $10 intervals, (the interval grouping we used

chart 2-6
Frequency Distribution (Histogram) of the Electricity Bills of 21 Households

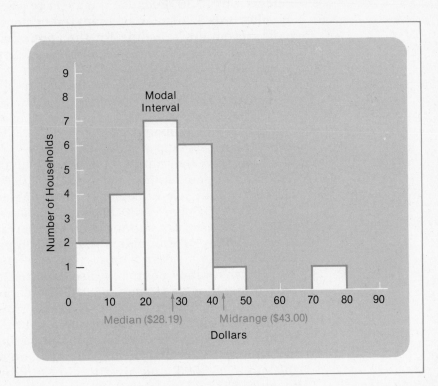

to create the histogram in Chart 2-6), the modal interval is $20 to $30, because of the 21 households more (7) received electricity bills in that interval than in any other interval.

(The median is a useful representative of any distribution of measurements. A distribution of measurements may not have a modal value, but it always has a median. Unlike the mid-range, the median is not pulled off center by an extremely high or low score. Still, the median is not the type of average that is reported most often when data are described. That honor goes to the type of average we will consider next—the arithmetic mean.)

the
mean

The management of an automobile insurance company must continually reevaluate its premium schedule to determine whether revenue from premiums adequately covers the claims against the company. In last year's report, one insurance company stated that its average claim from class-A collision policyholders was $130. Because the average premium for this type of policy was $240, management felt no need to revise the company's premium schedule for the coming year. (This decision was based on an average called the *arithmetic mean*, or simply, the mean.)

(MEAN   The sum of the observations in a distribution divided by the number of observations in that distribution.)

Last year the company carried 20,000 class-A collision policies and paid claims totaling $2,600,000. The mean of the claims from class-A collision policyholders therefore was $2,600,000/20,000 = $130 per policyholder. Of course, many policyholders did not submit claims to the company in the last year, so many of the claims that the company paid last year were much larger than $130.

The insurance company also reported the actual claims paid in each of its branch offices. The data from one branch office are tabulated in Chart 2-7.

We determine the mean of the amounts paid on these 13 claims by adding the individual claim payments together and dividing by 13. Adding the individual claim payments gives us the sum of $7,615, and dividing this sum by 13 results in the mean of $585.77 per claim.

chart 2-7

Claim Number	1	2	3	4	5	6	7	8	9	10	11	12	13	SUM	MEAN
Claim Payment (in dollars)	25	100	115	150	200	200	225	250	300	550	600	1,400	3,500	7,615	585.77

(A major reason for the popularity of the mean is that the mean is the only type of average that is linked arithmetically to the sum of *all* the observations in the distribution. Because the mean is defined as the sum of the observations divided by the number of observations, the sum can be found by multiplying the mean by the number of observations.)

For example, if we are told that the *mean* of the premiums paid to the company by the 13 claimants in Chart 2-7 is $285, we can determine that the *sum* of the premiums paid by these claimants is $285·13 = $3,705. This result reveals that the insurance company paid in claims an excess of $7,615 − $3,705 = $3,910 above the premiums it collected from these claimants.

The arithmetic linkage of the mean to the sum of the observations can sometimes be a disadvantage as well. First, if the exact values of the observations are unknown, it is impossible to calculate the sum and, therefore, the mean of all the observations.) As an illustration, suppose that the automobile insurance company wishes to determine the average amount of time that elapses between the issuance of a policy and the policyholder's first accident. If, say, 50 percent of the policyholders have an accident within 4 years, then the *median* is 4 years. But some policyholders may never have an accident. Because no number can be used to represent these policyholders, we cannot compute the mean.

Second, since the mean is based on all the observations in a distribution, it is sensitive to the pull of an extremely high or low score. The extremely high claim of $3,500 in Chart 2-7 has pulled the mean up to $585.77, making the mean larger than 10 of the 13 scores in the distribution. However, the mean is not nearly as off center as the midrange of $1,762.50, which is higher than all but the 13th score. (In general, extreme scores affect the mean less than they do the midrange, because the mean is based on the sum of *all* 13 observations whereas the midrange is based on the sum of only the smallest and largest scores. As before, the median is not affected by an extreme score, because, by definition, the median is simply the value of the middle score.) Here the median is $225, which is larger than six and smaller than six of the observations.

A pharmaceutical company was concerned with customer reaction to its new headache tablet ACT (Alleviates Cerebral Tension) and hired an advertising agency to survey ACT purchasers to determine how effectively they felt that the tablet relieved pain. The agency reported that 60 of the 80 customers questioned responded positively.

(The statistic used by the advertising agency to summarize the responses to its question was a *proportion* or *percentage*.) In this situation we can say that the proportion of positive (*yes*) responses was 60/80 = 0.75, or that 75 percent of the customers surveyed responded positively.

(Many problems requiring statistical solutions involve variables that have only two possible values. We call these variables *yes–no variables*. It can be helpful to assign a score of 1 to one of the two values and a score of 0 to the other value. Using this convention, we can represent a *yes–no* variable by *quantitative* observations, which enables us to make an arithmetic computation.)

In this example, a response that ACT has been effective could be assigned the score 1 (*yes*) and the response that ACT has not been effective could be assigned the score 0 (*no*). In these quantitative terms, 0.75 represents the *proportion* (and 75 percent represents the *percentage*) of 1 responses.

(Remember that the mean of a distribution is the sum of the observations divided by the number of observations. Accordingly, for a *yes–no* variable

$$\text{MEAN} = \frac{\text{Sum of Observations}}{\text{Number of Observations}} = \frac{\text{Number of 1 Responses}}{\text{Number of Responses}}$$

$$= \text{PROPORTION OF 1's}$$

PROPORTION   The number of observations within a certain category divided by the total number of observations. The proportion of 1's (or yes's) is the mean of a *yes–no* variable.)

(Other types of averages can be used to describe the distribution of a *yes–no* variable but rarely are. The median and modal values are usually identical. They will equal 0 if the majority of the observations are 0, and they will equal 1 if the majority of the observations are 1. The midrange is always 0.5 (midway between 0 and 1) and is therefore useless in representing a *yes–no* variable.)

(We will discuss *yes–no* variables in greater detail in later chapters, where we will learn how the convention of assigning 0's and 1's to observations expedites statistical analysis.)

EXERCISES

2-1   Lothario, a character in an eighteenth-century play, was reputed to have charmed women from the ages of 6 to 66. What can you report about the average age of Lothario's admirers? What type of average is this?

2-2   Refer to the array and the histogram of the lengths of service of Supreme Court justices given in Charts 1-3 and 1-7, pages 12 and 19 respectively.
(a) From the array, determine the mode of this distribution and interpret its value.
(b) From the histogram, determine the modal interval of this distribution and interpret its value.
(c) Occasionally, the mode is defined as the midpoint (midrange) of the modal interval. Based on this definition, what is the modal length of service of these justices?

2-3   An instructor recorded the following attendance rates (percentages of enrolled students who were present) in six classes:

84, 77, 53, 79, 80, 69

Determine the median attendance rate and interpret its value.

2-4   Students often write the *medium* of a distribution when they mean to say *median*—an understandable error, because the median actually occupies the medium or middle position in a distribution. In contrast, the *mean* can lie toward one end of the distribution. List any eight numbers, seven of which are larger than their mean. Then report the median of these numbers.

2-5   Referring to the planetary data set given in Chart 1-1 (page 6), determine:
(a) the midrange

(b) the mode
(c) the median
(d) the mean
of the number of satellites.

2-6   In a comparative price survey of seven supermarkets in a city, the following prices (in cents) were noted for a pound of coleslaw:

49, 79, 59, 69, 69, 89, 69

Determine:

(a) the midrange
(b) the mode
(c) the median
(d) the mean

of the prices for a pound of coleslaw.

2-7   A quarterback threw 17 passes during a game for a total gain of 204 yards. Calculate the average number of yards gained per pass. What type of average have you calculated?

*2-8   Referring to Exercise 1-8 (page 16):
(a) Determine the midrange and the mean of the distribution of annual unemployment rates for 1948–1974.
(b) Recompute the midrange and the mean, including the observations for 1975 and 1976 (1975: 8.5 percent; 1976: 7.7 percent).
(c) Was the midrange or the mean more sensitive to the inclusion of the extreme values for 1975 and 1976? Why?

2-9   A record of the number of push-ups performed by 100 new recruits at Fort Dix revealed
        A *mean* of 30 push-ups.
        A *median* of 24 push-ups.
        A *mode* of 22 push-ups.
        A *midrange* of 60 push-ups.
(a) What was the total number of push-ups completed by the group of 100 recruits?
(b) Did a majority of the recruits (51 or more) complete *less than* 23 push-ups?
(c) If the poorest performance by a recruit was 5 push-ups, what was the *best* performance by a recruit?

2-10 The following table summarizes the percentage of the world's total land area that lies within each of the seven continents:

CONTINENT	PERCENT OF LAND AREA	ACTUAL LAND AREA (Square Miles)
Asia	29.5	
Africa	20.0	
North America	16.3	
South America	11.8	
Antarctica	9.6	
Europe	6.5	
Australia	5.2	

If, on the average, a continent has a land area of 8,127,000 square miles (the mean), calculate the actual land area for any *one* of the individual continents.

2-11 Three of four registers in a supermarket are in use. The manager is considering whether to open the fourth register. A count of the number of carts in each waiting line reveals that there are

> 5 in Line 1
> 6 in Line 2
> at least 7 in Line 3

(a) Determine the median number of carts in line.
(b) Can you calculate the mean? Explain your answer.

2-12 Explain the meaning of a baseball player's batting average of .300. What type of average is a batting average?

2-13 Of 20 women involved in an experiment to evaluate the effectiveness of a new type of birth-control device, three became pregnant while using the device. If 1 denotes an instance of pregnancy and 0 denotes an instance of nonpregnancy, what is the mean of this *yes–no* variable (*yes* = pregnant; *no* = not pregnant)?

2.2 properties of the mean

the mean
in algebraic
notation

Ten students in an English composition class received the following grades on their first theme:

Student Number	1	2	3	4	5	6	7	8	9	10
Grade (10 = perfect)	9	9	6	7	8	10	8	8	9	4

In handing back the graded papers, the instructor reported that (the mean) grade was 7.8. When asked how she arrived at this figure, the instructor said she used (the equation

$$\overline{X} = \frac{\Sigma X}{n}$$ (2-1)

Of course, the students didn't feel they had received an adequate answer. But the instructor's answer was adequate; the students were simply not familiar with the notation she used. (The symbol Σ (upper case sigma in the Greek alphabet) conveys the instruction to "sum up" and is called the *summation sign*. The question of what scores are to be summed is answered by whatever follows Σ. In Equation (2-1), Σ is followed by the variable X. In our problem, X represents the variable *English grade*. So ΣX tells us to sum all 10 grades received on the theme. To complete the numerator in Equation (2-1), therefore, we write $\Sigma X = 78.$) (Many statistical equations contain the letter n. It is used to denote the number of observations in a distribution.) In the English class $n = 10$ because there are 10 students in the class. (The expression $\Sigma X/n$ therefore instructs us to divide the sum of the observations (the grades received on the theme) by the number of observations. By definition, we know that the result will be the mean grade:)

$$\text{Mean grade} = \frac{\Sigma X}{n} = \frac{78}{10} = 7.8$$

(The notation $\overline{X}$ ("X bar") is the accepted symbol for the mean of a distribution.) The statement $\overline{X} = 7.8$ therefore informs us that the mean grade on the English theme was 7.8.

We did not have to use the letter X to denote the variable. Any letter with an overbar represents the mean of the distribution of scores on that variable.

frequency and
relative frequency
equations for
the mean

(When the data are presented in a frequency distribution, it is more convenient to use a slightly different equation to compute $\overline{X}$.) Suppose, for example, that the grades received on the English theme are displayed in the frequency distribution and the relative frequency distribution given in Chart 2-8.

The first row of the table indicates the grades that could be

chart 2-8

English Grade X	0	1	2	3	4	5	6	7	8	9	10
Number of Students $f(X)$	0	0	0	0	1	0	1	1	3	3	1
Proportion of Students $p(X)$	0	0	0	0	$\frac{1}{10}$	0	$\frac{1}{10}$	$\frac{1}{10}$	$\frac{3}{10}$	$\frac{3}{10}$	$\frac{1}{10}$

received on the theme; the second row indicates the number of students who achieved each grade (the frequency); and the third row indicates the proportion of students who achieved each grade (the relative frequency).

If we let X denote the possible grades (0–10), then $\overline{X}$ can be found by using Equation (2-2) and the frequency information *or* by using Equation (2-3) and the relative frequency information.

$$\overline{X} = \frac{\Sigma[X \cdot f(X)]}{n} \qquad (2\text{-}2)$$

$$\overline{X} = \Sigma[X \cdot p(X)] \qquad (2\text{-}3)$$

Equation (2-2), which we will call the *frequency equation for the mean*, instructs us to multiply each value of X by the frequency with which that value occurs and then to add the products and divide by the number of observations n. The steps are shown in Chart 2-9.

chart 2-9
Calculation of $\overline{X}$ Using the Frequency Equation

ENGLISH GRADE X	NUMBER OF STUDENTS $f(X)$	$X \cdot f(X)$
0	0	$0 \cdot 0 = 0$
1	0	$1 \cdot 0 = 0$
2	0	$2 \cdot 0 = 0$
3	0	$3 \cdot 0 = 0$
4	1	$4 \cdot 1 = 4$
5	0	$5 \cdot 0 = 0$
6	1	$6 \cdot 1 = 6$
7	1	$7 \cdot 1 = 7$
8	3	$8 \cdot 3 = 24$
9	3	$9 \cdot 3 = 27$
10	1	$10 \cdot 1 = 10$
	$n = \Sigma f(X) = 10$	$\Sigma[X \cdot f(X)] = 78$

$$\overline{X} = \frac{\Sigma[X \cdot f(X)]}{n} = \frac{78}{10} = 7.8$$

chart 2-10
Calculation of $\overline{X}$
Using the Relative
Frequency
Equation

ENGLISH GRADE X	PROPORTION OF STUDENTS $p(X)$	$X \cdot p(X)$
4	$\frac{1}{10}$	$4 \cdot \frac{1}{10} = \frac{4}{10}$
5	$\frac{0}{10}$	$5 \cdot \frac{0}{10} = \frac{0}{10}$
6	$\frac{1}{10}$	$6 \cdot \frac{1}{10} = \frac{6}{10}$
7	$\frac{1}{10}$	$7 \cdot \frac{1}{10} = \frac{7}{10}$
8	$\frac{3}{10}$	$8 \cdot \frac{3}{10} = \frac{24}{10}$
9	$\frac{3}{10}$	$9 \cdot \frac{3}{10} = \frac{27}{10}$
10	$\frac{1}{10}$	$10 \cdot \frac{1}{10} = \frac{10}{10}$

$$\Sigma[X \cdot p(X)] = \frac{78}{10}$$

$$\overline{X} = \Sigma[X \cdot p(X)] = \frac{78}{10} = 7.8$$

(Equation (2-3), which we will call the *relative frequency equation for the mean,* instructs us to multiply each value of X by the proportion of times that value occurs. The steps are shown in Chart 2-10.)

the mean as the
balance point

(The ages of seven children playing in a pediatrician's waiting room are 2, 2, 2, 3, 6, 6, and 7 years old.

The average age of these children can be quoted as

2 (the mode), because 2 is the most frequently occurring age.

3 (the median), because 3 is the value of the age in the middle.

4 (the mean), because $\dfrac{\Sigma X}{n} = \dfrac{2 + 2 + 2 + 3 + 6 + 6 + 7}{7}$

$$= \frac{28}{7} = 4.$$

4.5 (the midrange), because 4.5 is midway between the smallest and largest ages.)

Chart 2-11 is a line graph of the age distribution of these children. In this graph, we can see that the mode is situated at the point of highest frequency, the median is situated at the

chart 2-11
Frequency Distri-
bution of
Children's Ages

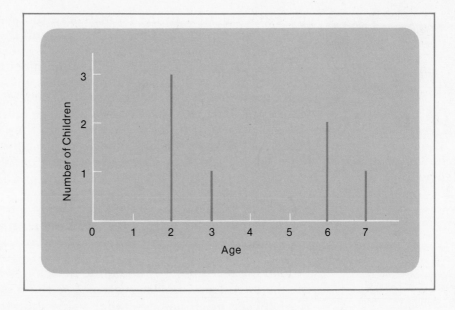

point that separates the scores into two equal groups (3 of the children are under 3 years old, and 3 of the children are over 3 years old), and that the midrange is situated at the point midway between the extreme scores. But at what point is the mean situated?

We can visualize the distribution of the children's ages by placing one disk for each child on a slab of cardboard (see Chart 2-12). If we set this slab on the edge of a fulcrum and

chart 2-12

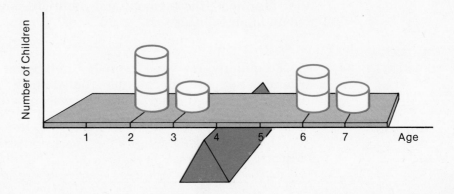

Illustration of a balance point.

move the fulcrum to the point where the slab doesn't tip to either the right or the left, we have found the balance point. *This point is always the* mean *of the distribution.*

But why is this true? You may remember the "playground problem" from childhood. If three friends who are all about the same weight want to use the seesaw, one child sitting on one side of the slide can balance two children on the other side by sitting twice as far from the center as they do. The seesaw balances because the distance from its center to the single child *equals* the sum of the distances from the center to the two children on the other side. The same principle works here:

(A point is a balance point if the sum of the distances to the right of the point *equals* the sum of the distances to the left.)

(Taking advantage of algebraic notation, we can symbolize the distance from an observation X to the mean $\overline{X}$ as $(X - \overline{X})$. If the observation has a smaller value than the mean (if X is less than $\overline{X}$), then the distance will be negative. If the observation has a larger value than the mean, then the distance will be positive. If the mean is to be the balance point, then the sum of the negative distances must *equal* the sum of the positive distances. This, in turn, implies that the sum of all distances, positive plus negative, must equal zero.)

(Symbolically, $\overline{X}$ will be a balance point if $\Sigma(X - \overline{X}) = 0$.)

Chart 2-13 lists the distances between each child's age and the mean age $\overline{X} = 4$. The sum of the negative distances is -7, and the sum of the positive distances is $+7$. Therefore, $\Sigma(X - \overline{X}) = 0$, and the mean is actually the balance point. You can verify that the mode, the median, and the midrange of the age distribution all fail this test.

the least squares criterion

Frequently, we have to predict individual scores in a distribution. Suppose, for example, that we had wanted to predict the age of the first child who is to be seen by the pediatrician.

chart 2-13

Age X	2	2	2	3	6	6	7
Distance from Mean $(X - \overline{X})$	-2	-2	-2	-1	$+2$	$+2$	$+3$

(We don't know the order of the children's arrival; we only know that their ages are 2, 2, 2, 3, 6, 6, and 7 years, respectively).

If our prediction is "2 years old" (the modal age — we stand the best chance of being correct) because more children are 2 years old than any other age. (*So our predictor is the mode of the distribution if our strategy is to maximize our chances of being correct.*)

But this strategy does not take into account the *size* of the prediction error we can make. If we predict "2 years old" and the 7 year old is called into the pediatrician's office next, then we've made a prediction error of 5 years. If instead our prediction is "4.5 years old" — the midrange of the age distribution — then the maximum prediction error we can make is 2.5 years. (*So our predictor is the midrange if our strategy is to minimize the maximum possible prediction error.*)

(Although the strategies leading to the choice of the mode and midrange as predictors have merit, the most widely employed prediction strategy leads us to choose the mean as the predictor. This strategy is to choose the predictor that satisfies the *least squares criterion.*

> **LEAST SQUARES CRITERION**   The best predictor of an individual score in a distribution is the predictor that keeps the sum of the possible squared prediction errors to a minimum.)

(Of all the single values that can be used as predictors (for example, the mode, midrange, median, or mean), the mean is the only value that *always* satisfies the least squares criterion.) We can see how this works by studying Chart 2-14.

Column (1) lists the individual scores in the age distribution. Column (2) assumes that the mean ($\overline{X} = 4$ years old) is chosen as the predictor. Column (3) lists the possible prediction errors. For example, if we predict "4 years old" and the next child called into the pediatrician's office is 2 years old, then we have made a prediction error of -2 years (the actual score minus the predicted value for the score). If we predict "4 years old" and the next child is 7 years old, we have made a prediction error of 3 years. Column (4) lists the square of the prediction errors that we can make using the mean as the predictor. These sum to 30; that is, $\Sigma(X - \overline{X})^2 = 30$.

chart 2-14

(1)	(2)	(3)	(4)
		PREDICTION ERROR	SQUARED PREDICTION ERROR
SCORES	PREDICTOR		
X	$\bar{X}$	$X - \bar{X}$	$(X - \bar{X})^2$
2	4	-2	4
2	4	-2	4
2	4	-2	4
3	4	-1	1
6	4	$+2$	4
6	4	$+2$	4
7	4	$+3$	9
			$\Sigma(X - \bar{X})^2 = 30$

If we had chosen the median as the predictor instead, we would enter the median of 3 years old in the entries in Column (2) and we would find that the quantity $\Sigma(X - \text{Median})^2 = 37$. If we had selected the mode as the predictor, we would obtain $\Sigma(X - \text{Mode})^2 = 58$. Finally, if we had chosen the midrange as the predictor, we would find that $\Sigma(X - \text{Midrange})^2 = 31.75$. Therefore, of the four types of averages, the mean is the best predictor on the basis of the least squares criterion: The mean keeps the sum of the possible squared prediction errors to a minimum.

The least squares criterion is satisfied by choosing the value k that keeps the quantity $\Sigma(X - k)^2$ as small as possible. It will always be true that the quantity $\Sigma(X - k)^2$ is at a minimum when $k = \bar{X}$.

EXERCISES

2-14   A fisherman weighing his catch for the morning recorded the following weights in pounds:

2.3, 1.4, 4.2, 2.0, 5.1

Letting X denote the weight of a fish, determine the value of
(a) ΣX (b) n (c) $\bar{X}$

2-15   (*Continuation of Exercise 2-14*) A second fisherman caught four fish, weighing an average of $\bar{X} = 3.5$ pounds. Which fisherman had the greater catch in terms of pounds of fish?

2-16 Based on the grading system A = 4 points, B = 3 points, C = 2 points, D = 1 point, F = 0 points, compute the mean of the final grade points in a class in which the distribution of final grades is

GRADE	POINTS X	PROPORTION OF STUDENTS $p(X)$
A	4	0.24
B	3	0.32
C	2	0.21
D	1	0.16
F	0	0.07

2-17 Referring to Chart 2-3, page 30, calculate the mean number of strokes taken to play the 15th hole.

2-18 Referring to Chart 1-9 (page 20), calculate the mean of the minimum ages to marry for males.

2-19 Referring to the data in Exercise 1-6 (page 15), calculate the mean number of syllables per word.

2-20 If you rolled a pair of dice over and over and recorded the result of each toss, you would expect the results to produce the following relative frequency distribution:

RESULT	PROPORTION OF ROLLS
2	1/36
3	2/36
4	3/36
5	4/36
6	5/36
7	6/36
8	5/36
9	4/36
10	3/36
11	2/36
12	1/36
	36/36

(a) Calculate the mean of this distribution.
(b) Determine the median, mode, and midrange of the distribution.

2-21   This is a frequency distribution of the *heights* of the college basketball players whose teams participated in a holiday tournament:

CLASS INTERVAL (Height in Inches)	FREQUENCY (Number of Players)
66 to 70	3
70 to 74	10
74 to 78	20
78 to 82	15
	48

Use Equation (2-2) to compute the mean height of these players. To do so, let X denote the midpoint of each class interval. For example, the midpoint of the interval 66 to 70 is 68, the value midway between 66 and 70.

*2-22   (a) Referring to your histogram of the relative frequency distribution of annual unemployment rates, 1948–1974, for Exercise 1-13 (page 22), use Equation 2-3) to calculate the mean of this distribution. Let X denote the midpoint of each class interval (for example, 2.5, 3.5, and so on).

(b) Comparing your result for $\overline{X}$ in (a) with the value of $\overline{X}$ you calculated in Exercise 2-8 from the definitional equation $\overline{X} = \Sigma X/n$, you notice a slight discrepancy. Why is some discrepancy to be expected? In other words, why does the use of a relative frequency equation (or, for that matter, a frequency equation) on *data grouped into class intervals* provide only an *approximate* value for $\overline{X}$?

2-23   A pretest administered to four newly enrolled students in the Reading Acceleration Institute produced the following reading scores, each measuring the number of words per minute (*wpm*):

STUDENT	1	2	3	4	
WPM X	270	340	195	395	$\Sigma X = 1,200$
$X - \overline{X}$	___	___	___	___	

(a) Calculate $\overline{X}$.

(b) For each student, calculate $X - \overline{X}$ (the deviation between the student's reading score and the group mean).

(c) Calculate $\Sigma(X - \overline{X})$, the sum of the deviations obtained in (b).

(d) Would $\Sigma(X - \overline{X})$ have the same value no matter what reading scores were achieved by the students? Explain your answer.

2-24 Five disks of equal weight are placed on a 10-foot beam in the positions illustrated in the following diagram:

No. of Feet from Left End of Beam: 0 1 2 3 4 5 6 7 8 9 10

(a) Where would you place a fulcrum ▲ under the beam so that it would balance (so that the beam would not tip over in either direction)?

(b) Show that neither the median nor the midrange of these distances on the beam is the balance point of the distribution. In other words, show that neither $\Sigma(X - \text{Median})$ nor $\Sigma(X - \text{Midrange})$ equals 0.

2-25 Consider a variable X whose observations are the four numbers, $[1, 3, 3, 9]$. Which of the following quantities would be the smallest and why?

(a) $\Sigma(X - 3)^2$ (b) $\Sigma(X - 4)^2$ (c) $\Sigma(X - 5)^2$

2-26 Refer to Chart 2-19, page 55. Suppose that you were asked to predict how many job offers one business graduate from the group had received. What prediction would you make if you wished to:

(a) Maximize your chances of being correct?

(b) Minimize the maximum prediction error you could make?

(c) Minimize the sum of the possible squared prediction errors?

2.3 comparison of averages: the effect of shape

(As we have noted, knowing whether a distribution has a few unusually high or low scores is important in determining which type of average to report. In more general terms, we can say that in choosing an average we should pay attention to the *shape of the distribution*.)

(The shape of a distribution can be classified broadly as either *symmetric* or *skewed*.)

symmetric
distributions

(A *symmetric distribution* is a distribution that can be divided into two identical halves—one half a mirror image of the other. The center of a symmetric distribution is called the *point of symmetry*. If you "fold" the distribution over at the point of symmetry, the two parts coincide.)

**chart 2-15
Symmetric
Distributions**

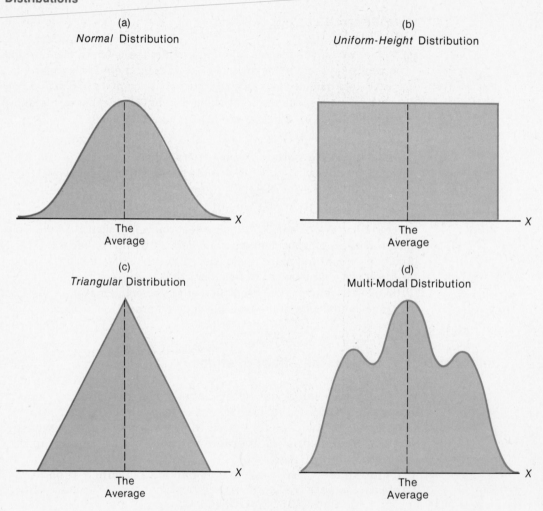

(a)
Normal Distribution

(b)
Uniform-Height Distribution

The
Average

The
Average

(c)
Triangular Distribution

(d)
Multi-Modal Distribution

The
Average

The
Average

(In a symmetric distribution, the point of symmetry is, at once, the mean, median, midrange, and mode (if one exists) of the distribution. In short, we may speak of the "average" of a symmetric distribution without specifying which average, because each average has the same value.)

Chart 2-15 depicts several examples of symmetric distributions that we will encounter in later chapters. Note that the point of symmetry, labeled "the average," in each graph:

1. Is midway between the lowest and the highest score, making it the *midrange*.
2. Divides the area under the frequency curve into two equal halves (so that half of the scores fall on either side of the average), making it the *median*.
3. Is the balance point of the distribution, making it the *mean*.
4. Is the point of maximum frequency in distributions (a), (c), and (d), making it the *mode*.

The bell-shaped curve of distribution (a) is a special form of symmetric distribution called the *normal distribution*, which we will examine in Chapter 4. Distribution (b) is called a *rectangular* or a *uniform-height* distribution. Although it is symmetric, this distribution has no modal value because all scores occur with equal frequency. We will encounter the *uniform-height* distribution as well as the *triangular* distribution (c) and the *multi-modal* distribution (d) in sampling experiments in Chapter 9.

skewed
distributions

A *skewed distribution* is a frequency distribution that is asymmetric. Chart 2-16 illustrates several skewed curves. Distributions (a) and (b) are said to be *skewed to the right*; that is, the longer tail extends to the right. When a distribution is skewed to the right, it contains a larger number of relatively low scores (the mode is a relatively low score) and a few extremely high scores.

When a distribution is skewed to the right, the modal value is generally smaller than the median value, which, in turn, is smaller than the mean value.

Distribution (b) is so sharply skewed to the right that the mode is the lowest value in the distribution.

In contrast, distributions (c) and (d) are said to be *skewed to the left,* revealing a large number of relatively high scores and a few extremely low scores.

chart 2-16
Skewed
Distributions

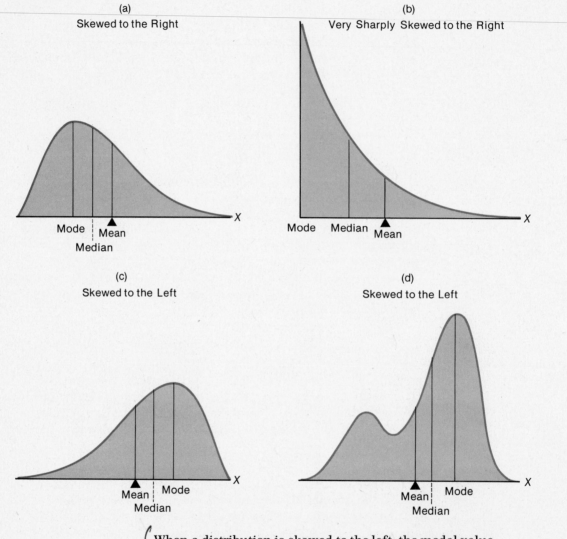

(a)
Skewed to the Right

(b)
Very Sharply Skewed to the Right

Mode ┊ Mean
Median

Mode   Median
Mean

(c)
Skewed to the Left

(d)
Skewed to the Left

Mean ┊ Mode
Median

Mean ┊ Mode
Median

⎛ When a distribution is skewed to the left, the modal value
is generally larger than the median, which, in turn, is
larger than the mean. ⎠

⎛ When a distribution is skewed, judgment must be exer-
cised in selecting an average to represent the distribution.
Let's consider a specific problem.⎠

a practical
problem:
selecting a career

The unemployment rate is so high today that many students are choosing courses of study based on their employment potential. One undergraduate had narrowed her choice of a major field to political science or psychology. The placement office gave her some summary information on the number of job offers received by recent college graduates in these two fields. This information is presented in Chart 2-17.

chart 2-17
Average Number
of Job Offers Per
Student by Field

	MEAN	MEDIAN	MODE	MIDRANGE
Political Science	2.3	1	0	4
Psychology	1.7	1	1	3

Although she could obtain no other information, the student was able to prepare the rough sketches shown in Chart 2-18 of the distribution of job offers in each field. In both dis-

chart 2-18

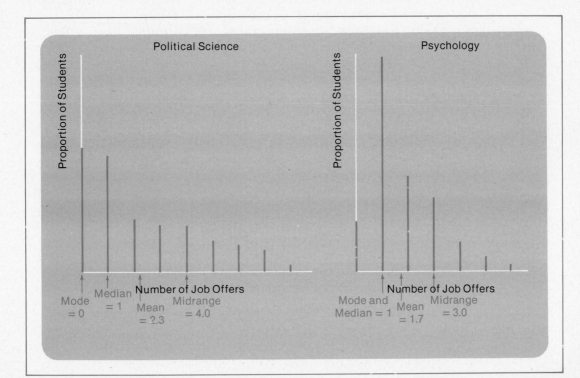

tributions, the median is less than the mean, which, in turn, is less than the midrange. This implies that a few students in each field received numerous job offers but most students in each field received only a few. Both distributions, therefore, are skewed to the right.

After examining the averages given in Chart 2-17, the student realized that these measures did not unanimously favor one field over the other. Compared to the distribution of job offers to psychology graduates, the mean number of job offers to political science graduates and the midrange of these job offers are higher, but the mode is lower. However, realizing that she was dealing with skewed distributions, the student decided to base her comparison on the *median* values.

(If the various types of averages differ substantially in value, the *median* is the best measure to consider for most purposes.)

No matter how asymmetric a distribution is, the median always divides the lower half of the observations from the upper half. Both the mean and, more noticeably, the midrange are affected by extreme observations. Therefore, if the distribution is skewed, these measures will be pulled away from the center in the direction of the skew. In our present example, the many job offers (up to 8)[1] received by at least one graduate have pulled the mean and midrange way up. Unless the student considers herself exceptionally capable, she will not want to base her comparision on the mean or the midrange.

Unfortunately, the medians in political science and psychology are identical. So the student cannot base her choice between the two fields on a comparison of the medians.

When she looked at the mode, she found it disconcerting that the most frequent number of job offers received by political science students was 0. Among the psychology graduates, at least more were receiving *one* job offer than any other number. The conservative choice of field, therefore, would seem to be psychology. Although the (presumably) most qualified political science majors received more job offers (8) than the most qualified psychology majors (6) and although 50 percent of the

[1] In political science, the midrange value is 4 and the minimum value is 0, so the maximum value must be 8 (4 is midway between 0 and 8). In psychology, the maximum number of job offers received was 6 (assuming that at least one psychology graduate has received 0 offers).

	MEAN	MEDIAN	MODE	MIDRANGE
Business Administration	4.6	5	6	4.0

students in both fields received at least one job offer, the risk of obtaining no job offers at all was higher in the field of political science.

This illustration has an epilogue. After checking with the placement service again, the student chose to enter the business administration program. It is easy to see why by looking at the summary statistics for recent business administration graduates given in Chart 2-19. Obviously, these students received a larger number of job offers than graduates in either of the other two fields. Moreover, the leftward skew of the distribution implies that there were relatively few business graduates who didn't receive at least three job offers, as Chart 2-20 shows.

chart 2-20
Business
Administration

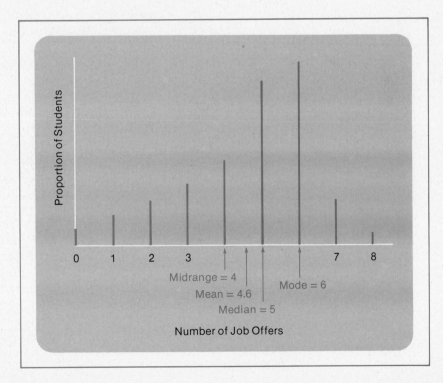

EXERCISES
2-27

A distribution contains a few extremely low scores and many relatively high scores.
(a) Is the shape of the distribution skewed to the right or to the left?
(b) Would a frequency curve of the distribution have a long tail extending to the right or to the left?
(c) Would the mean of the distribution be likely to be smaller or larger in value than the median of the distribution? Why?
(d) Would the mode of the distribution be likely to be smaller or larger in value than the median of the distribution? Why?

2-28

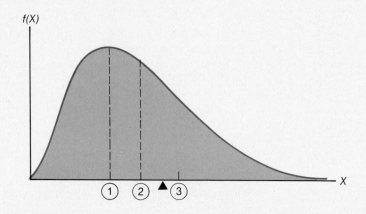

The fulcrum ▲ in this frequency curve indicates the location of the mean of the distribution.
(a) Points 1, 2, and 3 represent the three other types of averages. Label them appropriately.
(b) Characterize the shape of this distribution.

2-29

Is the relative frequency distribution of grades in Exercise 2-16 (page 47) (a) symmetric, (b) skewed to the right, or (c) skewed to the left? Sketch a line graph of this distribution to check your answer.

2-30

(a) Is the relative frequency distribution of the results from rolling a pair of dice in Exercise 2-20, page 47, (1) symmetric, (2) skewed to the right, or (3) skewed to the left?
(b) Suppose that you rolled only a single die over and over and that the

results were presented in a relative frequency distribution. Draw a line graph of the distribution you would expect to see and characterize the shape of the distribution.

2-31   (a) Referring to your histogram of the annual unemployment rates, 1948–1974, for Exercise 1-13 (page 22), state whether the distribution is (1) symmetric, (2) skewed to the right, or (3) skewed to the left.

(b) How is the shape of the distribution altered by the inclusion of the 1975 and 1976 unemployment rates (1975: 8.5 percent; 1976: 7.7 percent)?

Caveat: Variation from the Average

Statisticians tell stories of people who almost drowned by wading into a lake with an average depth of 3 feet, drop a fortune in a restaurant where the entrees average less than $5, and wipe out on a ski trail with a grade, or average angle of descent, rated at a mere 15°. These embarrassing incidents

Is variation important in choosing a ski trail?

occur because people sometimes forget that any *average* can provide seriously misleading information unless it is accompanied by some indication of the amount of *variation from the average.*

A lake with an average depth of 3 feet may be much deeper than 3 feet in certain places. A restaurant where the entrees average less than $5 may serve a few inexpensive meals but price most of its more desirable dishes well above $5. A ski trail with a grade of 15° may be extremely flat for most of the course but contain a few precipitous drops.

So it is clear that if we know the average level of some variable we do not necessarily know all we need to know about that variable. It is wise to determine some measure of the variation or spread of the observations we are examining as well. We will discuss measures of variation in Chapter 3.

COMING
TO
TERMS

Averages   Measures of the central or typical values of a distribution. When it is unqualified, "the average" refers to the mean.

Balance Point   A value has the property of being a balance point if the sum of the deviations between each score and that value equals 0. The mean is always the balance point of a distribution. The balance point is also referred to as the *fulcrum* or *center of gravity*.

Least Squares Criterion   The criterion that requires predictions to be made in a manner that minimizes the sum of the squared prediction errors.

Mean or Arithmetic Mean   The sum of the observations in a distribution divided by the number of observations in that distribution. The mean is always the balance point of a distribution.

Median   The value of the middle score in the distribution. If the number of observations is odd, then the median is the middle observation. If the number of observations is even, then the median is the average of the two middle observations.

Midrange   The value midway between the smallest and largest observations of a distribution. The midrange is calculated by adding the smallest and largest observations together and dividing the sum by 2.

Modal Interval   The class interval that includes the largest number of observations. Used when the data have been grouped into a frequency distribution.

Mode   The value that occurs more frequently than any other value in a distribution.

Proportion   The number of observations within a certain category divided by the total number of observations. The proportion of 1's (or yes's) is the mean of a *yes–no* variable.

Sigma Σ   The summation sign. The mathematical notation for the instruction, "sum up."

Skewed Distribution   A distribution in which the observations are concentrated at one end (tail) of the distribution, so that the most typical scores are relatively low (skewed to the right) or relatively high (skewed to the left).

Symmetric Distribution   A distribution that can be divided in two halves— one half the mirror image of the other. The center of a symmetric distribution is called the *point of symmetry* and simultaneously represents the mean, median, mode, and midrange of the distribution.

ISSUES FOR ANALYSIS

1. The following exhibit presents the actual averages (means) of the base (nine-month) salaries of U.S. business administration faculty for 1976–1977. (All figures are in thousands of dollars.)

Business Schools in U.S. Public Universities

RANK	MEAN	PROPORTION OF FACULTY
Professor	24.9	0.28
Associate Professor	19.7	0.28
Assistant Professor	16.5	0.34
Instructor	12.6	0.10

Business Schools in U.S. Private Universities

RANK	MEAN	PROPORTION OF FACULTY
Professor	25.8	0.27
Associate Professor	19.9	0.30
Assistant Professor	16.7	0.35
Instructor	13.8	0.08

Did business school faculties tend to earn higher base salaries in public or private institutions? Include in your analysis:

(a) A rank-by-rank comparison.

(b) An overall comparison of business salaries in public and private institutions.

(c) A discussion of the limitations of the data provided.

2. The following data (in billions of dollars) pertain to the annual budget of the U.S. government for 1970–1977. An excess of outlays over receipts constitutes a deficit.

YEAR	RECEIPTS	OUTLAYS	SURPLUS OR DEFICIT (−)
1970	193.7	196.6	−2.9
1971	188.4	211.4	−23.0
1972	208.6	232.0	−23.4
1973	232.2	247.1	−14.9
1974	264.9	269.6	−4.7
1975	281.0	326.1	−45.1
1976	300.0	366.5	−66.5
1977 (estimate)	354.0	411.2	−57.2

(a) Verify that the mean of the differences (receipts − outlays equals the difference between the means (mean receipts − mean outlays).

(b) In relation to receipts, was the deficit larger during the *last* four years or during the *first* four years of this time period?

3. The summary results of the Verbal Aptitude Test (VAT) administered to a large group of eight-year-old children in 1955 are provided in the exhibit below.

CATEGORY OF CHILD	MEDIAN VAT SCORE
1. Neither mother nor father had a college degree.	63
2. Mother did but father did not have a college degree.	72
3. Father did but mother did not have a college degree.	64
4. Both mother and father had college degrees.	73

Does the evidence provided indicate that the educational status of a child's parents had some influence on the child's VAT score? In your analysis, consider both the mother's and the father's status.

Heterogeneous heights (in contrast to homogeneous heights).

Homogeneous, consistent, predictable, and stable are a few of the adjectives used to describe distributions in which most of the scores have closely similar values.

Heterogeneous, inconsistent, unpredictable, and volatile, in contrast, are adjectives used to characterize distributions in which the scores are widely dissimilar.

Precise numeric measures of the extent to which scores in a distribution vary are called *measures of variation or spread*. Just as there are several types of averages of a distribution, there are several ways to measure variation in a distribution. We will begin this chapter by introducing three of the most common measures of variation: the *range*, the *interquartile range*, and the standard deviation. Each of these measures of variation is a logical companion to one of the types of averages we discussed in Chapter 2, and each shares the virtues and defects of its companion average.

	Batting Average Player A	Batting Average Player B
1970	.297	.285
1971	.296	.314
1972	.293	.267
1973	.300	.322
1974	.295	.358
1975	.298	.291
1976	.290	.246
1977	.292	.257
1978	.297	.333
	Predicted for 1979: ~ .290–.300	Predicted for 1979: ?
	Consistent Performance	Inconsistent Performance

variation

3.1 measures of variation

the
range

The measure of variation associated with the midrange of a distribution is the *range*.

RANGE   The distance from the smallest to the largest score.

After checking the lengths of 100 boards to be used in "do-it-yourself" bookcase kits, the manufacturer's inspector recorded a discrepancy of 3 in. between the shortest and the longest shelf. The "4-ft" boards varied in length from 3 ft 10 in. to 4 ft 1 in., a range of 3 in. A kit containing boards with this large a range of lengths could only lead to the disaster illustrated in Chart 3-1.

Like its logical companion the midrange, the great advantage of the range is that it is simple to calculate: All we need to do is determine the lowest and the highest scores in the distribution. But this "advantage" is also the major defect of the range, because it reveals nothing about the variation that occurs among the scores between the two extremes. In our example, it is important to know if only a few boards vary from

chart 3-1

the standard 4-ft length or if the board lengths are highly ir-regular in general. If only a few boards vary from the standard, they can be replaced. But if most of the boards are irregular, then something must be done to improve the cutting accuracy of the machinery.

The range therefore is a poor measure of how much *most* of the observations differ from one another—especially when we are dealing with a large set of data. After all, there is "one in every crowd," and it only takes one to make the range exces-sively wide.

How can we prevent extreme scores from producing an exaggerated picture of the spread that exists among most of the scores? One solution is to discard a certain portion of the highest and lowest scores and to report the range among the remaining scores. The *interquartile range* is a measure of variation that does just that.

the inter-quartile range

The interquartile range is based on measures called *per-centiles.* For example, the median is the 50th percentile, be-cause it divides the scores in a distribution into the lower 50 percent and the upper 50 percent. Other percentiles are defined similarly. The 25th percentile (or the *lower quartile*) divides

chart 3-2

Family Rank:	1	2	3	4	5	6	7	8	9	10	11	12
Number of Hours:	13	38	62	66	78	91	95	115	126	132	155	203

the distribution between the lower 25 percent and the upper 75 percent of the scores. The 75th percentile (the *upper quartile*) divides the lower 75 percent and the upper 25 percent of the scores.

(**INTERQUARTILE RANGE**   The distance from the 25th to the 75th percentile, usually reported as the distance that spans the *middle* 50 percent of the observations.)

(The interquartile range is the logical companion to the median. The median is the value of the middle score; the interquartile range represents the interval that contains the middle 50 percent of the scores. The lowest and the highest 25 percent of the scores lie outside the interquartile range)

Chart 3-2 displays the number of hours of television viewed by 12 individual families during a selected month. To calculate the interquartile range, we must locate the 25th and the 75th percentiles. Because the distribution has 12 scores, the 25th percentile must divide the distribution between the lower 3 and the upper 9 scores (25 percent of 12 is 3; 75 percent of 12 is 9). As shown in Chart 3-3, any value between 62 hours and

chart 3-3

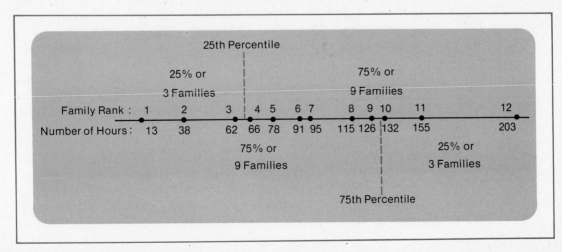

66 hours would divide the distribution this way. By convention, we call the value midway between these two scores the 25th percentile.

$$25\text{th percentile} = 64 \text{ hours}$$

The same logic allows us to identify the 75th percentile as the value midway between the 9th and 10th observations, or the value midway between 126 and 132 hours.

$$75\text{th percentile} = 129 \text{ hours}$$

The interquartile range is therefore the distance from 64 hours to 129 hours. If the families were arrayed in the order of the amount of television they watched, then the middle 50 percent of these families viewed between 64 and 129 hours of TV during the month.

This example illustrates how the 25th and the 75th percentiles and the interquartile range are calculated under the most straightforward conditions: when the number of scores can be divided evenly by 4. When the number of observations is not evenly divisible by 4, no value *exactly* separates the lower 25 percent from the upper 75 percent (or the lower 75 percent from the upper 25 percent). In such cases, we generally employ the following rule:

> The 25th percentile is the *smallest* score that is associated with a cumulative frequency of *at least* 25 percent. This score, together with all smaller scores, therefore comprises *at least* 25 percent of the scores in the distribution.

To illustrate, let's add to our list a 13th family with a score of 160 hours. The new array is shown in Chart 3-4. Following our rule, the 25th percentile is 66 hours (the 4th score), because the 4th score is the smallest score that has a cumulative frequency of *at least* 25 percent (actually $4/13 = 31$ percent). Note that the 3rd score, 62 hours, does not satisfy our rule, because 3 scores make up less than 25 percent of the 13 scores.

Similarly, the 75th percentile is 132 hours (the 10th score). The first 10 scores actually constitute 77 percent of the distribution $(10/13)$.

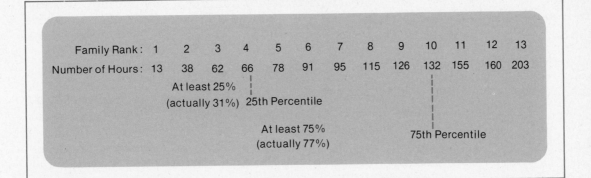

chart 3-4
Array of TV Hours Viewed by 13 Families

So the middle half (precisely, 77% − 31% = 46%) of the families viewed between 66 hours and 132 hours of television during the month.

(Compared to the range, the virtue of the interquartile range is that it excludes extreme observations. However, we do not have to discard the lowest and highest *quarters* of the distribution. Sometimes the interval from the 10th to the 90th percentile is reported—a measure called the *interdecile range.)* Another alternative is employed in judging Olympic diving events. Each of 7 judges gives a performance a numeric rating, and the lowest and the highest scores are then discarded. This amounts to excluding the lowest and the highest 14 percent (1/7) of the data.

(The interquartile range is a logical companion to the median. When instead we report the mean as the average, we report its companion measure of variation, the *standard deviation.)*

the standard deviation

(Although the standard deviation is the most frequently reported measure of variation, it is also the least intuitive. It is called the "standard" (average, usual) "deviation" (distance, difference), because it is a kind of "average" of the distances between the individual observations in a distribution and the mean of the distribution. The following example illustrates the calculation and meaning of the standard deviation.)

The water department of a city was monitoring the daily water usage of some of its largest customers. In particular, the city hospital seemed to show large fluctuations in water usage from day to day. The first row of Chart 3-5 presents the hospital's daily water usage in thousands of gallons over the past week.

The hospital's mean water usage was $140,000/7 = 20,000$ gallons per day, with a range of 10,000 to 36,000 gallons a day, suggesting considerable variation. The deviation (difference) between each day's water usage and the mean water usage over the week is given in the second row of Chart 3-5. The water department wanted a measure of variation in daily water usage to use in summarizing the day-to-day fluctuations from the weekly mean.

The department could not simply add together the daily deviations $X - \overline{X}$ to form a total or mean for the week. Because the mean $\overline{X}$ is the balance point of the scores, the sum of the deviations about the mean, $\Sigma(X - \overline{X})$, must be 0, no matter how large or how small the deviations might be. Accordingly, in the Weekly Total column of Chart 3-5, $\Sigma(X - \overline{X}) = 0$.

(A solution to this problem is to represent each deviation from the mean by the square of the deviation $(X - \overline{X})^2$. The square of either a positive or a negative quantity is positive. The average (mean) of the squared deviations then can be used as a measure of variation.)

The squared daily deviations in water usage appear in the third row of Chart 3-5. The sum of these squared deviations, indicated in the Weekly Total column, is 434,000 square gallons. Dividing this sum by 7 gives us the mean of the squared deviations, or 62,000 square gallons.

This measure of variation is important enough to be given a name. It is called the *variance* of a distribution.

chart 3-5

								WEEKLY TOTAL
Daily Water Usage (X):	14	24	20	36	15	21	10	$\Sigma X = 140$
Deviation ($X - \overline{X}$):	−6	4	0	16	−5	1	−10	$\Sigma(X - \overline{X}) = 0$
Squared Deviation ($X - \overline{X}$)²:	36	16	0	256	25	1	100	$\Sigma(X - \overline{X})^2 = 434$

(VARIANCE For each observation, calculate the
square of the deviation between the observation
and the mean. The *variance* is the average
(mean) of these squared deviations.)

The variance of daily water usage scores for the city hospital is 62,000 square gallons. This seems to be an enormous
amount of variation, but then what is a square gallon? The
units of variance are square gallons, because each daily deviation is squared. Squaring also makes the amount of variation
appear to be extremely large. So variance is not easy to interpret as a measure of variation. It appears that in solving one
arithmetic difficulty by squaring, we have created another.

But this problem can also be resolved. (If the variance is
measured in units of square gallons, then the *square root* of
the variance will be measured in units of gallons.) In our example, the square root of 62,000 square gallons is approximately 249 gallons. (We have just arrived at the measure called
the *standard deviation*.)

(STANDARD DEVIATION The square root of the
variance; literally, the square root of the average of
the squared deviations from the mean. The standard
deviation can be described as a measure of the
"average" spread of the observations about
the mean.)

So we can say that the city hospital's daily water usage
fluctuated by an "average" of 249 gallons from the hospital's
mean water usage over the week.

(Let's review the steps involved in the calculation of the
standard deviation:

1. For each observation, calculate the deviation: $X - \overline{X}$.
2. Square each deviation: $(X - \overline{X})^2$.
3. Find the mean of the squared deviations: $\dfrac{\Sigma(X - \overline{X})^2}{n}$. This

   is the variance.

4. Take the square root of the variance: $\sqrt{\dfrac{\Sigma(X - \overline{X})^2}{n}}$.)

Using the symbol s to denote standard deviation and s^2 to denote variance, we define these measures algebraically as[1]

$$s = \sqrt{\frac{\Sigma(X - \overline{X})^2}{n}} \qquad s^2 = \frac{\Sigma(X - \overline{X})^2}{n} \qquad (3\text{-}1)$$

We can remember the equation for the standard deviation by the letters RMSD, which stand for

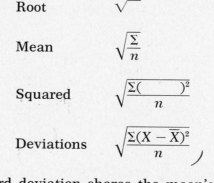

Root $\sqrt{}$

Mean $\sqrt{\dfrac{\Sigma}{n}}$

Squared $\sqrt{\dfrac{\Sigma()^2}{n}}$

Deviations $\sqrt{\dfrac{\Sigma(X - \overline{X})^2}{n}}$

The standard deviation shares the mean's advantage of using information from all the observations. However, this measure of variation also shares the mean's two major limitations:

1. Sometimes we cannot calculate the mean. Obviously, without a mean, we cannot calculate deviations from the mean and the standard deviation cannot be determined.
2. When extreme values in a highly skewed distribution make the mean unrepresentative of the majority of the observations, the deviations from these points to the mean will be large and the standard deviation will overstate the amount of variation among the scores in the distribution.

In both circumstances, it is common to report the median

[1] The variance and the standard deviation are frequently defined by using $n - 1$ in place of n in the denominator. The use of $n - 1$ is related to the role these measures play in statistical inference. We will discuss this point in more detail in Chapter 10.

and a companion measure of variation about the median, such as the interquartile range.)

(Each of the three measures of variation we have discussed here has its advantages and disadvantages. As a general rule, once we have chosen the appropriate average, determining which measure of variation we should use follows logically.)

EXERCISES

3-1   The pulse rates of eight contestants in the 100-meter dash, checked immediately after the race, were

99, 78, 96, 89, 98, 85, 107, 102

Determine:
(a) The range.
(b) The 25th percentile.
(c) The 75th percentile.
(d) The interquartile range.

3-2   Referring to the lengths of service of U.S. Supreme Court justices in Chart 1-3 (page 12), determine and interpret the value of:
(a) The range.
(b) The 25th percentile.
(c) The 75th percentile.
(d) The interquartile range.

3-3   Report the range of the distribution of the household electricity bills listed in Chart 2-5 (page 32). Why is the range a misleading measure of the amount of variation among most of the scores in this distribution?

3-4   The following information concerns the distribution of the number of new mopeds sold last year by salespersons in a dealer network:

10th percentile: 3 mopeds sold
25th percentile: 8 mopeds sold
50th percentile: 19 mopeds sold
75th percentile: 28 mopeds sold
90th percentile: 41 mopeds sold

(a) Determine and interpret the value of the interquartile range of this distribution.
(b) Determine and interpret the value of the interdecile range of this distribution.

(c) The value positioned midway between the 25th and the 75th percentiles is called the *midquartile*. Is the midquartile necessarily equal to the median?

(d) Under what assumption about the shape of a distribution would you have to answer the question in (c) affirmatively?

*3-5   Referring to your array for Exercise 1-8 (page 16), find the interquartile range of the annual unemployment rates for (a) 1948–1974; (b) 1948–1975; (c) 1947–1976. (Remember: 1975, 8.5 percent; 1976, 7.7 percent.)

3-6   (a) Use Equation (3-1) to calculate the standard deviation of the scores [1, 2, 3, 4, 5]. Report the result to three decimal places.

(b) Use Equation (3-1) to calculate the standard deviation of the scores [2, 4, 6, 8, 10]. Report the result to three decimal places.

(c) Interpret the results obtained in (a) and (b) in a comparative sense.

3-7   We have said that the standard deviation measures the "average" distance between the observations in a distribution and the mean of the distribution. *Literally,* the average (mean) distance between the scores and the mean would be measured as $\Sigma(X - \overline{X})/n$. Why is this measure useless as a measure of variation? How does squaring each deviation $(X - \overline{X})$ solve this problem?

3-8   What is the relationship between the *standard deviation* and the *variance* of a distribution? What advantage do we gain by reporting the standard deviation instead of the variance?

3-9   Which one of the two frequency distributions graphed below has the smaller standard deviation?

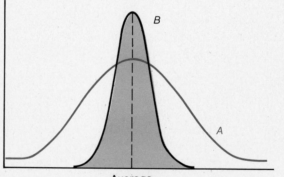

Average

3.2 computational equations for the standard deviation

The standard deviation is an important statistical measure, not only because it is the most frequently reported measure of variation in a distribution but also because the standard deviation is a component of so many other statistical measures, such as the correlation coefficient (Chapter 6) and the standard error of the mean (Chapter 9), as well as the coefficient of variation and the Z score — two measures that will be introduced later in this chapter.

Therefore, we will wish to calculate the standard deviation of a distribution on numerous occasions. Unfortunately, in many cases, the definitional equation, Equation (3-1), is not an efficient or even feasible method of calculating the standard deviation. For this reason, it is useful to know the following computational equations.

computation of *s*
from the basic
data

Whereas *s* is defined by Equation (3-1)

$$s = \sqrt{\frac{\Sigma(X - \overline{X})^2}{n}} \qquad (3\text{-}1)$$

by taking advantage of the algebraic fact that $\Sigma(X - \overline{X})^2 = \Sigma X^2 - n\overline{X}^2$, we obtain the more efficient computation equation

$$s = \sqrt{\frac{\Sigma X^2}{n} - \overline{X}^2} \qquad (3\text{-}2)$$

If we use Equation (3-2), we avoid computing the deviation of each observation from the mean, $(X - \overline{X})$, which is a time-consuming process. Specifically, Equation (3-2) tells us to:

1. Square each observation on X, obtaining X^2.
2. Find the mean of these squares $\Sigma X^2/n$.
3. Subtract the square of the mean, $\overline{X}^2$, from $\Sigma X^2/n$.
4. Take the square root of the remainder.

Chart 3-6 applies this computational equation to the water usage data presented in Chart 3-5.

chart 3-6

								WEEKLY TOTAL
Daily Usage X	14	24	20	36	15	21	10	$\Sigma X = $ 140
Find X^2:	196	576	400	1,296	225	441	100	$\Sigma X^2 = 3{,}234$

Compute $\dfrac{\Sigma X^2}{n}$: $\quad\dfrac{3{,}234}{7} = 462.$

Subtract $\overline{X}^2$: $\quad 462 - \left(\dfrac{140}{7}\right)^2 = 462 - 20^2 = 462 - 400 = 62$ (thousand) square gallons

Take the square root: $\quad \sqrt{62{,}000} \cong 249$ gallons.

X denotes daily water usage (in thousands of gallons).

We can confirm that $s = 249$ gallons is the same result we found using the definition of the standard deviation given in Equation (3-1).

frequency equations for s

When we introduced the mean $\overline{X}$ in Chapter 2, we provided two frequency equations for the calculation of $\overline{X}$ to be used when the data are presented in the form of a frequency distribution or a relative frequency distribution.

These frequency equations for the mean and the analagous equations for the standard deviation are given in Equations (3-3) and (3-4).

THE FREQUENCY EQUATIONS

$$\overline{X} = \frac{\Sigma[X \cdot f(X)]}{n} \qquad s = \sqrt{\frac{\Sigma[X^2 \cdot f(X)]}{n} - \overline{X}^2} \qquad (3\text{-}3)$$

where $f(X)$ denotes the *number* of observations.

THE RELATIVE FREQUENCY EQUATIONS

$$\overline{X} = \Sigma[X \cdot p(X)] \qquad s = \sqrt{\Sigma[X^2 \cdot p(X)] - \overline{X}^2} \qquad (3\text{-}4)$$

where $p(X)$ denotes the *proportion* of observations.

The frequency distribution of the golf scores given in Chart 2-3 is reproduced in Chart 3-7. Because the number of golfers with each score, $f(X)$, is provided instead of the proportion of golfers with each score, $p(X)$, we use Equations (3-3) to compute the standard deviation, as shown in the worksheet in Chart 3-7.

chart 3-7

NUMBER OF STROKES X	NUMBER OF GOLFERS $f(X)$
2 (Eagle)	1
3 (Birdie)	8
4 (Par)	29
5 (Bogie)	63
6 (Double Bogie)	21
7 (Triple Bogie)	3
	$n = \overline{125}$

WORKSHEET

Equations (3-3) $\overline{X} = \dfrac{\Sigma[X \cdot f(X)]}{n}$ $s = \sqrt{\dfrac{\Sigma[X^2 \cdot f(X)]}{n} - \overline{X}^2}$

X	$f(X)$	$[X \cdot f(X)]$	X^2	$[X^2 \cdot f(X)]$
2	1	$2 \cdot 1 = 2$	$2^2 = 4$	$4 \cdot 1 = 4$
3	8	$3 \cdot 8 = 24$	$3^2 = 9$	$9 \cdot 8 = 72$
4	29	$4 \cdot 29 = 116$	$4^2 = 16$	$16 \cdot 29 = 464$
5	63	$5 \cdot 63 = 315$	$5^2 = 25$	$25 \cdot 63 = 1,575$
6	21	$6 \cdot 21 = 126$	$6^2 = 36$	$36 \cdot 21 = 756$
7	3	$7 \cdot 3 = 21$	$7^2 = 49$	$49 \cdot 3 = 147$
		$\Sigma[X \cdot f(X)] = 604$		$\Sigma[X^2 \cdot f(X)] = 3,018$

$\overline{X} = \dfrac{\Sigma[X \cdot f(X)]}{n} = \dfrac{604}{125} = 4.832$, or approximately 4.8 strokes

$s = \sqrt{\dfrac{\Sigma[X^2 \cdot f(X)]}{n} - \overline{X}^2}$

$= \sqrt{\dfrac{3,018}{125} - (4.832)^2}$

$= \sqrt{24.144 - 23.348}$

$= \sqrt{0.796}$

$= 0.892$, or approximately 0.9 strokes

So the scores of the 125 golfers who played the hole varied by an "average" of 0.9 strokes from the mean of 4.8 strokes.

If the distribution of golf scores had been reported as the *relative* frequency distribution given in Chart 3-8, then we would have used Equations (3-4) to compute the standard deviation, as shown in the accompanying worksheet.

chart 3-8

NUMBER OF STROKES X	PROPORTION OF GOLFERS $p(X)$
2 (Eagle)	0.008
3 (Birdie)	0.064
4 (Par)	0.232
5 (Bogie)	0.504
6 (Double Bogie)	0.168
7 (Triple Bogie)	0.024
	1.000

WORKSHEET

Equations (3-4) $\overline{X} = \Sigma[X \cdot p(X)]$ $s = \sqrt{\Sigma[X^2 \cdot p(X)] - \overline{X}^2}$

X	$p(X)$	$[X \cdot p(X)]$	X^2	$[X^2 \cdot p(X)]$
2	0.008	$2 \cdot 0.008 = 0.016$	$2^2 = 4$	$4 \cdot 0.008 = 0.032$
3	0.064	$3 \cdot 0.064 = 0.192$	$3^2 = 9$	$9 \cdot 0.064 = 0.576$
4	0.232	$4 \cdot 0.232 = 0.928$	$4^2 = 16$	$16 \cdot 0.232 = 3.712$
5	0.504	$5 \cdot 0.504 = 2.520$	$5^2 = 25$	$25 \cdot 0.504 = 12.600$
6	0.168	$6 \cdot 0.168 = 1.008$	$6^2 = 36$	$36 \cdot 0.168 = 6.048$
7	0.024	$7 \cdot 0.024 = 0.168$	$7^2 = 49$	$49 \cdot 0.024 = 1.176$
		$\Sigma[X \cdot p(X)] = 4.832$		$\Sigma[X^2 \cdot p(X)] = 24.144$

$\overline{X} = \Sigma[X \cdot p(X)] = 4.832$, or approximately 4.8

$s = \sqrt{\Sigma[X^2 \cdot p(X)] - \overline{X}^2}$

$ = \sqrt{24.144 - (4.832)^2}$

$ = \sqrt{24.144 - 23.348}$

$ = \sqrt{0.796}$

$ = 0.892$, or approximately 0.9 strokes

EXERCISES

3-10   Referring to the data on the number of hours of television viewing in Chart 3-2 (page 65):
(a) Calculate the standard deviation, using the definitional Equation (3-1). Use four decimal places in your computations.
(b) Calculate the standard deviation, using the computational Equation (3-2). Confirm that you obtain the same result that you did in (a).
(c) Interpret the value you obtained for the standard deviation.

3-11   For the observations on a variable X, it is determined that

$$\Sigma(X^2) = 1{,}603 \qquad \Sigma X = 105 \qquad n = 7$$

Calculate s.

*3-12   Referring to the unemployment data given in Exercise 1-8 (page 16):
(a) Calculate the standard deviation of the distribution of annual unemployment rates for 1948–1974. (You should know the value of ΣX from Exercise 2-8 on page 38.) Save the results for $\Sigma(X^2)$.
(b) Would you expect the standard deviation to increase or decrease if the 1975 unemployment rate of 8.5 percent were included in the distribution?
(c) Recompute s after including the unemployment rate for 1975, building on the values you obtained earlier for ΣX and $\Sigma(X^2)$.

3-13   In a previous exercise, you determined that the following grade distribution has a mean of 2.5 points:

GRADE	POINTS X	PROPORTION OF STUDENTS $p(X)$
A	4	0.24
B	3	0.32
C	2	0.21
D	1	0.16
F	0	0.07

Use Equation (3-4) to compute the standard deviation of this distribution. Interpret the value you obtain.

3-14   In a previous exercise, you determined that the following distribution of basketball players' heights has a mean of 75.9 inches:

CLASS INTERVAL (Height in Inches)	MIDPOINT X	FREQUENCY f(X) (Number of Players)
66 to 70	68	3
70 to 74	72	10
74 to 78	76	20
78 to 82	80	15
		48

Use Equation (3-3) to calculate the standard deviation of these players' heights. Interpret the value you obtain.

3.3 comparing variation in two distributions: the coefficient of variation

A manufacturer employs two classes of workers: skilled laborers (machinists) and unskilled laborers (packers). The means and standard deviations of the hourly wages of these employees are presented in Chart 3-9.

We would expect a substantial difference to exist between the average wages paid to these two classes of employees, and we can see that the mean wage of the skilled workers is 5 times greater than the mean wage of the unskilled workers in this company.

We also may wish to compare the variation in the wages of the two classes of employees. Are the wages of the skilled workers more homogeneous (less variable) than the wages of the unskilled workers? Or do the wages of the unskilled workers concentrate more closely about their mean?

One perspective on the issue can be gained from a com-

chart 3-9
Hourly Wages

CLASS	MEAN $\bar{X}$	STANDARD DEVIATION s
Unskilled	$ 3.15	$0.29
Skilled	$15.75	$1.45

parison of the standard deviations of the two distributions of wages. For the unskilled workers, $s = \$0.29$, telling us that the wages of the unskilled workers vary by an "average" of $0.29 about the mean of $3.15. In contrast, $s = \$1.45$ for skilled workers, telling us that the wages of the skilled workers vary by an "average" of $1.45 about the mean of $15.75. Therefore, based on the standard deviations, it appears that the wages of the skilled workers are far more varied than the wages of the unskilled workers.

A different perspective on the relative variability of the scores in the two wage distributions can be gained by comparing the *coefficients of variation.*

COEFFICIENT OF VARIATION   The standard deviation of a distribution divided by the mean of the distribution.

Equation (3-5) defines the coefficient of variation algebraically as

$$\text{Coefficient of Variation} = \frac{s}{\bar{X}} \qquad (3\text{-}5)$$

Chart 3-10 illustrates the calculation of the coefficients of variation for the distributions of hourly wages for skilled and unskilled workers.

The interpretation of a coefficient of variation follows directly from the interpretation of a standard deviation, as shown in Chart 3-11.

So our perspective differs, depending on whether we compare standard deviations or coefficients of variation. The standard deviation provides a measure of the "average" *amount* of variation, expressed in the original units of mea-

chart 3-10

CLASS	COEFFICIENT OF VARIATION
Unskilled	$\dfrac{s}{\bar{X}} = \dfrac{0.29}{3.15} = 0.092 = 9.2\%$
Skilled	$\dfrac{s}{\bar{X}} = \dfrac{1.45}{15.75} = 0.092 = 9.2\%$

chart 3-11

CLASS	STANDARD DEVIATION	COEFFICIENT OF VARIATION
Unskilled	The hourly wage rates of the unskilled workers vary by an "average" of $0.29 about the mean of this distribution.	The hourly wage rates of the unskilled workers vary by an "average" of 9.2% about the mean of this distribution. (In other words, the standard deviation of $0.29 is equal to 9.2% of the mean of $3.15.)
Skilled	The hourly wage rates of the skilled workers vary by an "average" of $1.45 about the mean of this distribution.	The hourly wage rates of the skilled workers vary by an "average" of 9.2% about the mean of this distribution. (In other words, the standard deviation of $1.45 is equal to 9.2% of the mean of $15.75.)

surement (dollars in this example). In contrast, the coefficient of variation provides a measure of the "average" *degree* of variation, expressed as a percentage of the mean.)

Because both of the coefficients of variation in our illustrative example are equal to 9.2 percent, we can report that the wages of the skilled and unskilled workers exhibit the same *degree* of variation. This result implies that the larger *amount* of variation in wages among skilled workers is just proportional to the larger value for the mean wage of this class.

When we wish to compare the *degree* of variation in two distributions with average levels that differ sharply, it is wise to base this comparison on the coefficients of variation rather than on the standard deviations.)

Because the coefficient of variation is a percentage, coefficients of variation can be employed to compare the variability of scores that are expressed in different units of measurement — something that standard deviations and ranges cannot do.) For example, are the heights of females aged 20 through 29 more or less variable than the weights of females in this category? Suppose that the evidence to be examined is

	HEIGHT	WEIGHT
Mean	66 in.	125 lb
Standard Deviation	2 in.	10 lb

(We cannot directly compare the standard deviations of the two distributions, because they are recorded in different units of measurement: inches versus pounds. But we can compare the coefficients of variation, because they are percentages.) The coefficient of variation for the heights is 2/66, or 3.0 percent, and the coefficient of variation for the weights is 10/125, or 8.0 percent. Thus the heights varied from the mean height by an "average" of 3 percent, and weights varied from the mean weight by an "average" of 8 percent.

EXERCISES

3-15   The mean and the standard deviation of a distribution of scores are $\overline{X} = 50$ and $s = 20$, respectively. Calculate the coefficient of variation for this distribution.

3-16   The *period of revolution* for the nine planets in our solar system averages 60.17 years, with a standard deviation of 84.44 years. Complete the following sentence. "The periods of revolution vary by an 'average' of _____ percent from the mean of 60.17 years."

3-17   Which of these two sets of weights exhibits the greater *amount* of variation? Which set exhibits the greater *degree* of variation?

Weights of Dogs (lb): 10, 12, 14
Weights of Their Masters (lb): 100, 120, 140

3-18   Referring to the summary measures of the lifetimes and prices of a popular consumer item that follow:

	LIFETIME (Years)	PRICE
Mean	6.6	$125
Standard Deviation	0.2	10

(a) Are the lifetimes more or less homogeneous than the prices?

(b) Why would it be *in*appropriate to base your answer to (a) on a comparison of the *standard deviations*?

3-19   The scores in a distribution vary by an "average" of 25 percent from the mean of 100 units. Calculate and interpret the value of the standard deviation of the distribution.

3-20   *True or False:* If two distributions have equal coefficients of variation, then the ratio of their means will equal the ratio of their standard deviations. Support your answer.

3.4 measurement of relative standing: Z scores and the Chebyshev Inequality

Some statistical reports list only summary measures, such as means and standard deviations, without displaying the data set itself. This omission creates a problem when—as is often the case—we wish to determine the *relative standing* of a score in a distribution.

A sociologist earning $28,000 a year reads that the current salary distribution in her profession has a mean equal to $20,000 and a standard deviation equal to $4,000. She would like to know the relative standing of her salary. For example, does her $28,000 salary lie in the top quarter of the distribution?

A student receives grades of 78 and 92, respectively, on the first two exams in a mathematics course. He is given the class averages (means) and the standard deviations on each exam but not the individual grades of his classmates. The student would like to know whether he did better on Exam 2 than he did on Exam 1, relative to the class as a whole.

What can we say about the relative standing of a score in a distribution when we know only the mean and the standard deviation of the distribution? In certain cases, we can make crude estimates of relative standing by computing a Z score.

Relative standing.

Z scores

A Z *score*, also called a *standard score*, is a measure that indicates how far an individual score lies from the mean of the distribution *in number of standard deviations*. The Z score is defined by

$$Z \text{ score} = \frac{\text{Score} - \text{Mean}}{\text{Standard Deviation}} \qquad (3\text{-}6)$$

or, symbolically

$$Z_X = \frac{X - \overline{X}}{s} \ )$$

(Z_X is read as "the Z score of X" and represents the Z score of an observation on the variable X. A Z score of $+1.0$ indicates that the observation lies a distance of 1 standard deviation above the mean.)For example, if $\overline{X} = 10$ and $s = 2$, then the observation $X = 12$ has a Z score of

$$Z_X = \frac{X - \overline{X}}{s}$$

$$Z_{12} = \frac{12 - 10}{2} = 1.0$$

(A Z score of -2.0 indicates that the observation lies a distance of 2 standard deviations below the mean.)If $\overline{X} = 10$ and $s = 2$, then the Z score of $X = 6$ is

$$Z_6 = \frac{6 - 10}{2} = \frac{-4}{2} = -2.0$$

(A Z score of 0 indicates that the observation is equal to the mean of the distribution.)Thus, if $\overline{X} = 10$ and $s = 2$, the Z score of $X = 10$ is

$$Z_{10} = \frac{10 - 10}{2} = 0.0$$

The sociologist who earns \$28,000 knows that the salary distribution in her profession has a mean of $\overline{X} = \$20,000$ and a standard deviation of $s = \$4,000$. Accordingly, the Z score of her \$28,000 salary is

$$Z_{28,000} = \frac{28,000 - 20,000}{4,000} = \frac{8,000}{4,000} = +2.0$$

Her salary score is 2 standard deviations above the mean for her profession.

How does this result help the sociologist determine the relative standing of her salary in the salary distribution for her profession? One thing she can say is that her salary exceeds *at least 75 percent* of the salaries of her professional colleagues; that is, her salary is in the top quarter of the distribution. The sociologist can say this because of a theorem known as the Chebyshev Inequality.

> **CHEBYSHEV INEQUALITY   In any distribution (whatever its shape), the proportion of the observations whose *Z* scores are between the two values −*k* and +*k* must be *at least* $1 - 1/k^2$. This means that at least $1 - 1/k^2$ of the observations must lie within the interval $[\overline{X} - ks$ to $\overline{X} + ks]$.**

The sociologist's statement about the relative standing of her $28,000 salary was derived as follows:

1. She determined the Z score of $X = 28{,}000$.

$$Z_{28,000} = \frac{28{,}000 - 20{,}000}{4{,}000} = +2.0$$

Having found that her salary has a Z score of 2.0 (that her salary lies 2 standard deviations above the mean):

2. She stated the Chebyshev Inequality, substituting $k = 2$.

The proportion of observations (salaries) within a distance of 2 standard deviations from the mean must be *at least*

$$1 - \frac{1}{k^2}$$

$$= 1 - \frac{1}{2^2}$$

$$= 1 - \frac{1}{4}$$

$$= \frac{3}{4} \text{ (or 75 percent)}$$

So *at least* 75 *percent* (between 75 percent and 100 percent) of the salaries of professional sociologists lie within 2 standard deviations of the mean of this distribution. If a salary of $28,000 is 2 standard deviations above the mean ($Z_{28,000} = 2$), it must have a relative standing in the top quarter of the distribution.

As a second example of the Chebyshev Inequality, consider what can be said about the relative standing of a sociologist whose salary is $8,000.

1. We determine the Z score of $X = 8,000$.

$$Z_{8,000} = \frac{8,000 - 20,000}{4,000} = -3.0$$

2. We restate the Chebyshev Inequality, substituting $k = 3$.

The proportion of observations (salaries) within a distance of 3 standard deviations from the mean must be *at least*

$$1 - \frac{1}{k^2}$$

$$= 1 - \frac{1}{3^2}$$

$$= 1 - \frac{1}{9}$$

$$= \frac{8}{9} \text{ (or 89 percent)}$$

So the salaries of at least 89 percent of the sociologists lie within 3 standard deviations of the mean. If an $8,000 salary lies 3 standard deviations below the mean, it must have a relative standing in the bottom 11 percent of the distribution.

Chart 3-12 illustrates the implications of the Chebyshev Inequality for the proportion of observations that lie, respectively, within 3 ($k = 3$), 2 ($k = 2$), and 1 ($k = 1$) standard deviations of the mean.

Observe that the Chebyshev Inequality fails to supply any useful information about the proportion of scores within 1

chart 3-12
The Chebyshev
Inequality

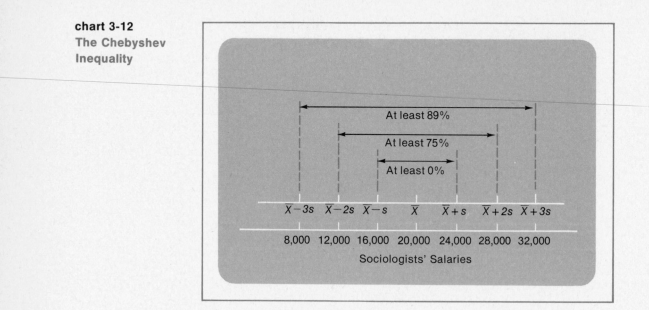

standard deviation of the mean. This is because when $k = 1$

$$1 - \frac{1}{k^2}$$

$$= 1 - \frac{1}{1^2}$$

$$= 1 - 1$$

$$= 0$$

This result verifies only the trivial fact that at least 0 percent (between 0 and 100 percent) of the observations lie within 1 standard deviation from the mean.

It is important to remember that the Chebyshev Inequality is an *inequality*, not an equality. This inequality only permits the sociologist who earns $28,000 to determine that her salary lies somewhere in the upper quarter of the distribution. Her salary may be in the upper fifth or tenth of the distribution, or it even may be the highest salary in the distribution. She can't tell from the Chebyshev Inequality.

However, if the sociologist knew more about the *shape of*

the salary distribution, she could determine the relative standing of her own salary more precisely. In Chapter 4, we will see what we can learn about the relative standings of scores in a distribution that is *normal* in shape.

comparison of
relative standing

Let's turn now to the case of the student whose grades on the first two exams were 78 and 92 and who wishes to determine whether his relative standing in the class on Exam 2 was superior to his rank in the class on Exam 1. The means and standard deviations of the grade distributions are

	EXAM 1	EXAM 2
Mean $\overline{X}$	70	80
Standard Deviation s	5	9

If we ignored the information provided by the standard deviations, we would conclude that the student's performance on Exam 2 was superior to his performance on Exam 1. Although both exam grades are above average, the 78 on Exam 1 is only 8 points above average but the 92 on Exam 2 is 12 points above average.

However, the standard deviations tell us that the class scores on Exam 2 were more variable than the class scores on Exam 1. So it is possible that scores at least 12 points above average on Exam 2 occurred more frequently than scores at least 8 points above average on Exam 1.

To adjust for differences in the variability of the two sets of exam scores, the student calculated the Z scores of his two grades. The results

$$\text{Exam 1: } Z_{78} = \frac{78 - 70}{5} = \frac{8}{5} = +1.60$$

$$\text{Exam 2: } Z_{92} = \frac{92 - 80}{9} = \frac{12}{9} = +1.33$$

indicate that his score on Exam 1 was 1.60 standard deviations above average but that his score on Exam 2 was only 1.33 standard deviations above average.

The student could then apply the following rule to his comparison of exam scores:

If two distributions of scores have roughly similar shapes (for example, if the frequency graphs of two distributions are both moderately skewed to the right or to the left), we can be reasonably sure that the larger (more positive or less negative) Z score implies a higher rank, or that equivalent Z scores imply equivalent ranks.

So the student could draw the provisional conclusion that if the shapes of the distributions of scores on Exams 1 and 2 were roughly similar, his class standing on Exam 2 (based on a Z score of 1.33) was inferior to his class standing on Exam 1 (based on a Z score of 1.60). Despite the higher grade on Exam 2 (92 versus a score of 78 on Exam 1), the student did not do as well on Exam 2 as he did on Exam 1 in relation to the rest of the class.

Too often, however, people compare the Z scores of observations in different distributions without paying attention to the shapes of the distributions. The caveat at the end of this chapter illustrates the danger of drawing conclusions about relative standing from comparisons of Z scores when the distributions have radically different shapes.

EXERCISES
3-21

A variable X has a mean of 20 units and a standard deviation of 8 units. Calculate the Z score of an observation on X equal to (a) 12 units; (b) 32 units; (c) 20 units.

3-22

A variable X has a mean of 30 units. The Z score of the observation $X = 15$ units is equal to -1.5. Determine the standard deviation of the variable X.

3-23

A variable X has a standard deviation of 5 units. The Z score of the observation $X = 20$ is equal to $+2.2$. Determine $\overline{X}$.

3-24

You have previously determined that the following grade distribution has a mean of 2.5 points and a standard deviation of 1.2 points:

GRADE	POINTS X	PROPORTION OF STUDENTS $p(X)$
A	4	0.24
B	3	0.32
C	2	0.21
D	1	0.16
F	0	0.07

Calculate the Z scores of *each* grade, and interpret the value you obtain.

3-25 Referring to the array of the lengths of service of U.S. Supreme Court justices in Chart 1-3 (page 12) and given that the distribution has a mean of 15.4 years and a standard deviation of 9.7 years:

(a) Calculate the Z scores of the lengths of the terms served by:
  (1) Fortas (Justice 8).
  (2) Burton (Justice 39).
  (3) Black (Justice 89).
(b) Name the justice whose term of service has a Z score that is virtually equal to 1.5.
(c) How many years would a justice have had to serve to earn a Z score of 3.0?

3-26 A state Department of Agriculture reports that the mean acreage of dairy farms within the state is 720 acres, with a standard deviation of 160 acres. Based on the Chebyshev Inequality, make a statement about the proportion of dairy farms whose size is:

(a) Between 400 and 1,040 acres.
(b) Above 1,040 acres.
(c) Between 240 and 1,200 acres.
(d) Below 240 acres.

3-27 In a recent bowling tournament, members of the Alley Cats rolled a mean of 160 points per game, with a standard deviation of 12 points. The shape of the distribution of points per game is unknown. What can you report about the relative standing of each of the following game scores?

(a) 184
(b) 136
(c) 196
(d) 124
(e) 160
(f) 160—assuming that the shape of the distribution is symmetric.

3-28   State the Chebyshev Inequality for the case in which $k = 4$.

3-29   Referring to the array of the lengths of service of U.S. Supreme Court justices in Chart 1-3 (page 12) and given $\overline{X} = 15.4$ years and $s = 9.7$ years:
(a) The Chebyshev Inequality would predict that at least 75 percent of the justices served terms lasting between _____ and _____ years.
(b) In fact, the actual percentage of justices serving terms within the interval indicated in the answer to (a) was _____ percent.

3-30   Referring to your array of the data given in Exercise 1-8 (page 16), verify the Chebyshev Inequality statement, "At least three-quarters of the observations in a distribution must lie within 2 standard deviations of the mean" by determining the actual proportion of years in which the annual unemployment rate fell between $\overline{X} - 2s$ and $\overline{X} + 2s$. (Remember from Exercise 3-12: $\overline{X} = 4.8$; $s = 1.1$.)

3-31   The average annual production of new gold and silver (in units of fine troy ounces) for 1950–1975 is summarized as follows:

	GOLD	SILVER
Mean	1,750 oz	38,300 oz
Standard Deviation	450 oz	2,200 oz

Assume that the distributions of annual gold production and annual silver production were similar in shape.
(a) During 1950, 2,400 oz of new gold and 42,500 oz of new silver were produced. Relative to usual production levels, which of the two minerals was produced in greater quantity that year?
(b) During 1975, 1,050 oz of new gold and 34,100 oz of new silver were produced. Relatively speaking, which of the two minerals was produced in a lesser quantity that year?

3-32   A study of housing costs in a community reported separate information on the cost of owning versus renting a housing unit. Defining the "monthly housing expense" as the sum of mortgage interest, insurance, property taxes, and utilities for homeowners and as gross rent including utilities for renters, the study revealed the following results:

	MONTHLY HOUSING EXPENSE	
	HOMEOWNERS	RENTERS
Mean	$350	$200
Standard Deviation	75	50

(a) Relatively speaking, which housing unit is more expensive: a rental for $350 a month or a homeowner expense of $450 a month?

(b) What assumption about the shapes of the two distributions was implicit in your answer to (a)?

3-33   Is a 90-degree day in Augusta, Maine, a more unusual occurrence than a 100-degree day in Phoenix, Arizona? Explain how you would obtain your answer to this question if you knew the means and the standard deviations of the daily temperature distributions.

3-34   A survey of restaurants listed in the *Yellow Pages* of a city telephone directory quoted these price statistics:

	STEAK DINNER (12-oz sirloin)	TURKEY DINNER	LOBSTER DINNER ($1\frac{1}{4}$–$1\frac{1}{2}$ lb)
Mean	$6.00	$4.00	$8.00
Standard Deviation	1.25	0.50	0.40

The menu in one "expensive" restaurant in the area contains the following entrees:

Steak: $7.75   Turkey: $4.95   Lobster: $8.25

(a) Relatively speaking, which dinner is the most expensive; that is, which dinner is the highest priced relative to the average price of that dinner in the area?

(b) What assumption must you make to justify your analysis in (a)?

Caveat: Consider How the Data Shape Up

The shape of a frequency distribution influences the way the average level and spread of the observations are reported. We know that when the shape of a distribution is highly skewed, the mean and the standard deviation will be distorted by the weights of a few extremely high or low scores. In this case, it is preferable to report the median as the average along with a companion measure of variation, such as the interquartile range.

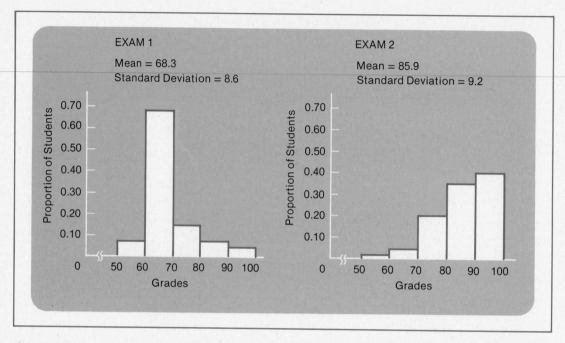

EXAM 1

Mean = 68.3
Standard Deviation = 8.6

EXAM 2

Mean = 85.9
Standard Deviation = 9.2

chart 3-13

When we wish to determine the relative standing of a score in a distribution but we do not know the shape of the distribution, we must rely on the Chebyshev Inequality. But we have seen that the Chebyshev Inequality affords at best *crude* estimates of relative standing and that it tells us nothing about the relative standings of scores that lie within 1 standard deviation of the mean.

Finally, the use of Z scores to compare the relative standings of scores in different distributions (for example, to compare the relative standings of a student's grades on two exams) is risky, unless the shapes of the distributions are roughly similar.

To illustrate this risk, consider the distributions of exam grades shown in Chart 3-13.

The shape of the grade distribution for Exam 1 is skewed to the right, and the distribution of grades on Exam 2 is skewed to the left. So these two distributions are radically dissimilar in shape.

Now assume that a student scores 70 on Exam 1 and 90 on Exam 2. The Z score of the grade of 70 on Exam 1 is

$$Z_{70} = \frac{70 - 68.3}{8.6} = +0.20$$

The Z score of the grade of 90 on Exam 2 is

$$Z_{90} = \frac{90 - 85.9}{9.2} = +0.45$$

Can we conclude from the result (Z_{90} on Exam 2 is greater than Z_{70} on Exam 1) that the student's relative standing in the class was higher on Exam 2? A glance at the histograms in Chart 3-13 tells us that we cannot.

The student's score of 70 on Exam 1 exceeded the grades of 75 percent of the class on Exam 1: A grade of 70 is at the 75th percentile. However, the student's score of 90 on Exam 2 exceeded the grades of only 60 percent of the class: A grade of 60 is at the 60th percentile. Despite the fact that Z_{90} on Exam 2 exceeded Z_{70} on Exam 1, the student's relative position in the class was higher on Exam 1 than on Exam 2.

(Clearly, an absence of knowledge about the shape of a distribution severely limits our ability to report statistical facts. In Chapter 4, we will learn the value of working with distributions that exhibit well-defined shapes, such as the bell shape of a *normal distribution*.)

COMING
TO
TERMS

Chebyshev Inequality   A theorem that allows us to make a crude estimate of the relative standing of an observation based on its Z score. For example, the Chebyshev Inequality states that an observation whose Z score is +2.0 must be positioned in the top quarter of the distribution.

Coefficient of Variation   The standard deviation divided by the mean. A measure of the *degree* of variation in a distribution.

Interquartile Range   The distance from the 25th to the 75th percentile. The distance that spans the middle 50 percent of the observations.

Percentile   The 25th percentile is the value that divides a distribution between the lower 25 percent and the upper 75 percent of the scores. The 75th percentile is the value that divides a distribution between the lower 75 percent and the upper 25 percent of the scores. Other percentiles are defined similarly.

Range   The distance from the smallest to the largest score in a distribution.

Standard Deviation   A measure of the amount of variation about the mean
of a distribution, defined by the equation $s = \sqrt{\Sigma(X - \overline{X})^2/n}$.

Variance   The square of the standard deviation, defined by the equation
$s^2 = \Sigma(X - \overline{X})^2/n$.

Z score   Also called a *standard score*. The distance of an individual score
from the mean in number of standard deviations. A Z score is used to
determine and compare the relative standings of scores.

**ISSUES
FOR
ANALYSIS**

1. A list of the number of home runs (HR) hit by Major League baseball
teams during the 1975 season follows:

AMERICAN LEAGUE	NO. OF HR	HR RANK	NATIONAL LEAGUE	NO. OF HR	HR RANK
Boston	134	4.5 (*tie*)	St. Louis	81	11
Minnesota	121	8	Cincinnati	124	3
New York	110	10	Philadelphia	125	2
Cleveland	153	1	Pittsburgh	138	1
Kansas City	118	9	Chicago	95	8
Texas	134	4.5 (*tie*)	San Francisco	84	9.5 (*tie*)
Chicago	94	11	New York	101	6
Oakland	151	2	Houston	84	9.5 (*tie*)
Baltimore	124	7	Los Angeles	118	4
Milwaukee	146	3	Montreal	98	7
Detroit	125	6	Atlanta	107	5
California	55	12	San Diego	78	12
Mean = 122.08			Mean = 102.75		
Standard Deviation = 27.16			Standard Deviation = 19.80		

(a) Compare the home-run performance of the two leagues in terms of
the average *and* in terms of variation from the average.

(b) Compare the home-run performance of the two Chicago teams—
White Sox (American League) versus Cubs (National League).

(c) The teams are listed in the order in which they finished in their
respective pennant races. Does there seem to be any connection
between a team's relative standing in its league pennant race and
its relative standing in terms of home-run performance?

2. A medical-school graduate wishes to compare her earnings potential
in individual practice and group practice before making a career

choice. The following information is available to her:

	MEDIAN ANNUAL INCOME	MEAN ANNUAL INCOME	90th PERCENTILE ANNUAL INCOME
Individual Practice	$41,000	$69,000	$120,000
Group Practice	56,000	59,000	85,000

Based on these figures, is it more profitable for the graduate to enter individual or group practice?

3. Two manufacturers of defense products are competing for a contract to be awarded by the Department of Defense (DOD) for the research and development of a new strategic weapons system. The previous performance records of the companies on DOD contracts follows:

GENERAL SYSTEMS, INC.	DYNAMETRICS COMPANY
1.13	1.06*
1.09	0.89*
1.08*	1.53
1.18*	1.30*
1.12	0.97*
Total: 5.60	1.45
(five contracts)	Total: 7.20
	(six contracts)

Each observation represents a "cost-overrun" index—the final cost incurred on a contract divided by the cost estimate made on the bid for the contract. (An observation of 1.00 indicates that the final cost equals the estimated cost.) All completed jobs were at least adequate in quality; asterisks indicate superior jobs.

Based on these data regarding previous performance records, which company merits the new DOD contract? In your answer, consider:

(a) The level of the cost-overrun to be expected.
(b) The consistency of each company's performance with respect to costs.
(c) The quality of each company's performance.
(d) The cost-overrun to be expected for superior work.

THE SCORING GAME

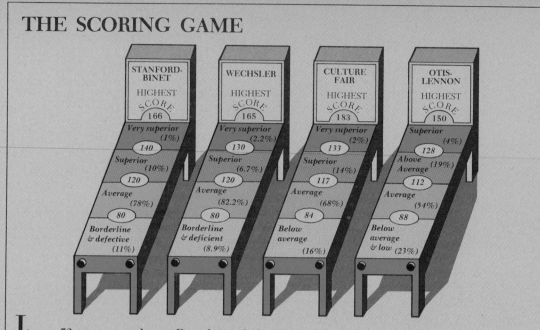

I was 72 years ago when a French psychologist named Alfred Binet first devised a test that attempted to measure a child's intelligence. . . . Today, close to 200 different tests are in use. [The four most widely known tests are shown here.]

. . . since scores were found to distribute themselves along a bell curve — centered at 100 — individual IQs are now measured in standard deviations along such a curve. In the tests, about 68% score between 85 and 115; less than 3% score below 70 — or above 130. Because scores fluctuate widely in the high IQ range, researchers have scrapped the designation genius (once defined at 140 level or above). Now they prefer more subtle terms like superior and very superior. Because terminology differs from one test to another, anyone with a 120 on the Wechsler test is designated superior, while the same score rates only above average on the Otis–Lennon.

Behind the Headline

Frequency distributions can assume many shapes, but one—the bell-shaped curve called the *normal distribution*—deserves our special attention. Of course, every frequency distribution is not normal in shape, but some important classes of distributions, like the IQ distributions illustrated here, do form bell-shaped curves. A knowledge of the structure of *normal* distributions therefore can be of considerable value in reporting facts, as well as in drawing inferences from the data at hand.

In this chapter, we will learn what it means to say that a frequency distribution is *normal* and how to describe data that are *normally* distributed. We also will see why the bell-shaped curve appears so frequently in nature.

the normal distribution

4

"You can imagine, Watson, with what eagerness I listened to this extraordinary sequence of events, and endeavored to piece them together, and to devise some common thread upon which they might all hang."

The Musgrave Ritual

At Coney Island, a popular amusement park in New York City, visitors are beckoned to the Test of Strength. There, for the price of a ticket, players can test their physical prowess on a scale that ranges from the puniest Weakling to the mightiest Powerhouse, as shown in Chart 4-1.

Everyone who played the game always asked how the other players had scored. One summer, the owner began keeping track of the scores achieved by adult males. By the end of the season, the owner had acquired enough data to state that, "The distribution of the scores of adult males is *normal* in shape, with a mean of 50 and a standard deviation of 15."

chart 4-1
Numeric Scale for the Test of Strength Scores

SCORE	STATUS
below 20	Weakling
20 to 40	Softy
40 to 60	Average Joe
60 to 80	Strongman
80 and above	Powerhouse

chart 4-2

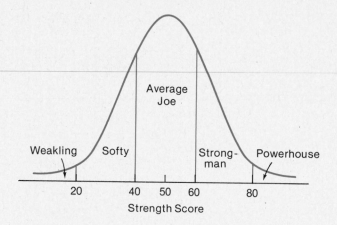

A frequency curve of this distribution appears in Chart 4-2. The graph permits us to see the major characteristics of a *normal* distribution at a glance.[1]

4.1   the characteristics of a normal distribution

> *In graphing the data they fell*
> *In line with a swoop and a swell*
> *Experientially normal*
> *Their pattern was formal*
> *Their shape emulating a bell*
> —From the uncollected works
> of Robert L. Lamborn

(When we say that a distribution is *normal*, we are asserting that it has the following properties:)

(1. *A normal distribution is symmetric.* If we were to fold the segment of the normal curve in Chart 4-2 that lies to the left of 50 over the segment of the curve that lies to the right of 50, the two halves would coincide perfectly)

---

[1] The bell-shaped frequency curve depicted in Chart 4-2 actually results from substituting the values of the mean (50) and the standard deviation (15) into a complex mathematical equation. We will examine the characteristics of a *normal* distribution by referring to the graph of this equation, rather than by employing the equation itself.

There are the same number of Weaklings and Power-houses and the same number of Softies and Strongmen. (We know that if a distribution is symmetric, the mean of that distribution must equal the median.) So half of the adult males who tried the Test of Strength have scored below 50 and half have scored above 50.

2. *In a normal distribution, the frequency of scores is greatest at the mean and declines as the distance from the mean increases.* In Chart 4-2 we can see that the frequency curve reaches its peak at the score of 50.)The largest number of participants were Average Joes (the *modal interval*).)A lesser number were Strongmen and Softies, and Weaklings and Powerhouses were seldom found.

3. *Although the range of a normal distribution is infinite[2] — in other words, the tails never quite reach the horizontal axis — almost all of the scores in a* normal *distribution lie within a range of 3 standard deviations on either side of the mean.*)

The frequency distribution of the scores achieved by adult males on the Test of Strength is graphed in Chart 4-3, where intervals having a width of 1 standard deviation are marked on the horizontal axis. Note that the score of 5 lies 3 standard deviations below the mean of 50 and that the score of 95 lies 3 standard deviations above the mean. (Practically the entire area under the curve lies between these scores. We will call the distance from the point 3 standard deviations below the mean $(\overline{X} - 3s)$ to the point 3 standard deviations above the mean $(\overline{X} + 3s)$ the *virtual range* of a *normal* distribution.)

4. *The normal distribution has points of inflection that lie a distance of 1 standard deviation above and 1 standard deviation below the mean.*)

(As we begin to move from the center or mean of a *normal* distribution, the path of the curve becomes increasingly steep (vertical). At a point 1 standard deviation on either side of the mean, the path of the curve begins to flatten out (becomes more horizontal). We call the point on a curve at which the rate of descent changes

---

[2] It is not clear from the frequency curve that the range is infinite. The mathematical equation of the *normal* distribution gives us this property.

chart 4-3

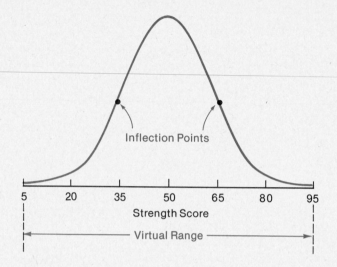

from increasingly vertical to increasingly horizontal an *inflection point.*)

   The inflection points on the curve in Chart 4-3 occur at the scores of 35 and 65, which are 15 points, or 1 standard deviation, below and above the mean of 50.

(5. *When a distribution is normal in shape, we can determine the* exact *relative standing of any observation in the distribution once we have computed the* Z *score of the observation.*)

   (Recall that the Chebyshev Inequality (Section 3.4, pages 84–87 permits us to make crude estimates of the relative standing of an observation in a distribution. These estimates are crude because the Chebyshev Inequality applies to distributions of all shapes. With the specific knowledge that a distribution has a *normal* shape, we can replace these inexact estimates with an exact description of the relative standings of individual observations.)

   (For example, the Chebyshev Inequality indicates that at least 75 percent of the observations will be within the interval extending from $\overline{X} - 2s$ to $\overline{X} + 2s$; that is, at least 75 percent of the observations will have Z scores between -2 and $+2$. If we know that a distribution is *normal*, we can state that *precisely* 95.4 percent of the observations will have Z scores between −2 and +2. This fact comes from the *normal* table, which we will examine in Section 4.2.)

EXERCISES

4-1 The distribution of the observations on a certain variable *X* has a *normal* shape, a mean of 10 units, and a standard deviation of 3 units. Draw a frequency curve of this distribution and enter on the graph the values of (a) the mean, (b) the inflection points, and (c) the lower and upper bounds of the virtual range.

4-2 The mean of a distribution that is *normal* in shape is equal to 150 units. Is it correct to report that half the scores are above 150 and that half the scores are below 150? Explain your answer.

4-3 The virtual range of a *normal* distribution is 20 through 90. Determine the value of:
(a) The mean.
(b) The standard deviation.

4-4 After being shown that the distribution of the size of single-family homes in the community is *normal* in shape, with a mean of 1,750 square feet and a standard deviation of 250 square feet, a housing official concludes that:
(a) Half of the single-family homes are smaller than 1,750 square feet and half are larger than 1,750 square feet.
(b) As many single-family homes occupy less than 1,500 square feet as occupy more than 2,000 square feet.
(c) Virtually every single-family home in the community has between 1,000 and 2,500 square feet.
(d) The proportion of single-family homes declines sharply between 1,750 and 2,000 square feet and continues to decline less sharply beyond 2,000 square feet.
Indicate the specific characteristic of a *normal* distribution that justifies each of the housing official's conclusions.

4.2 using the normal table

reading the table (The *normal* table is presented in Chart 4-4. (This table is also reproduced as Appendix Table B.) The Z scores in the table are expressed to two decimal places: The first decimal place is listed in the left-hand column, and the second decimal place is

chart 4-4
The Normal
Distribution

Z_U	\multicolumn{10}{c	}{SECOND DECIMAL PLACE OF Z_U}
	0	1	2	3	4	5	6	7	8	9
0.0	.0000	.0040	.0080	.0120	.0160	.0199	.0239	.0279	.0319	.0359
0.1	.0398	.0438	.0478	.0517	.0557	.0596	.0636	.0675	.0714	.0753
0.2	.0793	.0832	.0871	.0910	.0948	.0987	.1028	.1064	.1103	.1141
0.3	.1179	.1217	.1255	.1293	.1331	.1368	.1406	.1443	.1480	.1517
0.4	.1554	.1591	.1628	.1664	.1700	.1736	.1772	.1808	.1844	.1879
0.5	.1915	.1950	.1985	.2019	.2054	.2088	.2123	.2157	.2190	.2224
0.6	.2257	.2291	.2324	.2357	.2389	.2422	.2454	.2486	.2517	.2549
0.7	.2580	.2611	.2642	.2673	.2704	.2734	.2764	.2794	.2823	.2852
0.8	.2881	.2910	.2939	.2967	.2995	.3023	.3051	.3078	.3106	.3133
0.9	.3159	.3186	.3212	.3238	.3264	.3289	.3315	.3340	.3365	.3389
1.0	.3413	.3438	.3461	.3485	.3508	.3531	.3554	.3577	.3599	.3621
1.1	.3643	.3665	.3686	.3708	.3729	.3749	.3770	.3790	.3810	.3830
1.2	.3849	.3869	.3888	.3907	.3925	.3944	.3962	.3980	.3997	.4015
1.3	.4032	.4049	.4066	.4082	.4099	.4115	.4131	.4147	.4162	.4177
1.4	.4192	.4207	.4222	.4236	.4251	.4265	.4279	.4292	.4306	.4319
1.5	.4332	.4345	.4357	.4370	.4382	.4394	.4406	.4418	.4429	.4441
1.6	.4452	.4463	.4474	.4484	.4495	.4505	.4515	.4525	.4535	.4545
1.7	.4554	.4564	.4573	.4582	.4591	.4599	.4608	.4616	.4625	.4633
1.8	.4641	.4649	.4656	.4664	.4671	.4678	.4686	.4693	.4699	.4706
1.9	.4713	.4719	.4726	.4732	.4738	.4744	.4750	.4756	.4761	.4767
2.0	.4772	.4778	.4783	.4788	.4793	.4798	.4803	.4808	.4812	.4817
2.1	.4821	.4826	.4830	.4834	.4838	.4842	.4846	.4850	.4854	.4857
2.2	.4861	.4864	.4868	.4871	.4875	.4878	.4881	.4884	.4887	.4890
2.3	.4893	.4896	.4898	.4901	.4904	.4906	.4909	.4911	.4913	.4916
2.4	.4918	.4920	.4922	.4925	.4927	.4929	.4931	.4932	.4934	.4936
2.5	.4938	.4940	.4941	.4943	.4945	.4946	.4948	.4949	.4951	.4952
2.6	.4953	.4955	.4956	.4957	.4959	.4960	.4961	.4962	.4963	.4964
2.7	.4965	.4966	.4967	.4968	.4969	.4970	.4971	.4972	.4973	.4974
2.8	.4974	.4975	.4976	.4977	.4977	.4978	.4979	.4979	.4980	.4981
2.9	.4981	.4982	.4982	.4983	.4984	.4984	.4985	.4985	.4986	.4986
3.0	.4987									
3.5	.4998									
4.0	.49997									
4.5	.499997									
5.0	.4999997									

The entries in this table are the proportion of the observations from a *normal* distribution that have Z scores between 0 and Z_U. This proportion is represented by the colored area under the curve in the figure.
Reprinted with permission from *CRC Standard Mathematical Tables*, Fifteenth Edition (West Palm Beach, Fla.: Copyright The Chemical Rubber Co., CRC Press, Inc.).

listed across the top of the columns.) For example, a Z score of 2.81 is associated with the entry .4975, which lies at the intersection of the row labeled 2.8 and the column labeled 1.

(Each entry in the table indicates the proportion (relative frequency) of the observations in an interval whose lower bound L has a Z score of 0 ($Z_L = 0$) and whose upper bound U has the Z score identified as Z_U. This is the colored area under the *normal* curve in Chart 4-4.) For example, the entry .4975 tells us that 49.75 percent of the observations lie within the interval bounded by Z scores of 0 and 2.81, or within an interval from the mean to a point 2.81 standard deviations above the mean. (So the area under the *normal* curve between 0 and $Z_U = 2.81$ is 49.75 percent of the total area.)

(Only positive Z scores are listed in the table because a *normal* distribution is symmetric.) Therefore, the proportion of scores in an interval extending from 2.81 standard deviations below the mean to the mean ($Z_L = -2.81$, $Z_U = 0$) must equal the proportion of scores in an interval extending from the mean to 2.81 standard deviations above the mean ($Z_L = 0$, $Z_U = 2.81$), which we have determined to be 49.75 percent.

(Both when we report facts about a *normal* distribution and when we make inferential statements based on a *normal* distribution, two types of intervals are of particular interest:)

1. *An interval centered on the mean.* The bounds L and U are located at the same distance below and above the mean, respectively.) For example, the bounds of the Average Joe interval of the Test of Strength scores are $L = 40$ and $U = 60$, each of which lies 10 points from the mean of 50.
2. *An interval from a designated value to one end of the distribution.)* Again, using the Test of Strength scores as an example, the Powerhouse interval includes all scores above a low of 80. The Weakling interval includes all scores below a high of 20.

finding the proportion of scores within an interval centered on the mean

PROBLEM 1   What percentage of the adult males undertaking the Test of Strength scored between 40 and 60, classifying them as Average Joes?

chart 4-5

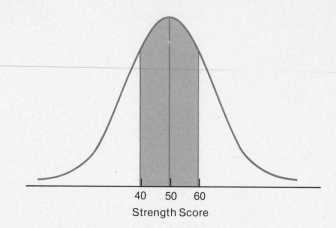

40 50 60
Strength Score

To answer this question, we must measure the blue area in Chart 4-5. The symmetry of the *normal* distribution implies that the area under the curve between 40 and 50 equals the area under the curve between 50 and 60. So we can answer the question posed in Problem 1 by calculating the area under the *normal* curve that lies between 50 and 60 and then doubling the value we obtain. We begin by converting the upper bound $U = 60$ to a Z score:

$$Z_U = \frac{U - \overline{X}}{s}$$

or

$$Z_{60} = \frac{60 - 50}{15} = +0.67$$

U therefore lies 0.67 standard deviations above the mean.

The entry corresponding to $Z_U = 0.67$ in the *normal* table in Chart 4-4 is .2486. This tells us that 24.86 percent of the scores in a *normal* distribution fall between the mean and a point 0.67 standard deviations above the mean. In Problem 1, 24.86 percent of the players scored between 50 and 60.

Doubling this percentage, we can state that 49.72 percent of the adult males trying the Test of Strength were Average Joes.

Generally, it is sufficient to report the final result to the nearest percentage point. In this case, we would say that essentially 50 percent of the players were Average Joes.

PROBLEM 2   The daily average temperatures in a region were reported to be *normally* distributed with a mean of 70°F and a standard deviation of 10°F. If we consider any temperature between 60°F and 80°F to be "moderate," during what percentage of the days was the average temperature level moderate?

Again we are concerned with an interval centered on the mean. Following the same procedure we used in Problem 1, we compute the Z score of the upper bound $U = 80$

$$Z_U = \frac{80 - 70}{10} = +1.00$$

U lies 1 standard deviation above the mean. The entry in the *normal* table for a Z score of +1.00 is .3413, and doubling this value gives us .6826. This tells us that approximately 68 percent of the scores in a *normal* distribution lie within 1 standard deviation on either side of the mean. Accordingly, the daily average temperature level was "moderate" during 68 percent of the days.

(The general rule for calculating the proportion of scores within an interval that is centered on the mean is: Determine the proportion (area) in the upper half of the interval and double it.)

finding the proportion of scores beyond a designated value

PROBLEM 1   To be rated a Powerhouse on the Test of Strength, an adult male must score at least 80. What percentage of the players achieved the status of Powerhouse?

Chart 4-6 depicts the area of interest in this problem. Because 50 percent of the scores lie above the mean of 50, we can measure the colored area by determining the percentage of scores that lie between 50 and 80 and subtracting this result from 50 percent.

The Z score of 80 is $(80 - 50)/15 = +2.00$, so that an adult male must score at least 2 standard deviations above the mean to be a Powerhouse. The entry in the *normal* table for a Z score of 2.00 is .4772, which tells us 47.72 percent of the players scored between 50 (the mean) and 80. Subtracting this result from 50 percent gives us the answer 2.28 percent, or 2 percent.

chart 4-6

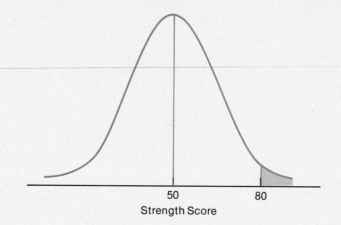

50   80
Strength Score

Thus, being a Powerhouse gives an adult male player a relative standing in the upper 2 percent of the distribution.

PROBLEM 2 Tests made on the actual weight of aspirin tablets rated at 5 g (grams) showed a *normal* distribution of weights with a mean of 5.20 g and a standard deviation of 0.16 g. What percentage of the aspirin tablets tested weighed less than the rated weight of 5 g?

In this problem, the area of concern is the colored portion of Chart 4-7. Since 50 percent of the scores lie below the mean

chart 4-7

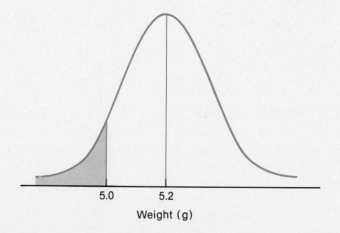

5.0   5.2
Weight (g)

of 5.2 g, we can measure the colored area by determining the percentage of scores that lie between 5.0 g and 5.2 g and sub-tracting the result from 50 percent.

The Z score of 5.0 is $(5.0 - 5.2)/0.16 = -1.25$, and the Z score of the mean is 0. Due to the symmetry of the *normal* distribution, we know that the relative frequency in the interval that lies between $Z = -1.25$ and $Z = 0$ is the same as the relative frequency in the interval that lies between $Z = 0$ and $Z = 1.25$. The entry in the *normal* table for a Z score of 1.25 is .3944, and

$$.5000 - .3944 = .1056, \text{ or } 10.56\%$$

Thus 11 percent of the aspirin tablets tested actually weighed less than 5 g.

(The general rule for calculating the proportion of scores that lie beyond (farther from the mean than) a designated value is: Determine the proportion (area) between the mean and the designated value and subtract it from 0.5.)

relative frequencies for noncentered intervals

Intervals that are centered on the mean or that span the distance from a designated value to one end of a distribution are most frequently considered in statistical inference. Obviously, not all intervals fall into one of these two categories, but the general two-step method for determining relative frequencies that we have discussed in this section can be applied in all cases.

1. Find the Z scores of the interval's lower and upper bounds, L and U.
2. Using the properties of a *normal* distribution, determine the desired relative frequencies from the *normal* table.)

Let's consider another problem. What percentage of the male adults who undertook the Test of Strength achieved the status of Strongman by scoring from 60 to 80? The area of concern is the colored portion of Chart 4-8.

The area from 50 to 80 and the area from 50 to 60 can both be determined from the *normal* table. The difference between these two areas is the proportion of adult males who scored

chart 4-8

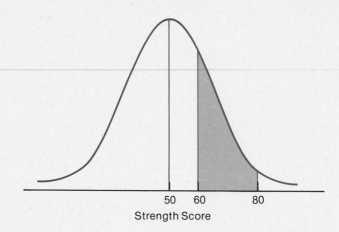

Strength Score

between 60 and 80 on the Test of Strength. To complete the problem, we calculate

$$Z_{60} = \frac{60 - 50}{15} = +0.67 \qquad Z_{80} = \frac{80 - 50}{15} = +2.00$$

From the *normal* table:

The entry for a Z score of 2.00 is .4772
The entry for a Z score of 0.67 is .2486
The difference between the entries is .2286 = 22.86%

Therefore, approximately 23 percent of the players who undertook the Test of Strength achieved the status of Strongman.

determining scores from relative frequencies

(In each of the examples in this section, we have used the *normal* table to determine the relative frequency of scores that lie within a designated interval. But the *normal* table can be used in quite a different way (backwards, in a sense). Consider the following two problems.)

PROBLEM 1   Find the interquartile range for the Test of Strength scores; that is, determine the bounds, L and U, for the interval that includes the middle 50 percent of these scores.

chart 4-9

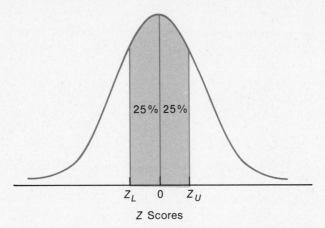

Z Scores

(To answer this question, we reverse the procedure we employed earlier to determine the proportion of scores in an interval centered on the mean. First, we find the Z scores that are associated with this proportion. Then knowing Z_L and Z_U, we calculate L and U. The interval of interest in this problem is represented by the blue area in Chart 4-9.

To find Z_U, we refer to the *entries* in the *normal* table. The relative frequency closest to 25 percent is .2486, and the Z score associated with this proportion is 0.67. So $Z_U = +0.67$ and $Z_L = -0.67$.

From the meaning of a Z score, we know that if $Z_U = 0.67$, then U lies 0.67 standard deviations above $\overline{X}$. In general, we state

$$U = \overline{X} + Z_U \cdot s \,\Big/$$

In this problem

$$U = 50 + 0.67 \cdot 15$$
$$= 50 + 10.05$$
$$= 60.05$$

Therefore, $U = 60.05$, a value that lies 10.05 points above the mean of 50. Given the symmetry property of the *normal* distribution, L must lie 10.05 points *below* the mean of 50. Thus, $L = 50 - 10.05 = 39.95$. Accordingly, the interquartile range extends from 39.95 to 60.05, or from 40 to 60.

chart 4-10

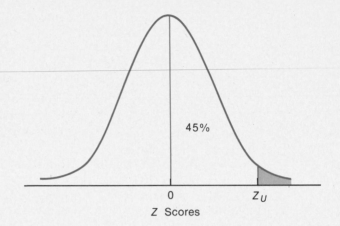

45%

0   Z_U

Z Scores

PROBLEM 2   How high a score on the Test of Strength would place an adult male player in the top 5 percent of the distribution?

As illustrated in Chart 4-10, we wish to determine the value of U that is required to place 5 percent of the area under the *normal* curve in the blue portion. Again, we begin by determining Z_U. The area from the mean to Z_U is 45 percent, or .4500.

The *normal* table contains two entries that are equally close to .4500: .4505 and .4495. We can select either entry. Choosing .4495 gives us $Z_U = 1.64$, and

$$U = \overline{X} + Z_U \cdot s$$
$$= 50 + 1.64 \cdot 15$$
$$= 74.6$$

Therefore a score of 75, a powerful Strongman, is required for status in the top 5 percent of the distribution.

comparing relative standings

(When we first introduced Z scores in Chapter 3, we said they could be used both to indicate the relative standing of a score in a distribution and to compare the relative standings of two scores from different distributions, if the shapes of the two distributions are similar.)

(If scores are being compared from two distributions that are *normal* in shape, then we can calculate the *exact* difference in their relative standing.)

PROBLEM 1 The Test of Strength scores for adult women are assumed to be *normally* distributed with a mean of 30 and a standard deviation of 10. Relative to the player's own sex, how does a woman whose score is 50 compare with a man whose score is 80? (Remember that for the men's Test of Strength scores, the mean is 50 and the standard deviation is 15.)

Because both male and female Test of Strength scores are *normally* distributed, we can directly compare the Z scores of the woman whose score is 50 and the man whose score is 80:

$$Z_{woman} = \frac{50 - 30}{10} = +2.00 \qquad Z_{man} = \frac{80 - 50}{15} = +2.00$$

The woman and the man occupy equivalent positions in their respective distributions. With Z scores of +2.00, both rank approximately in the top 2 percent.

PROBLEM 2 The following information pertains to the distributions of scores on career aptitude tests in graphic arts and creative writing:

	GRAPHIC ARTS	CREATIVE WRITING
Mean	45	55
Standard Deviation	4	8
Shape	Normal	Normal

One student who took both tests received a rating of 55 in graphic arts and 63 in creative writing. In which career area does her comparative strength lie?

The Z scores of the student's ratings are

$$Z_{graphic} = \frac{55 - 45}{4} = +2.50 \qquad Z_{creative} = \frac{63 - 55}{8} = +1.00$$

This indicates that although the student scored well above the average on both tests, her relative standing in the graphic arts test distribution is higher. Reference to the *normal* table reveals that she ranked in the top 1 percent of students taking the graphic arts test but only at about the 84th percentile on the creative writing test.

EXERCISES

4-5   Using the *normal* table in Chart 4-4, report the percentage of the area under a *normal* curve to two decimal places (for example, 82.55 percent) between the two points that lie:
(a) ±1.00 standard deviation from the mean.
(b) ±2.00 standard deviations from the mean.
(c) ±3.00 standard deviations from the mean.
(d) ±1.96 standard deviations from the mean.
(e) ±1.64 standard deviations from the mean.
(f) ±2.57 standard deviations from the mean.

4-6   According to the records of a large health insurance company, the lengths of newborn children last year averaged 21.50 in., with a standard deviation of 1.25 in. The distribution is essentially *normal* in shape.
(a) Draw a frequency curve of the distribution of the lengths of newborns, and enter the values of the mean and the standard deviation on the graph.
(b) Mark off the interval from 21 to 22 in. What percentage of the newborns were between 21 and 22 in. long at birth?
(c) Mark off the interval from 20 to 23 in. What percentage of the newborns were between 20 and 23 in. long at birth?
(d) Based on your answer to (c), what percentage of the newborns exceeded 23 in. in length at birth?

4-7   The distribution of scores on an IQ test is *normal* in shape, with a mean of 100 and a standard deviation of 15.
(a) Draw a frequency curve of the distribution of IQ scores, and enter the values of the mean and the standard deviation.
(b) Determine the percentage of IQ scores between 95 and 105.
(c) Determine the percentage of IQ scores between 85 and 115.
(d) Determine the percentage of IQ scores between 75 and 125.

4-8   A manufacturer offers three types of "100-yard tape measures": steel tape, cloth tape, and rubberized cloth tape. Inevitably, the actual length of any individual tape measure differs slightly from the exact specification of 100.00 yards. Information from the company's quality control division reveals that although the mean lengths of the tape measures of each type almost exactly meet specifications, the standard deviations are:

TYPE	STANDARD DEVIATION
Steel	4.40 in.
Cloth	8.00 in.
Rubberized Cloth	6.61 in.

If the shapes of the distributions are *normal,* what percentage of each type of tape measure will be within 6 in. of specification?

4-9   Financial data pertaining to U.S. commercial banks indicate that during a recent year, the banks' rates of return (profit rates) were *normally* distributed, with a mean of 8.4 percent and a standard deviation of 1.6 percent. Determine the proportion of commercial banks for which the rate of return:
(a) Was more than 8.4 percent.
(b) Was more than 10.0 percent.
(c) Was more than 11.6 percent.
(d) Was less than 6.8 percent.
(e) Was less than 5.2 percent.

4-10   Referring to the information on the lengths of newborn children given in Exercise 4-6, determine the percentage of newborns who were:
(a) More than 22.5 in. long at birth.
(b) More than 25 in. long at birth.
(c) Less than 19 in. long at birth.
(d) Less than 17.75 in. long at birth.

4-11   Referring to the information regarding IQ scores given in Exercise 4-7, determine the percentage of scores that are:
(a) Above 110 and then below 110.
(b) Above 120 and then below 120.
(c) Above 130 and then below 130.

4-12   Interviews of numerous individuals attending the annual gathering of a religious sect indicated a *normal* distribution of ages, with a mean of 32 years and a standard deviation of 7 years. What percentage of the individuals were:
(a) Between 25 and 39 years old?
(b) Over 40 years old?
(c) Under 35 years old?
(d) Between 20 and 30 years old?
(e) Between 45 and 53 years old?

4-13   (*Continuation of Exercise 4-12*)

(a) Determine the interquartile range of the age distribution and interpret your result.

(b) Is the interquartile range of a *normal* distribution wider or narrower than the distance between the two points that lie ±1 standard deviation from the mean?

4-14   Again referring to the information on the lengths of newborn children given in Exercise 4-6:

(a) Determine and interpret the interquartile range of the lengths of newborns.

(b) To be positioned in the upper 10 percent of the distribution, the length of a newborn would have had to be *at least* how many inches?

4-15   Again referring to the information regarding IQ scores given in Exercise 4-7:

(a) Determine the IQ score at the 75th percentile.

(b) Determine the IQ score at the 95th percentile.

4-16   A university gymnasium is used for student-sponsored concerts as well as for home basketball games. Some attendance measures for each type of event follow:

	CONCERTS	BASKETBALL GAMES
Seating Capacity	4,200	2,500
Mean Attendance	3,800	1,600
Standard Deviation	280	400

(a) Assuming that both attendance distributions are *normal* in shape, determine whether a capacity crowd is a less frequent occurrence at basketball games or at concerts.

(b) Based on the attendance measures provided, do you think either distribution is *normal?* Support your answer.

4-17   An archeological research team received endowments from two research foundations to assist in financing a proposed dig. Foundation A granted $10,000; Foundation B awarded $25,000. Summary measures of the distributions of awards by the two foundations during the past three years are:

	FOUNDATION A	FOUNDATION B
Mean	$5,000	$15,400
Standard Deviation	3,000	6,000

Assuming that both awards distributions are *normal* in shape, is the $10,000 from Foundation A or the $25,000 from Foundation B the more generous gift, relatively speaking?

4.3 the occurrence of normality

In this chapter, we have described the properties of a *normal* distribution and have examined ways of determining relative frequencies under a *normal* curve. Now we'll discuss why *normal* distributions occur so frequently in daily life. What characteristics of a variable give the distribution of its observations a bell-shaped curve?

(In general, if each observation on a variable reflects the combined action of numerous, individually minor factors, each exerting a positive or a negative force, then the distribution of these observations will be approximately *normal* in shape.)

An individual player's score on the Test of Strength can be attributed to a great many factors, none of which is overriding. Factors of heredity, diet, exercise, and motivation at time of performance all combine to produce the strength score we observe. Each factor may be viewed as exerting a small force that increases or decreases a player's strength.

Nature infrequently stacks the entire deck for or against an individual. So many factors influence a player's strength that it would be highly unusual to find that every factor exerts a force in the same direction. Accordingly, Powerhouses and Weaklings seldom occur.

Individuals who have many positive characteristics and a few defects, or vice versa, are much more likely to occur. This is why there are more Strongmen than Powerhouses, and more Softies than Weaklings. In the Average Joe, the positive and negative influences on strength approximately balance out.

If a Powerhouse has *rare* strength, this fellow must have *raw* strength.

Thus the Average Joe may be a highly motivated, diligent exerciser who weighs only 135 points or a pot-bellied brute of 210 pounds who hasn't tensed a muscle in 25 years.

(When the observations measure a physical dimension of one sex of a species, then the distribution is apt to be *normal*. Heights, hat sizes, and heart diameters, as well as strength scores tend to form *normal* distributions.)

⸨The actual measurements of items manufactured to specification also meet the conditions that describe a *normal* distribution. Each of the many factors involved in the production process can induce slight variations in the final product.⸩If we weighed several 1-lb boxes of cereal, most of the boxes would weigh close to 1 lb, a few boxes would be slightly lighter or heavier than 1 lb, and an occasional box would be rejected because it deviated radically from the specified weight. So the actual weight measurements of 1-lb boxes of cereal would appear as a *normal* distribution with a mean of 1 lb.

⸨Finally, distributions of *mean values* from different groups of observations on a variable tend to be *normal.*⸩For example, the distribution of class averages on a test administered to many large classes will appear as a bell-shaped curve. After all, the average of a large group is made up of numerous, individually minor factors, the scores that are being averaged.

⸨The fact that group averages tend to be *normally* distributed is the basis for many techniques used in statistical inference. We will return to this phenomenon in Chapter 9, where we will examine a formal law of normality that applies to group means.⸩

Caveat: Not All Distributions Are Normal

⸨Because we can report the facts about a *normally* distributed variable with precision, it is frequently tempting to assume that a variable possesses a *normal* distribution even when it does not. Historically, in fact, attempts have been made to present *normality* as a universal state of nature for numeric measurements. Understandably, these attempts have failed, because many types of distributions bear little resemblance to the bell-shaped curve.⸩

⸨The distributions of variables that measure economic status—notably, income (what you earn), wealth (what you own), and debt (what you owe)—are typically skewed to the right.⸩Chart 4-11 provides an illustration of the distribution of income levels among U.S. households in 1973, which obviously

chart 4-11
Income Levels in
U.S. Households,
1973

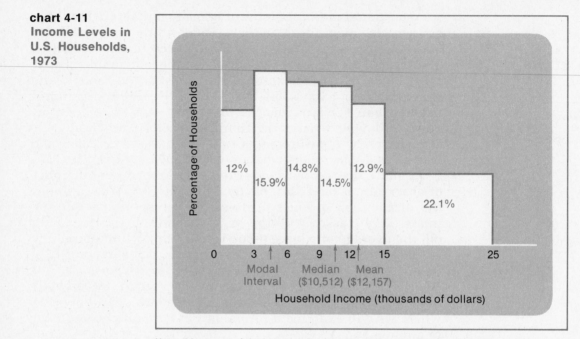

Note: 7.8 percent of the households earned an income greater than $25,000.

is not a *normal* curve. The median income of $10,512 was almost $1,500 less than the mean income of U.S. households. In fact, roughly 55 percent of the households earned less than the mean.

A primary reason that the income distribution is skewed is that a few factors, such as inherited wealth and educational background, exert a profound influence on a household's earning power. In a great many families, workers are hampered by limited inheritance and limited schooling; but in a few families, workers begin their careers with sizeable endowments or are offered high-paying jobs due to their extensive education.

(When we record the number of accidents that occur during a fixed period of time, we often find that the distribution is sharply skewed to the right.) Numerous factors combine to set the scene for an accident, but we cannot say that each factor contributes a small positive or negative force to the occurrence of the accident. Instead each factor is crucial; if one is missing, then the accident may not occur. An automobile collision may occur because the driver is tired, the visibility is poor, the road is slippery, and the approaching vehicle is speeding. If any one of these factors weren't present, the driver might be able to react fast enough to avoid a collision.

Accidents are examples of rare events. They are rare because so many conditions have to be present for the accident to occur. (Therefore, the distribution of the number of occurrences of a rare event is not apt to be *normal*.)

(Finally, we note that the basic outcomes of many gambling devices — dice, the roulette wheel, a deck of cards — assume distributions that are not *normal* or skewed but are uniform in height.) When a fairly balanced die is tossed a large number of times, each of the six faces will turn up with the same frequency (1/6 of the time), as shown in Chart 4-12.

Here, too, many factors influence the value (face) observed, including the initial position of the die, the way it is tossed, and any irregularities in the playing surface on which the die is thrown. But not one of these factors affects the outcome of the throw in a systematic way, either positively or negatively. Without positive and negative forces to average out, each outcome is apt to occur with the same frequency.

(The preceding list of *non-normal* distributions is hardly a complete one. The main point to be made here is that *normality* is not a way of life: The relative frequencies that characterize areas under a *normal* curve do not apply universally. We must check the shape of a distribution and consider the conditions

chart 4-12
Outcome of
Die Toss

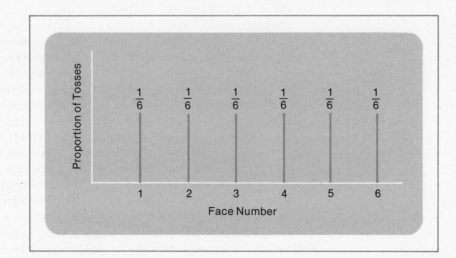

for *normality* before we report the statistical facts that would apply only if the distribution were *normal.*)

1. Criticize the use of the term *abnormal* to characterize a distribution whose shape does not conform to the bell shape of a *normal* distribution.
2. Which of the following distributions is apt to appear more nearly bell shaped? Why?
   (a) The distribution of income levels among all individuals within a specific profession.
   (b) The distribution of the mean income levels among all professions.
3. Referring to the frequency curve of the lengths of service of U.S. Supreme Court justices given in Chart 1-8 (page 16), you can see that the shape of this distribution is skewed to the right. Present a possible rationale for the *non-normality* of this distribution.
4. Would you expect the distribution of the ages of individuals attending a rock concert on a college campus that is open to the public to be *normal* in shape?

COMING
TO
TERMS

Area Under a *Normal* Curve   Geometric representation of the relative frequency of scores in a *normal* distribution.

Centered Interval in a *Normal* Distribution   An interval between two values *L* and *U*, where *L* is the same distance below the mean as *U* is above the mean.

Inflection Point   The point at which the path of a curve changes from increasingly steep to increasingly flat, or vice versa. Inflection points on a *normal* curve occur at distances of 1 standard deviation below and above the mean.

Normal Distributions   A class of distributions that occur frequently in nature and that are represented graphically as "bell-shaped" curves. It is characteristic of *normal* distributions that the relative frequency within an interval can be determined exactly by knowing the *Z* scores for the lower and upper bounds of that interval.

Normal Table   A table based on *Z* scores from which the relative frequencies of a *normal* distribution can be determined.

Virtual Range of a *Normal* Distribution   The distance from the point 3 standard deviations below the mean ($\overline{X} - 3s$) to the point 3 standard deviations above the mean ($\overline{X} + 3s$) in a *normal* distribution. Called the virtual range because virtually all the scores from a *normal* distribution will lie within this range.

ISSUES
FOR
ANALYSIS

1. Given the distributions of the heights of adult males and adult females shown here, would the composite distribution of the heights of both males and females also be *normal* in shape? How would you characterize the shape of the composite distribution? (*Hint:* Sketch a composite distribution by adding the heights of the two curves at 64 in., 67 in., and 70 in.)

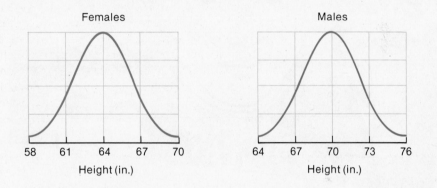

2. Use an ordinary deck of playing cards, counting *Ace* = 1, *Jack* = 11, *Queen* = 12, and *King* = 13.
   (a) Shuffle well and deal yourself a hand of six cards. Calculate the mean of the face numbers and record the value you obtain.
   (b) Repeat this process four more times, recording each mean value that results.
   (c) Assume that if you continued this process 20 more times, you would obtain the following mean values:

   8.33   6.44   7.02   7.58   4.90   6.26   8.60   7.16   8.01   6.95
   6.81   7.73   5.34   6.68   7.29   5.56   9.35   7.43   6.02   6.70

   Calculate the mean, midrange, and median of the 25 mean values resulting from this experiment.
   (d) Prepare a histogram of the distribution of the 25 $\bar{X}$'s. Use class intervals having a width of 1 unit, such as 6.50 to 7.50.
   (e) Characterize the shape of the distribution of $\bar{X}$'s, and present a rationale for the perceived shape.

2 investigating relationship: the tools of correlation and regression

1978 Gas Mileage Guide

U.S. Environmental Protection Agency

U.S. Department of Energy

How To Use This Guide

This Gas Mileage Guide gives information on the relative fuel economy performance of 1978 model year cars, station wagons, and light trucks. The estimates are expressed in terms of miles per gallon measured by standardized EPA fuel economy tests. **These estimates allow you to compare the relative fuel economy efficiency of 1978 model year cars; these estimates DO NOT MEAN that you will get the same mileage in these cars.** The mileage that you will get will depend to a large degree on where you drive—city versus country, mountains versus flat terrain, cold versus mild climate—and your personal driving habits.

Fuel Economy and Fuel Cost Estimates

City fuel economy reflects trips for local errands, driving to work, and general stop-and-go driving in urban and suburban areas. **Highway** fuel economy reflects non-stop driving on rural roads at a speed averaging about 50 mph. The **combined** fuel economy estimate is a weighted average of city and highway estimates. It assumes slightly over half city and under half highway driving, which is about the average U.S. driving pattern according to the Federal Highway Administration.

All values reflect the performance of a well-maintained car in warm weather driving on dry level roads after the car has been broken in.

relationship 5

5.1   evidence of a relationship

Due to sharp increases in gasoline prices in addition to general
anxiety over the energy crisis, drivers are paying increasing
attention to automobile fuel economy ratings and to the factors
that affect gasoline mileage. Chart 5-1 presents United States
Environmental Protection Agency (EPA) data on the gasoline
mileage achieved by various 1978 automobiles.

We can see that each automobile achieved better gasoline
mileage in *highway* travel than in *city* driving. In fact, the 12
cars listed in Chart 5-1 averaged close to 30 miles per gallon
(mpg) on the highway, compared to barely 20 mpg in city driving. In *combined* highway and city driving, these 12 automobiles averaged 24 mpg.

We also can see that gas mileage varies considerably from
model to model. In combined driving, for example, performance ranged from a low of 11 mpg to a high of 38 mpg. What
causes such variation?

Intuitively, we would expect an automobile's weight to be
one variable that has an important bearing on the automobile's
fuel economy. It should require more gasoline to drive a heavier
car than to drive a lighter car the same distance, all other
factors being equal.

chart 5-1
U.S. EPA Gasoline
Mileage Data, 1978

MODEL	CITY mpg	HIGHWAY mpg	COMBINED mpg
Renault LeCar	26	41	31
Chevette Scooter	30	40	34
Dodge Colt	34	45	38
Datsun 510	25	35	29
Chevy Monza	24	34	28
Ford Mustang II	23	33	26
AMC Concord	19	26	22
Pontiac Grand Prix	18	25	20
Buick Riviera	15	22	18
Ford LTD II	13	21	16
Mercury Marquis	13	20	15
Cadillac Eldorado	10	15	11

An easy way to demonstrate the effect of the variable *weight* on the variable *mpg* is to compare the mean gasoline mileage of a group of lighter cars to the mean gasoline mileage of a group of heavier cars. In Chart 5-2, the 12 automobiles listed in Chart 5-1 are divided into two groups, according to weight. The pairs of observations on the two variables *weight* and *mpg (combined)* are shown for each automobile.

Calculating each group's mean *mpg,* we find that the six lighter cars (Group 1) averaged 31 mpg and the six heavier cars (Group 2) averaged 17 mpg in combined driving. Driving one of the heavier cars involves a sacrifice of 14 mpg on the average, compared to driving one of the lighter cars.

But suppose that we need to obtain more specific information about the relationship between *weight* and *mpg.* For example, can we determine *how much* of a sacrifice in gas mileage we should anticipate for each 100 lb that is added to an automobile's weight? Can we evaluate how useful it is to know the weight of an automobile in attempting to predict its gas mileage?

Such questions are the focal point of an investigation into the statistical relationship between variables. In this chapter, we will give you a systematic list of the types of questions that have led statisticians to create measures of relationship. Then in Chapters 6 and 7, we will discuss some important measures of relationship between variables and learn how these measures are used to report statistical facts.

chart 5-2

	WEIGHT (lb)	MPG (Combined)
GROUP 1: Lighter Automobiles		
Renault LeCar	1,819	31
Chevette Scooter	1,993	34
Dodge Colt	2,068	38
Datsun 510	2,380	29
Chevy Monza	2,652	28
Ford Mustang II	2,712	26
		Mean *mpg* = 31
GROUP 2: Heavier Automobiles		
AMC Concord	3,052	22
Pontiac Grand Prix	3,309	20
Buick Riviera	3,873	18
Ford LTD II	4,123	16
Mercury Marquis	4,511	15
Cadillac Eldorado	5,101	11
		Mean *mpg* = 17

In presenting the questions of relationship, it will be helpful to refer to a graph of the information displayed in Chart 5-2. Just as a histogram or a frequency curve graphically represents the frequency distribution of a *single* variable, the *scatter diagram* graphically represents the relationship between *two* variables.

5.2   the graph of a relationship: the scatter diagram

Scattered diagrams?

In Chart 5-3, *weight*, denoted by X, is measured along the horizontal axis, and *mpg*, denoted by Y, is measured along the vertical axis. Each point on the graph represents the *weight* and *mpg* scores of an individual automobile. As a group, the points are scattered in the (X, Y) plane, which is why this type of graph is called a *scatter diagram.*

Computer printouts of scatter diagrams frequently "collapse" the axes to the approximate range of the data, as shown in Chart 5-4.

We have drawn a straight line on the (X, Y) plane in Chart 5-4 to depict the directional path of the scatter. Think of this line as one that passes through the "center" of the scatter. This line will provide us with a tool for illustrating the questions of relationship that we pose in the next section.

chart 5-3

chart 5-4

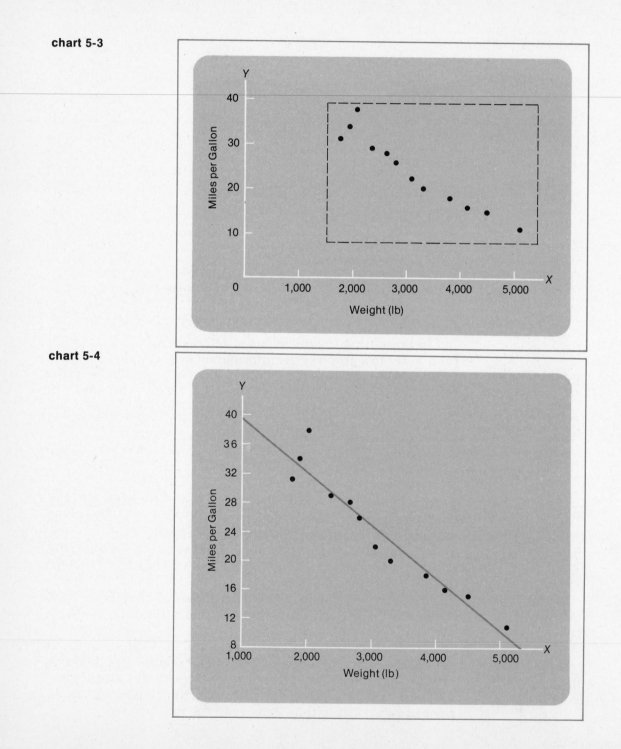

5.3   the questions of relationship

The nature of a statistical relationship between two variables can be characterized by answering three questions:

1. What is the *direction* of the relationship?
2. How *close* or, equivalently, how *strong* is the relationship?
3. If the value of one variable changes, how large a change can we expect to occur in the other variable? Equivalently, what is the *rate of response* of the second variable to changes in the first variable?

the question of
the direction of
the relationship

Are the two variables related positively (directly) or negatively (inversely)? As automobile weight increases, does gas mileage tend to increase or decrease?

The direction of the relationship is *positive* when two variables *tend to increase or decrease together*. As one variable assumes higher values, the other variable tends to assume higher values also. Analogously, as one variable assumes lower values, the other variable tends to follow suit.

A *positive* relationship is revealed on a scatter diagram by a scatter of points that moves *upward* from left to right. A line drawn through the center of this scatter has a *positive slope*.

The scatter diagram in Chart 5-4 reveals a negative relationship between the two variables, *weight* and *mpg*. The direction of the relationship is negative when two variables tend to move in opposite directions. As the variable *weight* assumes higher values, the variable *mpg* tends to assume lower values. Consequently, the scatter of points is in a *downward* direction, and the line drawn through the center of the scatter has a *negative slope*.

the question of
the closeness of
the relationship

How closely can we predict the values of one variable from the observed values of the other variable? How useful is it to know the weight of an automobile in predicting the gas mileage the car will obtain?

The *closeness* of a relationship is revealed on a scatter diagram by the *proximity* of the data points to a straight line. How closely, we ask, do the data points follow the straight and narrow path?

At one extreme, the points in a scatter diagram may line up exactly. This is illustrated in Chart 5-5(a) for a positive relationship and in Chart 5-5(b) for a negative relationship. In either circumstance, we say that there exists *an exact linear relationship* between the two variables: This means that all data points can be linked by a straight line. Thus, for any value of one variable, the equation of the straight line will predict the corresponding value of the other variable without error.

In statistical investigations, we rarely see an *exact* linear relationship. Logically, such an occurrence would mean that the value of one variable could be exactly predicted by knowing *one* other variable. You do not have to be an automotive engineer to realize that weight is not the sole factor that influences the gas mileage of an automobile. For example, two cars of the same weight can still obtain different gas mileage ratings if they differ in engine displacement or in type of transmission.

An extreme contrast to an *exact* linear relationship is illustrated in Chart 5-5(c), where the data points are scattered so randomly across the (X, Y) plane that it is impossible to delineate a directional path for the scatter. In this case, we say that

chart 5-5

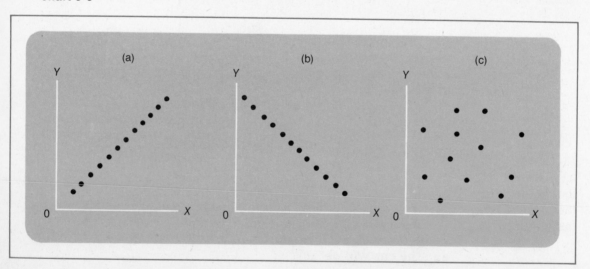

(a)   (b)   (c)

the two variables are *unrelated*. Neither variable would be of any value in predicting the behavior of the other.

Our scatter diagram of the relationship between *weight* and *mpg* (Chart 5-4) clearly falls between the two extremes illustrated in Chart 5-5. The data points do not line up exactly in a straight line, but a directional path is quite evident. This indicates that automobile weight is a useful variable but is not the only variable to consider in predicting gas mileage.

Thus the scatter diagram gives us a general picture of the closeness of the relationship between two variables. In Chapter 6, we will introduce a measure called the *correlation coefficient* that will permit us to report the closeness of a relationship in precise numeric terms.

the question of
the rate of
response

How much of a change should we expect in one variable if we change the value of the other variable? If we increase the weight of an automobile 100 lb, how much of a sacrifice in gas mileage should we anticipate? More generally, based on a car's weight, what gas mileage would we expect the automobile to obtain?

The question of the rate of response can be answered if we know the equation of the straight line that extends through the center of the point scatter.

The equation of the straight line can be expressed symbolically as $Y = a + bX$. The coefficient a, the *intercept*, denotes the value that the variable Y is expected to assume when the variable X is 0. In our present example, a would represent the expected *mpg* of a weightless car. The coefficient b, the *slope*, indicates the rate of response in the variable Y to a change in X. In our example, b would indicate the expected rate of change in *mpg* as *weight* increases.

Once the numeric values for a and b are known, the equation $Y = a + bX$ will supply a predicted value for Y (*mpg*) for any specified value of X (*weight*). We will discuss the method for determining these coefficients in Chapter 7.

recap

We have characterized the nature of a relationship between two variables on the basis of *direction*, *closeness*, and *rate of response*. Each of these features can be easily envisioned on the graphic picture of a relationship between variables—the scatter diagram. In reference to a straight line drawn through the center of a point scatter:

1. The question of *direction* asks us to determine whether the line assumes a positive or a negative slope.
2. The question of *closeness* asks us to judge how closely the points scatter about the straight line.
3. The question of *rate of response* asks us to find the equation of the line.

5.4   recognition of a nonlinear relationship

At this point, we must mention a complication that may arise in the investigation of a relationship: The relationship between two variables may be of a *nonlinear* form.

When we first introduced the scatter diagram, we depicted the directional path of the point scatter by a *straight* line. In some relationships, however, the flow of the data points is better represented by a *curved* line. Chart 5-6 illustrates two of a countless number of curved or *nonlinear relationships.*

The scatter diagram in Chart 5-6(a) could represent data pertaining to the relationship between

> X: the annual income of a family
> and
> Y: the amount of that family's annual expenditure on housing

As the annual income level of a family X increases, that family's expenditures on housing tend to grow also—but at an increasingly slow rate. Alternatively stated, richer families tend to spend a smaller proportion of their income on housing than poorer families do.

The scatter diagram in Chart 5-6(b) could depict a relationship between

> X: the age of a person
> and
> Y: the amount of leisure time available to that person

People normally have greater amounts of leisure time when

chart 5-6

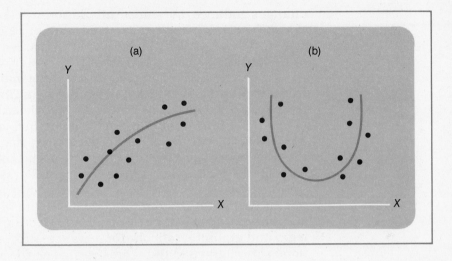

they are quite young and quite old than they do during the years in between when they must earn a living.

The statistical investigation of a nonlinear relationship is more complex than the investigation of a linear relationship, because the shape for the curve must be chosen from among many candidates. Fortunately, the linear relationship has been found to be favorably suited to a wide range of applications. For this reason, the remainder of our examination of relationship between variables will be limited to relationships of linear form.

5.5 statistical investigation of a relationship: toward what goals?

We frequently hear the statement, "There is a definite correlation (relationship) between. . . ." This is a valid, if somewhat imprecise, way of suggesting that an observed relationship between two variables seems too close to be dismissed as accidental or coincidental. It is a way of stating that *further research into the reasons why a statistical relationship exists is warranted.*

Indeed, one of the primary functions of statistical measures of relationship is to aid in the identification of important *linkages* between events. Finding a close statistical relationship between cigarette smoking and respiratory disease, for example, has stimulated research into the physiological changes induced by smoke inhalation. Close statistical relationships established between the rate of price increases and the growth rate of money and credit have led economists to a more careful investigation of the financial roots of inflation. Historians probed more deeply into the motivations of the framers of our Constitution after historian Charles Beard suggested that a statistical relationship existed between voting behavior at the Constitutional Convention of 1787 and the wealth of the participants. In summary, the investigation of the statistical relationship between variables frequently has served as a preliminary phase of scientific research.

Investigation of statistical relationship also plays a key role in efforts to evaluate the dependability of one variable as an *indicator* for another variable. In selecting students for colleges and graduate schools, for instance, substantial consideration is often given to the scores earned on the aptitude tests administered by the Educational Testing Service. These College Boards are regarded as an indicator of how applicants will perform in the college curriculum. However, studies have shown that College Board scores are not sufficiently dependable to warrant the exclusion of other factors when considering an applicant for admission to college.

In a similar vein, law enforcement officials commonly employ a breath test to measure the concentration of alcohol in the blood of a person suspected of driving while intoxicated. Studies have shown that a close correspondence exists between individuals' scores on breath tests and their blood alcohol levels measured from actual blood samples. As a result, the breath test has been judged sufficiently reliable to be used in prosecuting people for drunken driving.

Finally, statistical investigation of relationships has become an essential part of forecasting. From economic forecasting to weather forecasting, evidence of the relationship between variables based on past data is used to project the behavior of variables in the future. For example, economists have demonstrated their success in forecasting the sales of goods and services based on factors that have affected sales historically.

Whatever the ultimate intent of the relationship investigation, however, the three basic characteristics of statistical relationship—*direction, closeness,* and *rate of response*—must be examined.

EXERCISES
*5-1

In the following table, the variable *income* denotes the disposable (after-tax) personal income of all U.S. households and the variable *food* denotes the expenditures of all U.S. households on food products. All observations are measured in units of hundreds of billions of dollars (for example, 7.43 = 743 billion dollars).

YEAR	INCOME	FOOD
1971	7.43	1.41
1972	8.01	1.50
1973	9.02	1.68
1974	9.82	1.90
1975	10.81	2.10
1976	11.82	2.25

(a) Measuring *income* on the horizontal axis and *food* on the vertical axis, plot a scatter diagram of the relationship between *income* and *food* during the years 1971–1976. Follow the format of the diagram in Chart 5-3, where the origin is assigned the value of 0. After plotting the data points, draw a line through the center of the scatter.

(b) What is the direction of the relationship between *food* and *income*? What does the direction of this relationship tell you about the response of *food* to increases in *income* during 1971–1976?

(c) Does it appear that the annual level of food expenditures during 1971–1976 can be predicted fairly closely from knowledge of the annual level of disposable personal income during this time period? Why or why not?

*5-2

The following data are based on a state's rate schedule for long distance, intrastate telephone calls. The variable *distance* denotes the approximate distance of the call in units of miles, and the variable *cost* denotes the first-minute cost of a weekday call in units of cents.

DISTANCE (Miles)	COST (Cents)
5	21
12	27
16	33
21	39
27	45
35	51
48	57
66	63
88	69
130	75

(a) Measuring *distance* on the horizontal axis and *cost* on the vertical axis, plot a scatter diagram of the relationship between *cost* and *distance*. Follow the format of the diagram in Chart 5-3, where the origin is assigned the value of 0.

(b) Does the relationship between *cost* and *distance* appear to be linear or nonlinear? Explain your answer. Make a statement about the response of *cost* to increases in *distance*.

*5-3   The following data pertain to single-family homes in a suburban community. The variable *age* denotes the age of a house, and *ratio* denotes the structure's assessment ratio (the ratio of assessed value—the value subject to property tax—to actual sales value).

AGE (Years)	RATIO
1	0.45
3	0.40
7	0.35
12	0.37
20	0.30
26	0.32

(a) Measuring *age* on the horizontal axis and *ratio* on the vertical axis, plot a scatter diagram of the relationship between *age* and *ratio* for these six homes. Draw a line through the center of the scatter.

(b) What is the direction of the relationship between *age* and *ratio* among the six houses? What does the direction of this relationship imply about the likely assessment ratio for single-family homes to be completed in the near future?

*5-4   Referring to the planetary data given in Chart 1-1 on page 6, plot a scatter diagram of the relationship between the variables *distance from sun index* (horizontal axis) and *relative size* (vertical axis). Characterize the direction and strength of this relationship.

5-5   Temperature in degrees Fahrenheit (F°) can be determined exactly from temperature in degrees Celsius (C°) by the equation $F° = 32 + 1.8C°$.

(a) Are the variables F° and C° related positively or negatively?

(b) How close is the relationship between F° and C°?

(c) What is the rate of response of F° to changes in C°?

(d) What meaning does the intercept, a value of 32, have?

5-6   The annual sales of Brite Beer (*beer sales*) were found to be:
1. Positively related to the amount spent on advertising by the company (*ads*).
2. Negatively related to the relative price of Brite Beer (*price*) (that is, the price of Brite compared to the average price charged by competitors).
3. Negatively related to the rate of unemployment (*u-rate*) in the region where Brite distributes its product.

In terms of the variable *ads*, *price*, and *u-rate*, characterize a year in which *beer sales* would be relatively high.

COMING
TO
TERMS

Closeness of a Linear Relationship   The proximity of the data points to a straight line. The less scattered the data points are about the line, the closer the relationship is said to be.

Direction of a Relationship   Two variables have a positive (direct) relationship if they tend to increase and decrease together. The direction of relationship between two variables is negative (inverse) if they tend to change in opposite directions.

Exact Linear Relationship   All data points fall along a straight line.

Indicator   A variable whose behavior is indicative of the likely behavior of another variable.

Nonlinear Relationship   A relationship where the flow of the data points is best represented by a curved line.

Rate of Response   The amount of change we should expect in one variable for each unit change in the value of the other variable. Technically, rate of response is measured by the slope of the line that goes through the center of the point scatter.

Scatter Diagram   A graph of the pairs of observations on two variables. A scatter diagram is used to show the relationship between two variables.

ISSUE
FOR
ANALYSIS

Suppose that we recorded
X: the number of representatives in the U.S. House of Representatives
Y: the number of U.S. Senators
for each of the 50 states.

(a) If you were to plot a scatter diagram, placing X on the horizontal axis and Y on the vertical axis, what would be the path of the points? If you switched axes labels, then what would be the path of the points?

(b) Do your results in (a) indicate that (1) there is an exact linear relationship between X and Y or that (2) there is no relationship between X and Y? Explain your answer.

NIXON WINS ELECTION WITH 60% OF VOTE

Republicans Victors in Only 44% of House Seats

CARTER WINS ELECTION WITH ONLY 50.5% OF VOTE

Democrats Win 67% of House Seats

Behind the Headlines

In November 1972, Richard Nixon (Republican) was re-elected President of the United States with more than 60 percent of the popular vote—the third largest vote percentage in a U.S. presidential election in this century to date. At the same time, the Republican Party succeeded in winning only 192 (44 percent) of the seats in the House of Representatives. In contrast, in 1976, Jimmy Carter (Democrat) squeaked into the Presidency by receiving 50.5 percent of the popular vote, and the Democratic Party garnered 292 (67 percent) of the House seats.

How indicative are these results of the historical relationship between the presidential and the congressional vote? Has it been true that the more popular the

president-elect, the *less* successful his party has been in House elections? Or were the 1972 and 1976 elections atypical of general voting patterns? What, in fact, is the direction of the relationship between these two variables and how close is the relationship?

In place of the *direction and closeness of relationship,* we can speak simply of the *correlation* between two variables. A correlation is positive or negative as the relationship between variables is positive or negative, and a correlation is stronger or weaker, respectively, as the data points cling more or less closely to a straight line. Our main objective in this chapter is to derive and show applications of the most frequently used measure of correlation—the *correlation coefficient.*

correlation 6

"Indeed, I have found that it is usually in unimportant matters that there is a field for observation, and for the quick analysis of cause and effect which gives charm to an investigation. The larger crimes are apt to be the simpler, for the bigger the crime, the more obvious, as a rule, is the motive."

Holmes to Watson
in *A Case of Identity*

6.1 organizing and graphing the data

For each of the 16 U.S. presidential elections between 1916 and 1976 Chart 6-1 lists scores for two variables:

X: The percentage of the popular vote received by the president-elect (abbreviated *President's vote %*).

Y: The number of seats in the House of Representatives won by members of the president-elect's party (abbreviated *House support*).

For the purposes of investigating the correlation between two variables, we must regard the chronological organization of the information in Chart 6-1 as unstructured. It is difficult, if not impossible, to determine from this chart whether the two variables X and Y tend to move in the same direction or in opposite directions.

To help reveal the direction of correlation, we can list *one* of the variables (either one) in an array, either from the variable's lowest to its highest observation or conversely. Chart 6-2 presents the election results in increasing order of variable X, *President's vote %*.

chart 6-1

ELECTION YEAR	PRESIDENT-ELECT (Party)		X PRESIDENT'S VOTE %	Y HOUSE SUPPORT* (Seats)
1916	Wilson	(D)	49.8	216
1920	Harding	(R)	60.4	301
1924	Coolidge	(R)	54.0	247
1928	Hoover	(R)	58.1	267
1932	Roosevelt	(D)	57.4	310
1936	Roosevelt	(D)	60.8	331
1940	Roosevelt	(D)	54.7	268
1944	Roosevelt	(D)	53.4	242
1948	Truman	(D)	49.6	263
1952	Eisenhower	(R)	55.1	221
1956	Eisenhower	(R)	57.4	200
1960	Kennedy	(D)	49.7	263
1964	Johnson	(D)	61.1	295
1968	Nixon	(R)	43.3	192
1972	Nixon	(R)	60.7	192
1976	Carter	(D)	50.5	292

* Throughout this period, a majority in the U.S. House of Representatives required 218 of the 435 seats.

chart 6-2

PRESIDENT-ELECT	X PRESIDENT'S VOTE %	Y HOUSE SUPPORT (Seats)
1. Nixon (1968)	43.4	192
2. Truman (1948)	49.6	263
3. Kennedy (1960)	49.7	263
4. Wilson (1916)	49.8	216
5. Carter (1976)	50.5	292
6. Roosevelt (1944)	53.4	242
7. Coolidge (1924)	54.0	247
8. Roosevelt (1940)	54.7	268
9. Eisenhower (1952)	55.1	221
10. Roosevelt (1932)	57.4	310
11. Eisenhower (1956)	57.4	200
12. Hoover (1928)	58.1	267
13. Harding (1920)	60.4	301
14. Nixon (1972)	60.7	192
15. Roosevelt (1936)	60.8	331
16. Johnson (1964)	61.1	295
Mean = 54.8		256
Standard Deviation = 4.97		42.0

If you look down the column of *House support* scores, you may be able to sense that a *positive* correlation exists between the two variables: Values of variable Y tend to grow larger as we proceed down this column. For example, excluding the 1972 election, *House support* for the four top presidential vote-getters (12, 13, 15, and 16) ranges from 267 to 331 seats. On the other hand, *House support* associated with the four weakest presidential vote-getters ranges from 192 to 263 seats. Chart 6-2 seems to provide evidence of a positive correlation between X and Y.

The direction of correlation can be visualized more clearly if the data are presented graphically. Chart 6-3 shows the scatter diagram of the election results. The directional path of the data points is obviously positive, as revealed by the straight line drawn through the center of the scatter. Moreover, we can

chart 6-3

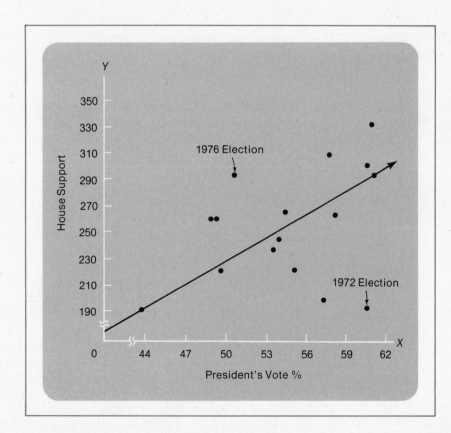

see that the 1972 and 1976 election points are somewhat atypical: The data points for these elections are positioned well away from the main path of the scatter. Evidently, *President's vote %* and *House support* are positively correlated variables, and the 1972 and 1976 election results are departures from this historical tendency.

We have now seen how the direction of correlation can be indicated in tables and graphs. But the strength of the correlation between two variables can be determined with precision only on the basis of a numeric measure.

The numeric measure we are about to introduce was created by the English statistician Karl Pearson (1857–1936). In some textbooks, this measure is referred to as "Pearson's product-moment correlation coefficient." However, because it is the single, most widely reported statistic on correlation and a prominent statistical tool, we will refer to this measure simply as the *correlation coefficient.*

6.2 the correlation coefficient

the correlation
coefficient
described

Equation (6-1) defines the correlation coefficient between two variables X and Y, denoted by $r(X, Y)$.

$$r(X, Y) = \frac{(1/n)\Sigma(X - \overline{X})\cdot(Y - \overline{Y})}{s_X \cdot s_Y} \tag{6-1}$$

where
$\overline{X}$ and $\overline{Y}$ are the means of variables X and Y
s_X and s_Y are the standard deviations of variables X and Y
n is the number of data points

In the next part of this section, we will learn how the correlation coefficient is derived by studying the logic underlying Equation (6-1). But first, let's examine the meaning of this measure.

Property 1: The sign (+ or −) of the correlation coefficient reveals the direction of the relationship between two variables. A positive r indicates that both variables tend to increase or decrease together. A negative r suggests that as one variable increases in value, the other tends to decrease in value. Therefore the sign of r indicates whether a straight line drawn through the center of a scatter diagram of data points will have a positive or negative slope.

Property 2: The numeric value of r *must* lie within the range from -1.0 to $+1.0$. A value of -1.0 indicates an exact *negative* linear relationship: All data points line up along a negatively sloped line. A value of $+1.0$ indicates an exact *positive* linear relationship: All data points line up along a positively sloped line. If the value of the correlation coefficient is unity ($+$ or $-$), the two variables are said to be *perfectly correlated.*

Property 3: A value for r midway between the extremes of ± 1.0 (the value $r = 0$) signals the absence of a linear relationship between the two variables. When $r = 0$, the variables are said to be linearly *uncorrelated,* although this does not necessarily mean that they are unrelated. (See Caveat 1 at the end of this chapter.)

These three properties of the correlation coefficient are illustrated in Chart 6-4.

Finally, note that when we speak about the correlation coefficient between two variables, it does not matter which variable we name first: $r(X, Y)$ and $r(Y, X)$ are interchangeable.

chart 6-4

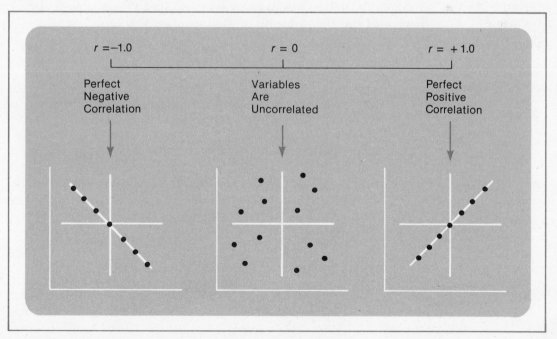

The correlation coefficient between *President's vote %* (X) and *House support* (Y) during the 16 election years under study turns out to be

$$r(X, Y) = 0.391$$

Because the possible values for $r(X, Y)$ range from -1.0 to $+1.0$, the value $r(X, Y) = 0.395$ lies less than halfway across the positive segment of the scale. From this result, we can conclude that although the two variables did *tend* to increase and decrease together, the correlation between X and Y was not especially strong. In itself, *President's vote %* was not a particularly reliable indicator of how successful the president-elect's party was in its battle for House seats. Conversely, a party's success in the House elections was not a highly dependable indicator of the percentage of the vote that was achieved by the elected head of that party's ticket. It is apparent that party loyalty was not the sole factor explaining voter behavior during these presidential elections.

the correlation
coefficient
derived

In our attempt to measure the direction and closeness of a linear relationship between two variables X and Y, how do we arrive at the formulation of Equation (6-1)?

$$r(X, Y) = \frac{(1/n)\Sigma(X - \overline{X}) \cdot (Y - \overline{Y})}{s_X \cdot s_Y} \tag{6-1}$$

If we consider the numerator and denominator of the correlation coefficient equation separately, we can show that

1. The numerator determines the *direction* of correlation (whether the correlation is positive or negative).
2. The denominator establishes the *scale* of correlation by adjusting the computed value of the numerator to insure that $r(X, Y)$ ranges from -1.0 to $+1.0$.

Covariance: The Numerator of the Correlation Coefficient

Each data point under study (each election in our illustrative example) can be located on a graph by specifying its distance (deviation) from the point at the mean of both variables $(\overline{X}, \overline{Y})$. Then each point can be expressed by the coordinates

$$(X - \overline{X}), (Y - \overline{Y})$$

where $(X - \overline{X})$ is the deviation of each score on X from $\overline{X}$ and $(Y - \overline{Y})$ is the deviation of each score on Y from $\overline{Y}$.

If the horizontal deviation $(X - \overline{X})$ is multiplied by the vertical deviation $(Y - \overline{Y})$, the result is the deviation product

$$(X - \overline{X}) \cdot (Y - \overline{Y})$$

The numerator of Equation (6-1) is the mean value of these deviation products, referred to as the *covariance* of X and Y and abbreviated cov(X, Y). Symbolically

$$\text{cov}(X, Y) = \frac{1}{n} \Sigma (X - \overline{X}) \cdot (Y - \overline{Y}) \tag{6-2}$$

The sign of cov(X, Y) conveniently reveals whether the direction of correlation between two variables is positive (+) or negative (−).

To illustrate, we reproduce the scatter diagram of *President's vote %* (X) and *House support* (Y) in Chart 6-5. First, we locate the mean point $(\overline{X}, \overline{Y})$. Over the 16 elections under study, the mean value of the *President's vote %* (X) is 54.8 percent and the mean value of *House support* (Y) is 256.3 seats.

Using the point $(\overline{X}, \overline{Y})$ as an origin (reference point), we draw vertical and horizontal lines through this point to divide the scatter diagram into four quadrants. Each quadrant is labeled with a Roman numeral. The sign of the deviation product, $(X - \overline{X}) \cdot (Y - \overline{Y})$, for points in each quadrant is shown in Chart 6-6.

Any point located in Quadrant I (for example, the point representing the Hoover election) will yield a positive deviation product. Hoover received 58.1 percent of the vote (3.3 percentage points *above* $\overline{X}$, and his House support numbered 267 seats (10.7 seats *above* $\overline{Y}$). The product of +3.3 and +10.7 is +36.4 units, a *positive* quantity.

In similar fashion, every point in Quadrant III will yield a positive deviation product. On the other hand, the points in Quadrants II and IV will yield negative products.

chart 6-5

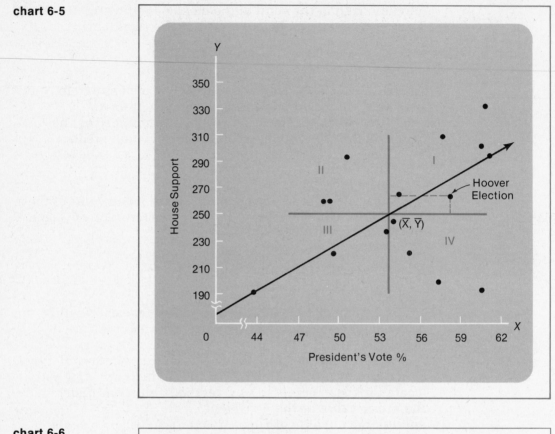

chart 6-6

LOCATION OF DATA POINT	SIGN OF $(X - \overline{X})$	SIGN OF $(Y - \overline{Y})$	SIGN OF $(X - \overline{X}) \cdot (Y - \overline{Y})$
Quadrant I	+	+	$+ \cdot + = +$
Quadrant II	−	+	$- \cdot + = -$
Quadrant III	−	−	$- \cdot - = +$
Quadrant IV	+	−	$+ \cdot - = -$

The average of all these products—the *covariance*—will be positive if the large majority of data points lie in Quadrants I and III and will be negative if the data points are concentrated mainly in Quadrants II and IV.

This result is intuitively appealing. If most points lie in Quadrants I and III, then higher (lower) than average observations on X tend to correspond to higher (lower) than average observations on Y (a positive correlation). If Quadrants II and IV are the crowded areas, then higher than average observations on one variable tend to correspond to lower than average observations on the other variable (a negative correlation).

Although Equation (6-2) is the definitional equation for covariance, we find—as we did in the case of the standard deviation—that another equation facilitates computation. The computational equation for covariance is

$$\text{cov}(X, Y) = \frac{\Sigma(X \cdot Y)}{n} - (\overline{X} \cdot \overline{Y}) \qquad (6\text{-}3)$$

Chart 6-7 illustrates the calculation of the covariance between the *President's vote %* and *House support*, based on

chart 6-7

X PRESIDENT'S VOTE %	Y HOUSE SUPPORT (Seats)	X·Y
43.4	192	8,332.8
49.6	263	13,044.8
49.7	263	13,071.1
49.8	216	10,756.8
50.5	292	14,746.0
53.4	242	12,922.8
54.0	247	13,338.0
54.7	268	14,659.6
55.1	221	12,177.1
57.4	310	17,794.0
57.4	200	11,480.0
58.1	267	15,512.7
60.4	301	18,180.4
60.7	192	11,654.4
60.8	331	20,124.8
61.1	295	18,024.5
$\Sigma X = 876.1$	$\Sigma Y = 4,100$	$\Sigma(X \cdot Y) = 225,819.8$
$\overline{X} = \frac{\Sigma X}{n} = 54.76$	$\overline{Y} = \frac{\Sigma Y}{n} = 256.25$	$\frac{\Sigma(X \cdot Y)}{n} = 14,113.74$

$$\text{cov}(X, Y) = \frac{\Sigma(X \cdot Y)}{n} - (\overline{X} \cdot \overline{Y})$$
$$= 14,113.74 - (54.76 \cdot 256.25)$$
$$= 81.49$$

Equation (6-3). The covariance of *President's vote %* and *House support* is +81.49 units, revealing that the correlation between the two variables is positive. The meaning of the numeric portion (81.49 units) of the measure remains to be explained.

Adjusting cov(*X, Y*) to a Standard Scale

Presumably, the numeric value of the covariance should indicate the closeness of the relationship between the variables. To give this numeric value of cov(*X, Y*) meaning, however, we must know its range of possible values. What is the numeric size (ignoring sign) of cov(*X, Y*) in the two extreme cases:

1. *X* and *Y* are uncorrelated?
2. *X* and *Y* are perfectly correlated?

X and *Y* Are Uncorrelated

Two variables are uncorrelated if their data points are scattered with essentially equal frequency throughout all four quadrants of the scatter diagram. In such cases, the positive products $(X - \overline{X}) \cdot (Y - \overline{Y})$ from the data points in Quadrants I and III plus the negative products from the data points in Quadrants II and IV sum to a value near 0. Consequently, cov(*X, Y*) will be close to 0. So we can designate the result cov(*X, Y*) = 0 as the *low* end of the numeric scale for covariance.

X and *Y* Are Perfectly Correlated

But what value marks the *high* end of the range for covariance—the value that reflects an exact linear relationship between *X* and *Y*? Unfortunately, this question is unanswerable as stated, because the maximum value for the covariance of a pair of variables depends on the units in which the variables are measured.

Consider the following simple illustration. Let *X* be a variable whose observations are 1, 2, and 3, respectively. Let *Y* be a variable whose corresponding observations are 2, 4, and 6. Obviously, *X* and *Y* are perfectly correlated: The data points lie on the straight line $Y = 2X$. If we compute the covariance of these two variables from Equation (6-2), we find that $cov(X, Y)$ = 1.333.

Now we'll change the units in which X and Y are measured by a factor of 10; that is, we will multiply each observation on X and Y by 10 and label the new variables X' and Y'. The observations on X' are 10, 20, and 30; the observations on Y' are 20, 40, and 60. Substituting these values into Equation (6-2), we obtain $cov(X', Y') = 133.3$.

$Cov(X', Y')$ is 100 times larger than $cov(X, Y)$. Yet X' and Y' are no more closely correlated than X and Y; both pairs of variables are perfectly correlated. The difference in the two covariance values is due to the fact that the change in units made the deviations of the X' values from their mean and the deviations of the Y' values from their mean larger than the corresponding deviations for the X and Y distributions. In other words, the distributions of X' and Y' are more variable than the distributions of X and of Y.

When we want to eliminate the effect of differences in variability in the case of one variable, we divide by the standard deviation to form a Z score. We use a similar procedure here. We divide $cov(X, Y)$ by the product $s_X \cdot s_Y$, adjusting our measurement of the deviation of a data point from $(\overline{X}, \overline{Y})$ to allow for the variability of both the X and Y observations. By definition, $cov(X, Y)/(s_X \cdot s_Y)$ is the correlation coefficient. Thus the correlation coefficient is a standardized measure of closeness; its value does not depend on the units in which the variables are expressed.

To demonstrate that this standardization process does work, let's continue our simple example. The standard deviation of X with observations (1, 2, 3) is $s_X = 0.817$. The standard deviation of Y with observations (2, 4, 6) is $s_Y = 1.634$. $s_X \cdot s_Y = 1.333$. Since $cov(X, Y) = 1.333$, the division of $cov(X, Y)$ by $s_X \cdot s_Y$ results in a quotient of $+1.0$. When $cov(X', Y')$ is divided by $s_{X'} \cdot s_{Y'}$, a quotient of $+1.0$ also results.

If two variables possess an exact linear relationship, the division of the covariance of these variables by the product of their standard deviations results in a quotient of unity ($+1.0$ if the variables are positively related; -1.0 if the variables are negatively related). This result holds no matter what units are chosen to express the values of the variables.

recap

We have developed the logic underlying the properties of the correlation coefficient. The equation defining the correlation coefficient is divided into two components:

The numerator, $\operatorname{cov}(X, Y)$, which establishes the direction of correlation.

The denominator, $s_X \cdot s_Y$, which standardizes the numeric value of covariance to a scale with a minimum value of 0 and a maximum value of ($+$ or $-$) unity.

In these terms, we can define the correlation coefficient by

$$r(X, Y) = \frac{\operatorname{cov}(X, Y)}{s_X \cdot s_Y} \qquad (6\text{-}4)$$

For our election example, we find that the value of the correlation coefficient is

$$\frac{\operatorname{cov}(X, Y)}{s_X \cdot s_Y} = \frac{81.49}{(4.971) \cdot (41.97)} = \frac{81.49}{208.632} = 0.391$$

as we reported earlier.

EXERCISES

6-1 Scatter diagrams (a), (b), and (c) illustrate the relationships between three pairs of variables.

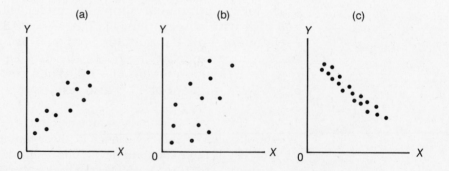

(a) Which relationship exhibits the strongest correlation?
(b) For which relationship would the correlation coefficient be closest to +1.0?

(c) Which relationship exhibits the weakest correlation?

(d) In which relationship would below-average values of Y be associated with above-average values of X?

*6-2 Based on the data given in Exercise 5-1 (page 135), compute the correlation coefficient between *food* and *income.* Is the value you obtain for $r(X, Y)$ consistent with the judgment you made about the closeness of this relationship in your answer to Exercise 5-1 (c)?

6-3 Compute the correlation coefficient for the data on *age* and *ratio* in Exercise 5-3 (page 136).

*6-4 Referring to the data on *weight* and *mpg* given in Chart 5-2 (page 127), calculate the correlation coefficient between these two variables. Comment on the closeness of the relationship between automobile weight and gas mileage.

6-5 Referring to Chart 6-3 (page 141), would the correlation coefficient between *President's vote* % and *House support* be higher or lower if the 1972 and 1976 elections were excluded?

6-6 Referring to Chart 6-1 (page 140), recompute the correlation coefficient between *President's vote* % and *House support,* excluding the data points representing the 1972 and 1976 elections.

6-7 Compute the correlation coefficient between *distance from sun index* and *relative size* (see Chart 1-1, page 6). Does your result confirm the characterization of the relationship you made based on the scatter diagram in Exercise 5-4, page 136?

6-8 A variable X has four observations: 1, 2, 4, 5.
A variable Y has four observations: 2, 4, 8, 10.

Match the observations on Y with the observations on X so that the resulting correlation coefficient $r(X, Y)$ is

(a) Equal to 1.0:   (b) Equal to −1.0:   (c) Equal to 0.0:

X	Y	X	Y	X	Y
1	___	1	___	1	___
2	___	2	___	2	___
4	___	4	___	4	___
5	___	5	___	5	___

6-9   Two students measured the correlation between the heights and weights of the same group of people. Both students expressed *weight* in pounds. One student expressed *height* in inches; the other, in feet (for example, 63 in. versus 5.25 ft). Did both students obtain the same numeric value for:

(a) The covariance of *height* and *weight*?

(b) The correlation coefficient between *height* and *weight*?

6-10   (a) You find that $\text{cov}(X, Y) = 0$. Can you conclude that $r(X, Y) = 0$, or do you need more information?

(b) You find that $\text{cov}(X, Y) = +1.0$. Can you conclude that X and Y are perfectly correlated, or do you need more information?

(c) You find that $\text{cov}(X, Y) > \text{cov}(X, Z)$. Can you conclude that $r(X, Y) > r(X, Z)$, or do you need more information?

6.3   the correlation matrix

Until now, we have measured the president-elect's success at the polls by the results of the *popular* vote. We might have chosen to use as alternatives the *electoral* vote (votes cast by the electoral college) or the *proportion of states* won by the president-elect.

Chart 6-8 displays the three versions of *President's vote %*, labeled X_1 (popular vote), X_2 (electoral vote), and X_3 (vote by state). The observations are arrayed in increasing size of *House support* (Y).

Based on these data, we will:

1. Compare the correlations between *House support* and each variant of *President's vote %*. This will enable us to identify the measure of *President's vote %* that is the best indicator of *House support*.
2. Examine the correlations *among* the three measures of *President's vote %* to see how closely these variables match over the 16 elections under study.

A convenient way to display a set of correlation coefficients is in a *correlation matrix*. (A matrix is a formal arrangement.) Chart 6-9 presents the correlation matrix for the election data.

Each entry in the matrix represents the coefficient of cor-

chart 6-8

Y HOUSE SUPPORT (Seats)	X_1 % POPULAR VOTE	X_2 % ELECTORAL VOTE	X_3 % OF STATES
192	60.7	98.5	96.8
192	43.4	62.7	56.1
200	57.4	85.4	86.1
216	49.8	62.5	52.2
221	55.1	81.3	83.2
242	53.4	75.0	81.4
247	54.0	72.9	71.9
263	49.7	46.0	56.6
263	49.6	58.3	57.1
267	58.1	83.3	83.6
268	54.7	79.2	84.6
292	50.7	55.2	47.1
295	61.1	88.2	90.3
301	60.4	77.1	76.1
310	57.4	87.5	88.9
331	60.8	95.8	98.5

relation between the variable at its row heading (to the left) and the variable at its column heading (across the top). First, notice that each entry along the diagonal has a value of unity: This reflects the unexceptionable result that any variable is perfectly correlated with itself.

If we look down the first column of the correlation matrix, we see that

$$r(Y, X_1) = 0.391$$
$$r(Y, X_2) = 0.097$$
$$r(Y, X_3) = 0.172$$

chart 6-9
A Correlation Matrix

	Y	X_1	X_2	X_3
Y	1.000			
X_1	0.391	1.000		
X_2	0.097	0.841	1.000	
X_3	0.172	0.855	0.949	1.000

chart 6-10

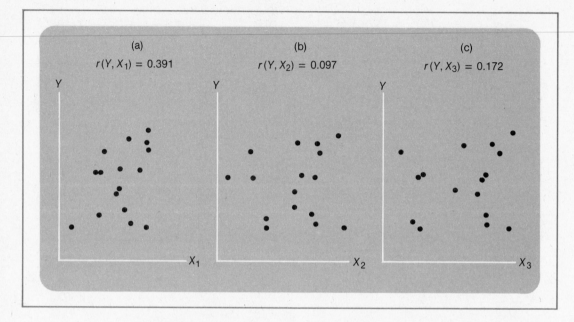

Chart 6-10 presents the scatter diagrams for each of these relationships. Although none of these relationships is especially strong, it is clear that the popular vote percentage X_1 is more highly correlated with *House support* Y than are the electoral vote percentage X_2 or the percentage of states X_3. So among the three variables X_1, X_2, and X_3, X_1 is the best indicator of *House support*.

We would expect the correlations among the three measures of *President's vote %* to be quite high, because the variables X_1, X_2, and X_3 are simply alternative measures of the same phenomenon. Our expectations are fulfilled. From the correlation matrix, we find that

$$r(X_1, X_2) = 0.841$$
$$r(X_1, X_3) = 0.855$$
$$r(X_2, X_3) = 0.949$$

The scatter diagrams reflecting these correlations appear in Chart 6-11. In contrast to the scatter diagrams in Chart 6-10, each of the scatters in Chart 6-11 conforms much more closely

chart 6-11

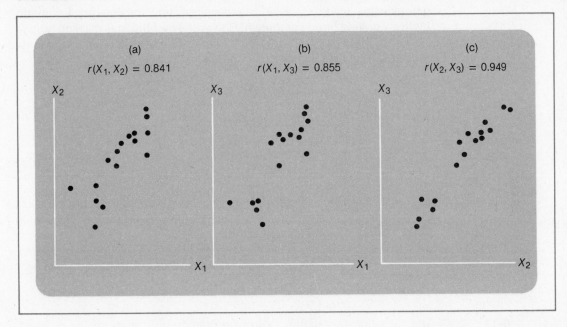

to a straight-line path. In particular, the correlation between variable X_2 (electoral vote) and X_3 (vote by state) is nearly perfect. After all, the electoral votes of each state are given in their entirety to the candidate who wins the state.

EXERCISES

6-11   The correlation matrix of the planetary data set given in Chart 1-1 is presented here. Note that POR denotes *period of revolution*, DIST denotes *distance from sun index*, SAT denotes *number of satellites*, and SIZE denotes *relative size*.

	POR	DIST	SAT	SIZE
POR	1.000			
DIST	0.989	1.000		
SAT	−0.182	−0.085	1.000	
SIZE	−0.143	−0.018	0.842	1.000

(a) What can you conclude from the 1.000 values along the diagonal?

(b) Of the six pairs of variables, only two pairs seem to be linearly related. Which pairs are they, and what is the direction of the relationship in each case?

6-12   In Chapter 5, we studied the relationship between automobile weight and gas mileage, using the EPA gas-mileage rating for combined city and highway driving (*mpg combined*). However, two other gas mileage ratings were presented: *mpg city* and *mpg highway*. The correlation coefficients for the six possible variable pairs are:

$$
\begin{aligned}
r(\textit{weight, mpg combined}) &= -0.954 \\
r(\textit{weight, mpg city}) &= -0.939 \\
r(\textit{weight, mpg highway}) &= -0.959 \\
r(\textit{mpg city, mpg highway}) &= 0.984 \\
r(\textit{mpg city, mpg combined}) &= 0.997 \\
r(\textit{mpg highway, mpg combined}) &= 0.991
\end{aligned}
$$

(a) From this information form a correlation matrix for the four variables.

(b) Does the strength of the relationship between *weight* and *mpg* differ substantially with the type of driving?

(c) Does a car that achieves superior gas mileage in one type of driving tend to achieve superior gas mileage in the other types of driving?

6-13   The following correlation matrix resulted from a study of the costs of damage in student dormitories at a state university. The variables are defined as

damage: Dollars of damage in a dorm per student.

gpa: The mean grade-point average of dorm residents.

% male: The percentage of dorm residents who are male.

% in-state: The percentage of dorm residents whose families reside in state.

	DAMAGE	GPA	% MALE	% IN-STATE
DAMAGE	1.000			
GPA	-0.741	1.000		
% MALE	0.683	-0.656	1.000	
% IN-STATE	0.002	0.014	-0.027	1.000

(a) What are the likely characteristics of a dorm in which a relatively large damage cost per student was incurred?

(b) How does the percentage of males in a dormitory relate to the quality of grades received by the dorm's residents?

6-14   A study of the relationship between average annual temperature (*temp*), latitude (*lat*), longitude (*long*), and elevation (*elev*) among selected cities in the United States resulted in the following correlation matrix:

	TEMP	LAT	LONG	ELEV
TEMP	1.000			
LAT	−0.937	1.000		
LONG	−0.169	0.330	1.000	
ELEV	−0.027	−0.125	0.315	1.000

(a) Which of the locational variables *lat, long*, or *elev* is the best indicator of *temp*?

(b) Which pair of *location* variables has the weakest correlation?

Caveats: Interpreting the Correlation Coefficient

The correlation coefficient is a measure of the *closeness* of a *statistical relationship* to a *linear path*. The italicized terms in this statement provide the key to interpreting the correlation coefficient, because they establish the boundaries that separate the conclusions about a relationship that are fact from those that are speculation. We will examine these key terms in this order:

1. *Linear* relationship.
2. *Statistical* relationship.
3. *Closeness* of relationship.

CAVEAT 1:   Do not overlook the possibility of a nonlinear relationship.

A value for $r(X, Y)$ that is close to 0 indicates that the statistical relationship between the two variables is poorly represented by a *straight line*, but does not rule out the possibility that the two variables are closely related in a *nonlinear way*. Chart 6-12 illustrates a hypothetical situation in which an extremely close

chart 6-12

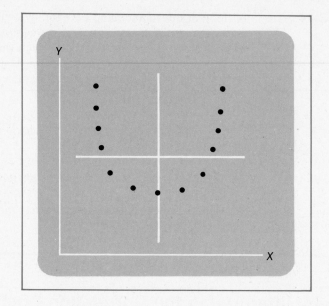

but nonlinear relationship would have a correlation coefficient close to 0.

Despite the apparent closeness of the points to the path of a smooth curve, the points are spread evenly throughout the four quadrants. So $r(X, Y)$ is approximately 0.

Before quoting the correlation coefficient, therefore, we must check the scatter diagram. If the path of the scatter has a marked curvature, a more sophisticated measure of the closeness of the relationship is required.

CAVEAT 2:   Do not assign *cause* and *effect* labels based on the correlation coefficient.

When the correlation between two variables is particularly strong—that is, when $r(X, Y)$ is close to ± 1.0—it is tempting to say that the behavior of one variable causes the behavior of the other variable.

In such cases, one variable is labeled *cause* and the other variable is labeled *effect. However, the assignment of cause and effect labels is speculation—not fact.*

It is true that if variable X exerts an important influence on the behavior of variable Y, then we *will* observe a close statistical relationship between X and Y. For example, engineers confirm that the weight of an automobile is a primary determinant (cause) of its gasoline mileage. We can observe that a close statistical relationship exists between weight and fuel economy under given driving conditions, because variation in weight causes changes in fuel economy.

However, a close statistical relationship can develop between two variables for other reasons. Two possibilities in particular should be considered:

1. Two variables can be highly correlated because they mutually influence each other; that is, each variable is at once a cause and an effect of the other.

"Robins must like baseball. They always come back when the season starts."

The dollars contributed to the campaigns of political candidates (X) and candidates' expenditures for television spots (Y) are interdependent. Media exposure requires money but may stimulate new contributions.

The relationship we have observed between *President's vote %* and *House support* probably represents variable interaction. Just as it is reasonable to suppose that the president-elect's popularity has positively affected his party's success in House elections, it is also reasonable to suppose that the party's popularity has spilled over on the president-elect.

2. Two variables can be highly correlated because the behavior of both variables is influenced proportionally by the behavior of an outside force. In such cases, neither variable is a direct cause or an effect of the other. Instead, both variables are effects of an *outside cause*.

For example, suppose that we record a group of elementary school children's heights (X) and scores on a reading ability examination (Y). We will probably find that these two variables are positively and highly correlated. However, it would be foolish to conclude that height affects reading ability, or, for that matter, that reading ability affects height. Elementary school children normally range in age from 5 to 13 years, and the older children are undoubtedly both taller and more able readers. Age, the outside force, influences both height and reading ability.

CAVEAT 3:   Do not use the correlation coefficient to measure rate of response.

Each scatter diagram in Chart 6-13 depicts an exact linear relationship; for each of these relationships, $r(X, Y) = 1.0$. But the *slopes* of the straight lines all differ. This can happen because correlation refers to the proximity of a scatter about a straight line and not to the slope of that line.

A correlation coefficient indicates how closely two variables are correlated but provides no evidence whatsoever as to the rate of response. The correlation coefficient does not reveal the amount by which one variable will change per unit change in the other variable.

chart 6-13

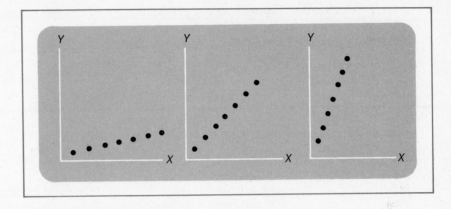

To measure rate of response, we must determine the equation of the straight line that represents the scatter of data points. Such a line is called a *regression line*. We will turn to the subject of regression in Chapter 7.

EXERCISES

1. Comment on the following statements:
   (a) If the correlation coefficient between variables X and Y is very close to ± 1.0, we can safely conclude that variation in Y is caused by variation in X.
   (b) If $r(X, Y)$ is very close to 0, we can safely conclude that no form of relationship exists between X and Y.

2. A study investigating the relationship between the wealth of families in a school district and the district's expenditures per pupil on public education (*spend*) measured wealth in two ways:
   (1) Median family income in the town (*income*).
   (2) Median value of the homes owned by families in the town (*property*). The study revealed that

$$r(spend, income) = +0.387$$
$$r(spend, property) = +0.546$$

   (a) Can we report that *property* is a better predictor of *spend* than *income* is?
   (b) Can we report that a $1,000 increase in *property* will raise *spend* more than a $1,000 increase in *income* will?

3. (*Continuation of Exercise 5-2, page 135*) The correlation coefficient relating *distance* and *cost* is r(cost, distance) = 0.927. Is the correlation coefficient a suitable measure of the closeness of the relationship in Exercise 5-2?

COMING TO TERMS

Correlation A synonym for *linear relationship*. More accurately stated *linear correlation*.

Correlation Coefficient r(X, Y) A standardized measure of the direction and closeness of a linear relationship.

Correlation Matrix A display of a set of correlation coefficients. Each entry in the matrix represents the correlation coefficient of the variable at its row heading and the variable at its column heading.

Covariance of X and Y, cov(X, Y) The numerator of the correlation coefficient. Technically, the mean of the deviation products.

Perfect Correlation Exists when data points line up along a positively sloped line (perfect *positive* correlation r(X, Y) = +1.0) or along a negatively sloped line (perfect *negative* correlation r(X, Y) = −1.0).

Uncorrelated Lacking a *linear* relationship, r(X, Y) = 0. Does not necessarily imply the absence of a relationship.

ISSUES FOR ANALYSIS

1. (a) Referring to Equation (6-1) on page 142 for the correlation coefficient r(X, Y) and to Equation (3-6) on page 83 for the Z score, express the equation for r(X, Y) in terms of Z_X, Z_Y, and n.
   (b) State the condition required for perfect positive correlation in terms of the equation you generated in (a).

2. A nation's annual *trade balance* is defined loosely as the difference between the *exports* (the value of domestic goods sold to foreign countries) and the *imports* (the value of foreign goods sold domestically) of the nation during a year. For each of the suppositions given below, indicate whether or not:
   (a) The supposition implies that r(exports, imports) is equal to 1.0.
   (b) The supposition implies that the line extending through the data points has a slope equal to 1.0.
   (c) The supposition implies that the line extending through the data points has an intercept equal to 0.

Supposition 1: Suppose that between each of the past 5 years, the increase in imports was proportional to the increase in exports.

Supposition 2: Suppose that between each of the past 5 years, the increase in imports was equal to the increase in exports.
Supposition 3: Suppose that in each of the past 5 years, the nation's trade balance was 0 (although levels of exports and imports varied from year to year).

A Theoretical View of Economic Activity

Our economic system consists of many markets. Every commodity, service, and financial asset is viewed as constituting an individual market in which a particular item is traded and a price is determined. All of these markets are linked together in varying degrees, since prices in one market influence decisions made in other markets.

The . . . dependent and independent market variables are summarized in Exhibit 1. The dependent variables are determined by the interplay of market forces which results from changes in the independent variables. Market-determined variables include prices and quantities of goods and services, prices and quantities of factors of production, prices (interest rates) and quantities of financial assets, and expectations. Independent variables consist of slowly changing factors, forces from outside our economy, random events, and forces subject to control by fiscal and monetary authorities. A change in an independent variable (for example, a fiscal or a monetary action) causes changes in many of the market-determined (dependent) variables.

EXHIBIT I

Classification of Market Variables

DEPENDENT VARIABLES

Prices and quantities of goods and services.
Prices and quantities of factors of production.
Prices (interest rates) and quantities of financial assets.
Expectations based on:
  (a) Movements in dependent variables.
  (b) Expected results of random events.
  (c) Expected changes in fiscal and monetary policy.

INDEPENDENT VARIABLES

Slowly changing factors:
  (a) Preferences.
  (b) Technology.
  (c) Resources.
  (d) Institutional and legal framework.
Events outside the domestic economy:
  (a) Change in total world trade.
  (b) Movements in foreign prices and interest rates.
Random events:
  (a) Outbreak of war.
  (b) Major strikes.
  (c) Weather.
Forces subject to control by:
  (a) Fiscal actions.
  (b) Monetary actions.

From Leonall C. Andersen and Jerry L. Jordan, "Monetary and Fiscal Actions: A Test of Their Relative Importance in Economic Stabilization," Federal Reserve Bank of St. Louis *Review*, Vol. 50, No. 11 (November 1968), 11–12.

simple and multiple regression 7

"A dog reflects the family life. Whoever saw a frisky dog in a gloomy family, or a sad dog in a happy one? Snarling people have snarling dogs, dangerous people have dangerous ones. And their passing moods reflect the passing moods of others."

"Surely, Holmes, this is a little far-fetched," said I.

"The practical application of what I have said is very close to the problem which I am investigating. One possible loose end lies in the question: Why does Professor Presbury's faithful wolfhound, Roy, endeavor to bite him?"

I sat back in my chair in some disappointment. "The dog is ill."

The Adventure of the Creeping Man

7.1 the language of regression

The two most basic terms of regression are:

1. The *dependent variable*—A variable whose behavior (response) is the subject of investigation; sometimes called the *response variable*.
2. The *independent variable* or *predictor*—A variable that is believed to influence the dependent variable's behavior.

In the fuel economy illustration in Chapter 5, we considered how the behavior of gas mileage *mpg* is influenced by automobile *weight*. Now we can say that *mpg* is the dependent variable and *weight* is the independent variable or predictor.

An investigation of the response of *mpg* to automobile *weight* is an example of *simple regression*. A simple regression investigation involves only *one* independent variable. In contrast, a *multiple regression* investigation includes *two or more* independent variables.

A multiple regression investigation of *mpg*, for example, might include *displacement, horsepower, final drive ratio,* and *type of transmission* as predictors, in addition to *weight.*

A few years ago, *The New York Times* reported the results of a major study of crime rates in New York City police precincts. The two primary purposes of the study were (1) to provide guidance in the deployment of police officers by identifying the precincts in which crimes occur most frequently and (2) to determine the relationship between precinct crime rates and certain socioeconomic characteristics of precinct residents.

The data collected for this study represented conditions in New York City during 1970. In that year, crime rates varied widely from precinct to precinct. In one precinct, fewer than 10 robberies were committed per 10,000 residents, while another was beseiged by more than 620 robberies per 10,000 residents. In nearly 50 percent of the precincts, more than 100 robberies were committed per 10,000 residents.

Chart 7-1 presents the robbery rates of the 69 regular police precincts of New York City. We will give these observations the variable name *robbery rate,* denoted by Y. *Robbery rate* (Y) will serve as the *dependent* variable in our illustration of regression analysis.

Sociologists and criminologists in general believe that the prevalence of crime in a geographic area is at least partly related to the socioeconomic conditions of that area, including income levels, employment opportunities, population density, and age composition of residents. A noted sociologist once observed that, "Throughout history, whenever there has been a great discrepancy between groups in terms of income and opportunity, the group at the bottom has had higher levels of crime." Therefore, we may expect the dependent variable *robbery rate* to respond to changes in levels of income and opportunity, measures of which may serve as predictors.

Initially, we select one of the numerous socioeconomic variables tabulated in the New York City crime study to serve as a predictor of *robbery rate.* This variable, called *% low income* and denoted by X, measures the percentage of precinct families whose 1970 incomes were below $4,000. The observations on *% low income* are also listed in Chart 7-1.

chart 7-1

New York City Precincts

Y ROBBERY RATE (No. of Robberies per 10,000 Residents)	X % LOW INCOME	Y ROBBERY RATE (No. of Robberies per 10,000 Residents)	X % LOW INCOME
621.9	31.1	85.6	27.3
395.7	12.8	79.3	14.9
275.6	35.1	78.6	6.5
264.0	26.2	65.3	6.9
237.0	20.2	64.2	16.2
231.6	16.1	59.9	13.5
204.4	17.4	59.3	13.5
203.0	27.1	51.6	11.6
200.5	24.6	50.8	12.5
194.2	15.0	44.9	11.4
186.1	29.3	44.1	19.6
181.1	9.1	43.0	21.7
178.0	31.7	42.1	9.3
171.4	24.4	41.2	7.1
169.0	34.5	40.3	11.9
167.6	27.6	39.8	6.6
167.5	10.1	39.5	8.7
165.0	7.3	39.5	9.2
161.4	25.3	35.3	16.2
161.3	30.5	33.8	11.6
159.1	15.6	31.7	6.9
154.6	35.1	31.5	10.0
154.4	14.5	30.9	11.7
152.8	39.5	30.4	10.0
149.9	19.1	27.7	11.1
134.7	19.8	25.4	7.3
131.8	29.8	21.6	6.9
117.5	14.8	21.4	12.5
115.5	24.6	19.5	15.2
111.1	5.3	19.5	4.9
109.6	17.6	17.4	9.7
107.8	30.2	15.3	13.3
103.2	23.2	10.4	5.2
88.5	11.8	9.3	6.9
87.3	19.6		

Mean $\overline{Y} = 112.5$ $\overline{X} = 16.9$

Standard Deviation $s_Y = 100.6$ $s_X = 8.9$

chart 7-2

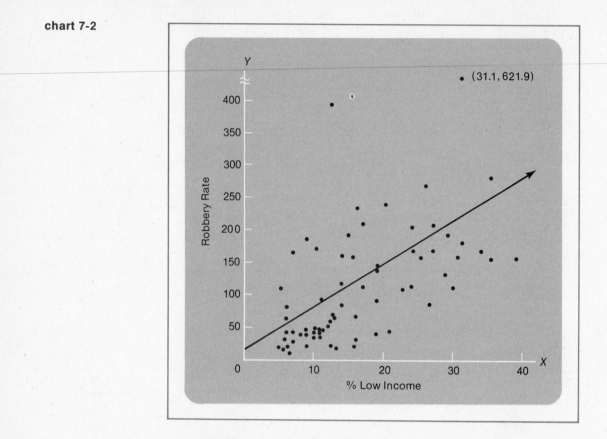

Chart 7-2 presents the scatter diagram of the observed values of *robbery rate* and *% low income* for each New York City police precinct. When plotting a scatter diagram as part of a regression investigation, it is general practice to position the independent variable on the horizontal axis and the dependent variable on the vertical axis. This is why the symbol *Y* is characteristically saved for the dependent variable: So we will remember to place it on the *Y* axis. The directional path of the scatter is represented by a straight line extending through the center of the scatter.

The direction of the relationship between the variables *robbery rate* and *% low income* is clearly positive: *Robbery rate* tends to increase in value as we advance from precincts with a small *% low income* to precincts with a high *% low income*. But how sensitive are precinct differences in robbery rates to precinct differences in percentages of low-income families? At what rate does *robbery rate* tend to increase for each one-point increase in *% low income?* To measure the rate of response of a dependent variable to changes in a predictor, we must determine the equation of the line extending through the center of the point scatter—a line we call the *regression line.*

EXERCISES

7-1   Suggest one or two plausible independent variables for each of the following dependent variables, and indicate the expected direction of the response in the dependent variable to an increase in each predictor.

(a) Graduating grade-point averages of students.

(b) Lifetimes of light bulbs.

(c) Income levels of physicians.

(d) Numbers of rebounds of college basketball players.

(e) Inches of snowfall in cities.

7-2   For each of the following variable pairs, indicate which variable is more likely to be the dependent variable and then state the question of *rate of response* in terms of the names of the dependent variable and the predictor.

(a) Food expenditures of families versus income levels of families.

(b) Severity of crimes committed versus lengths of jail sentences imposed.

(c) Number of construction workers employed during a year versus number of new housing starts during a year.

7.2   the regression line

We are used to seeing the equation of a line written

$$Y = a + bX \qquad (7\text{-}1)$$

where the coefficients a and b denote, respectively, the intercept and the slope of the line. The equation says that for each observation on X, there is a corresponding observation on Y that is equal to $a + bX$. Thus, the equation depicts an *exact* linear relationship between X and Y. Obviously, no exact relationship exists between *robbery rate* and *% low income.* If it did, the points in Chart 7-2 would line up rather than form a scatter.

When we write the equation of a *regression* line, we want to make it clear that a line drawn through the center of the scatter represents an *inexact* relationship. So we rewrite Equation (7-1), replacing Y with $\hat{Y}$ (Y-hat):

$$\hat{Y} = a + bX \qquad (7\text{-}2)$$

Y denotes *observed values* of the dependent variable, whereas $\hat{Y}$ denotes *predicted values* of the dependent variable. Equation (7-2) says that for each observation on X, there is a corresponding *point on the line* $\hat{Y}$ that is equal to $a + bX$. That point represents the value of Y that *would be predicted* by the equation of the line but that is not necessarily the actual value observed for Y. In our example, Equation (7-2) asserts that if a precinct's *% low income* $= X$, we *would predict* that it suffers $\hat{Y}$ ($\hat{Y} = a + bX$) robberies per 10,000 people.

Recognizing that $\hat{Y}$ represents the predicted values rather than the actual values of the dependent variable, we can interpret the intercept and slope coefficients in Equation (7-2) this way:

> a, the intercept: The value we *would predict* for the dependent variable when the independent variable is equal to 0. In our example, a represents the value of *robbery rate* that is predicted to occur in a precinct containing no low income families (*% low income* $= 0$).
>
> b, the slope: The *predicted* rate of response of the dependent variable to changes in the independent variable. In our example, b is the change we would predict in *robbery rate* if *% low income* increased by 1 point.

the least squares criterion

In choosing our regression line, we want to pick the line that yields the most accurate predictions of the dependent variable that the data permit. In statistical terms, we want to select *the line of best fit*. How do we find this line?

Many lines can be drawn through a scatter of data points. Visual inspection can eliminate some lines as candidates for the regression line. For example, line AA in Chart 7-3 would not serve our purpose: Its height (intercept) is too high, and its rate of ascent (slope) is too low. It does not go through the center of the scatter.

chart 7-3

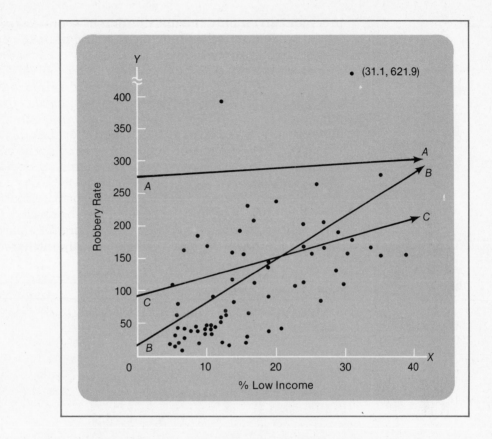

However, visual inspection cannot distinguish between lines BB and CC as candidates for the regression line. Both lines seem to be plausible paths through the center of the scatter. For this reason, we use a mathematical criterion, called the *least squares criterion,* to select the regression line.

LEAST SQUARES CRITERION Choose as the regression line the line that keeps the sum of the squared prediction errors to a minimum.

We were introduced to the least squares criterion in Chapter 2 (pages 44–46). There we learned that the *mean* is the best *single value* to choose for predicting individual scores on a variable, because the mean *always* satisfies the least squares criterion. If we wished to predict the robbery rates of individual precincts given only the distribution of scores on *robbery rate* Y, our prediction for each precinct would be $\overline{Y}$, the mean of the distribution of robbery rates.

In contrast to the situation in Chapter 2 involving the least squares criterion, now we have obtained information about a second variable X that we wish to use to improve the quality of our predictions. So instead of using a single value $\overline{Y}$ as a predictor, we seek the twin values a and b of the *line* $\hat{Y} = a + bX$, which satisfy the criterion of minimizing the sum of the squared prediction errors. In other words, we seek the *least squares regression line*.

To see how we use the least squares criterion to choose a regression line, look at Chart 7-4, where the circled data point Y represents the observed *robbery rate* score for a particular precinct. Then move vertically down to the line, where the circled

chart 7-4

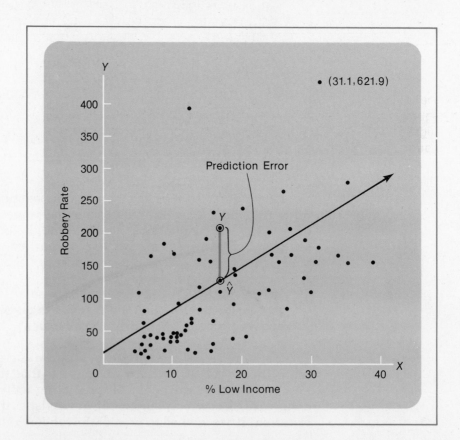

point $\hat{Y}$ represents the predicted *robbery rate* for the same precinct. The vertical distance between point Y and the line depicts the size of the *prediction error* for that precinct—the observed minus the predicted *robbery rate* values.

The prediction error for every other precinct also can be represented by the vertical distance between the precinct's data point and the line. In algebraic notation, each vertical distance—each prediction error—has a length $Y - \hat{Y}$ (Y = observed value; $\hat{Y}$ = predicted value). The line that satisfies the least squares criterion is the line that minimizes the sum of the squares of these vertical distances.

> Least squares criterion in algebraic notation: Find the line that minimizes $\Sigma(Y - \hat{Y})^2$.

Now we can precisely define "the straight line drawn through the center of the scatter" as the line that satisfies the least squares criterion. Technically, this line is called the *least squares regression line*.

the least
squares
regression
coefficients

The slope and intercept of the least squares regression line, called the *least squares regression coefficients*, are defined by the following equations:[1]

$$b = \frac{\Sigma(X - \overline{X}) \cdot (Y - \overline{Y})/n}{\Sigma(X - \overline{X})^2/n} = \frac{\text{cov}(X, Y)}{s^2_X} \qquad (7\text{-}3)$$

$$a = \overline{Y} - b\overline{X} \qquad (7\text{-}4)$$

The slope b turns out to be the ratio of cov(X, Y), the covariance of the two variables, to s^2_X, the square of the standard deviation of the independent variable (also called the variance of X). Because the denominator must be a positive quantity, the sign of the numerator cov(X, Y) determines the sign of b. This makes intuitive sense: If two variables are positively (negatively) related, then the regression line fit through their scatter of data points will have a positive (negative) slope.

---

[1] The derivation of these equations requires the use of calculus and will not be discussed here.

According to Equation (7-4), the intercept a is the mean of the dependent variable minus the product of b and the mean of the independent variable.

We will use these equations to compute the coefficients of the least squares regression line relating *robbery rate* to *% low income*. In doing this, we will assume that the raw materials — the means of X and Y, the standard deviations of X and Y, and $\text{cov}(X, Y)$ — have already been determined.

$$\begin{aligned}
\overline{X} &= 16.865 \\
\overline{Y} &= 112.525 \\
s_X &= 8.918 \\
s_Y &= 100.614 \\
\text{cov}(X, Y) &= 480.207
\end{aligned}$$

Substituting the appropriate values into Equations (7-3) and (7-4), we obtain

$$b = \frac{\text{cov}(X, Y)}{s^2_X} = \frac{480.207}{(8.918)^2} = 6.038$$

$$a = \overline{Y} - b\overline{X} = 112.525 - 6.038 \cdot 16.865 = 10.694$$

Rounding the results to a single decimal place, which is consistent with the way in which the data are reported, the regression equation is

$$\hat{Y} = 10.7 + 6.0X \tag{7-5}$$

or

$$\widehat{robbery\ rate} = 10.7 + 6.0 \cdot \%\ low\ income$$

interpreting
regression
results

On the surface, we can draw three principal conclusions from the computed regression line given by Equation (7-5):

1. (*From the intercept, $a = 10.7$*)
   In a precinct with no low-income families (*% low income* $= 0$), we would predict that 10.7 robberies occurred per 10,000 residents. However, see Caveat 1 (page 201) regarding the frequent need to qualify the interpretation of the intercept.

2. (*From the slope,* $b = 6.0$)
   We would predict that an increase of 6.0 robberies per 10,000 residents would occur for each 1 point rise in *% low income*. This is the predicted rate of response of *robbery rate* to changes in *% low income*.
3. (*From the equation as a whole,* $\hat{Y} = 10.7 + 6.0X$)
   Given the observed value of *% low income*, X, for any precinct, we would predict that the *robbery rate* of that precinct would be $10.7 + 6.0X$ robberies per 10,000 residents.

At this point, the wise investigator sits back and says, "Do the results that I obtained by putting numbers into these equations seem plausible? Do they make sense? I'll stop and check." Such a reaction has saved many a red face. Implausible results may indicate either that a gross error has been made in the computation or that simple linear regression is an inadequate tool for investigating the problem at hand. Thus, it becomes necessary to backtrack to identify computation errors or to investigate the data further before reporting the statistical results.

How can we check our results to see if they are reasonable? In many instances, our background knowledge and intuition are sufficient to specify the sign (+ or −) that each coefficient should have. Occasionally it is also possible to set upper and/or lower bounds on the numeric size of the coefficients.

In our present investigation, we would have expected to find that:

1. The slope coefficient b is positive: Increases in *% low income* result in *increases* in *robbery rate*. The scatter diagram in Chart 7-2 reveals a positive relationship, and the slope coefficient should confirm this. Our computed value $b = 6.0$ does have the expected sign.
2. The intercept a is *below* $\overline{Y}$ in size: A precinct with few or no low-income residents would be expected to have a below-average *robbery rate*. Our intercept $a = 10.7$ is actually well below the mean *robbery rate* of 112.5.

Having satisfied ourselves that our results are not implausible, we want to present them in a way that clearly illustrates the response rate of the dependent variable to changes in the

chart 7-5

Precinct with % *low income* = 10%   (X = 10):
$$\hat{Y} = 10.7 + 6.0 \cdot 10 = \ \ 70.7 \text{ robberies per 10,000 residents}$$

Precinct with % *low income* = 20%   (X = 20):
$$\hat{Y} = 10.7 + 6.0 \cdot 20 = 130.7 \text{ robberies per 10,000 residents}$$

Precinct with % *low income* = 30%   (X = 30):
$$\hat{Y} = 10.7 + 6.0 \cdot 30 = 190.7 \text{ robberies per 10,000 residents}$$

predictor. A good approach is to perform some illustrative predictions and then to summarize these predictions in a chart.

Chart 7-5 shows predicted *robbery rates* for precincts whose scores on *% low income* are, respectively, 10 percent, 20 percent, and 30 percent—all of which are within the range of values observed for *% low income*.

Clearly, differences in the percentage of low-income families are associated with sizable differences in precinct robbery rates. Predictions indicate that only 70.7 robberies per 10,000 residents should occur in a precinct with *% low income* = 10 percent, but that 190.7 robberies per 10,000 residents should occur in a precinct with *% low income* = 30 percent. Stated more generally, an increase of 10 points in *% low income* is predicted to result in an additional 60 robberies per 10,000 residents.

EXERCISES

7-3   Explain the essential distinction between the meanings of the two equations

$$Y = a + bX \qquad \text{versus} \qquad \hat{Y} = a + bX$$

Under what special circumstance will $\hat{Y}$ always equal Y?

7-4   Interpret the values of the regression coefficients (slope and intercept) in each of the following equations:
(a) *book price* (in dollars) = $2.80 + 0.03 \cdot$ *number of pages*
(b) *salesperson's weekly salary* (in dollars) = $100.00 + 0.25 \cdot$ *value of orders written* (in dollars)
(c) *daily fuel oil use* (in gallons) = $2.0 + 0.285 \cdot$ *number of degree-days*
(d) *number of reports written* = $0 + 2.4 \cdot$ *number of years on the job*

7-5 Referring to the scatter diagram drawn here, explain what the least squares regression line $\hat{Y} = a + bX$ accomplishes in terms of the distances denoted by the e's.

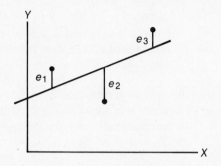

7-6 Referring to the equations defining (1) the slope coefficient of the least squares regression line $\hat{Y} = a + bX$, and (2) the correlation coefficient $r(X, Y)$ given by Equation (6-4) on p. 150:
(a) What statistic forms the numerators of both b and r?
(b) Are the denominators identical? If not, in what respect do they differ?
(c) If you switch the dependent and the independent variables, so that the regression line is $X = a' + b'Y$, will $b' = b$?

7-7 (a) What characteristic of a relationship between two variables can be obtained equally well from the correlation coefficient or the regression slope coefficient?
(b) What characteristic of a relationship between two variables can be obtained from the correlation coefficient but not from the regression slope coefficient?
(c) What characteristic of a relationship between two variables can be obtained from the regression slope coefficient but not from the correlation coefficient?

7-8 (a) Calculate the slope of the regression line relating *food* to *income* from the data given in Exercise 5-1 (page 135). (Note that you have already computed the numerator and part of the denominator of this coefficient in Exercise 6-2 on p. 151.)
(b) Using Equation (7-4) on page 173, obtain the intercept of this regression line.
(c) Report the values you obtain in (a) and (b) in equation form; that is, substitute into $\hat{Y} = a + bX$ the variable names for Y and X as well as the computed values for a and b.

(d) Interpret the slope and the intercept values in terms of what they tell you about the response of *food* to *income*.

(e) Using the regression line, predict *food* for each year from 1971 through 1976.

(f) Define a prediction $\hat{Y}$ to be *an accurate prediction of Y if $\hat{Y}$ is within 5 percent of Y*; that is, if $|Y - \hat{Y}|/Y \leq 5$ percent. Does the regression line relating *food* to *income* provide an accurate prediction of *food* during each year, 1971 through 1976?

*7-9   The following table contains data on (1) the distance between Montpelier, Vermont, and five other Vermont cities, and (2) the New England Telephone Company charge for the first minute of a weekday station-to-station call.

EXHIBIT I

CITY	DISTANCE X (Miles)	COST Y (Cents)
Morrisville	21	39
Saint Johnsbury	30	45
Burlington	35	51
Saint Albans	45	57
Alburg	63	63

(a) Assume that over this range of distances, the relationship between *cost* and *distance* is approximately linear in form. Given the information in the worksheet in Exhibit II, calculate the slope and the intercept of the regression line $cost = a + b \cdot distance$.

EXHIBIT II

	X	Y	X²	Y²	X·Y
	21	39	441	1,521	819
	30	45	900	2,025	1,350
	35	51	1,225	2,601	1,785
	45	57	2,025	3,249	2,565
	63	63	3,969	3,969	3,969
Σ	194	255	8,560	13,365	10,488

(b) Interpret the values you obtain for the slope and the intercept.

(c) Use the regression line to predict the *cost* from Montpelier to each city.

(d) Define a prediction $\hat{Y}$ to be *an accurate prediction* of Y if $\hat{Y}$ is within 5 percent of Y; that is, if $|Y - \hat{Y}|/Y \leq 5$ percent. Does the regression line relating *cost* to *distance* provide an accurate prediction of *cost* for each of the five cities?

7-10   (a) Referring to your results in Exercise 6-4 (page 151), calculate the slope and intercept of the regression line relating *weight* to *mpg combined*.

(b) Predict the number of miles per gallon that would be lost by a 100-pound increase in an automobile's weight.

7-11   A 12-month follow-up study on a group of students who had completed the Fleet School speed-reading course revealed that:

The group's mean reading speed at the time the course was completed was 720 words per minute (wpm).

During each month that elapsed following the completion of the course the group's mean reading speed decreased by an average of 20 wpm.

(a) Express these results in the form of a regression equation, where *wpm* denotes the group's mean reading speed and *time* denotes the number of months following the completion of the course.

(b) After learning the results of the study, a mathematician in the group pointed out that three years (36 months) after completion of the course, the group's mean reading speed would be 0 wpm. Is this a mathematical implication of the regression equation? Is this a fair criticism of the equation as a representation of the study's findings? (See Caveat 1, pages 201–203.)

7.3   evaluating predictive ability: the R^2 statistic

Our conclusions about the rate of response of *robbery rate* to changes in *% low income* must be considered highly tentative because they are based on results *predicted* by the regression line. We still don't know how the predicted values of *robbery rate* will compare to the values actually observed.

Our findings also must be considered tentative because we have not considered what effect variables other than *% low income* will have on *robbery rate*. Our conclusion about the rate of response of *robbery rate* to *% low income* may be altered if we include other variables in the regression investigation.

We will explore this possibility in Section 7.4. First we will learn how to evaluate the predictive ability of a simple regression line.

examination
of prediction
errors

Robbery rate varies in value from precinct to precinct. We have presumed that one reason for this is the variation in *% low income* from precinct to precinct. Logically, if variation in *% low income* were the *only source* of variation in *robbery rate*, a regression line including *% low income* as the sole predictor would provide extremely accurate, if not perfect, predictions of precinct robbery rates. On the other hand, if the differences observed among precinct robbery rates were unrelated to the differences in *% low income*, then the robbery rates predicted from a regression line that included *% low income* as the sole predictor would conform very poorly to the observed robbery rates, and the equation would produce large prediction errors.

Chart 7-6 is a partial list of the prediction errors that result if the regression line $\widehat{robbery\ rate} = 10.7 + 6.0 \cdot \%\ low\ income$ is used to predict precinct robbery rates. Column (1) lists the *% low income* observation X for each precinct, from smallest to largest. Column (2) identifies each precinct's observed robbery rate, Y. Column (3) contains the precinct robbery rates predicted by the regression line, $\hat{Y}$. Finally, column (4) lists the prediction errors as the observed minus the predicted robbery rate $(Y - \hat{Y})$.

The fact that column (4) is not a column of zeros — that prediction errors have been made — tells us that *% low income* is not the only variable that affects *robbery rate*. The omission of some other factor(s) resulted in a regression line that under-

chart 7-6
Table of Prediction Errors for 12 of the 69 Precincts

(1) % LOW INCOME X	(2) OBSERVED ROBBERY RATE Y	(3) PREDICTED ROBBERY RATE Ŷ	(4) PREDICTION ERROR (Y − Ŷ)
4.9	19.5	40.1	−20.6
7.1	41.2	53.3	−12.1
10.1	167.5	71.3	96.2
11.8	88.5	81.5	7.0
13.5	59.3	91.7	−32.4
14.8	117.5	99.5	18.0
16.2	64.2	107.9	−43.7
17.6	109.6	116.3	−6.7
24.4	171.4	157.1	14.3
26.2	264.0	167.9	96.1
31.7	178.0	200.9	−22.9
35.1	275.6	221.3	54.3

estimated the actual robbery rate in certain precincts—those with positive entries in column (4)—and that overestimated the actual robbery rate in other precincts—those with negative entries in column (4).

It would be helpful if we had a measure based on all these prediction errors that would provide an indication of how successful the regression line is as a predictor of *robbery rate*. In short, we need a *predictive ability score* for the regression line.

a measure of
predictive ability

The logical basis for scoring the predictive ability of the regression line can be outlined in four steps:

1. Measure the scope of the prediction errors that would result if we ignored the independent variable (that is, if we predicted the behavior of the dependent variable *without using the regression line*).
2. Measure the scope of the prediction errors that result *using the regression line* to predict the behavior of the dependent variable.
3. Subtract the second measure from the first. The difference represents the *reduction* in the scope of the prediction errors accomplished by using the regression line.
4. Express this reduction as a percentage to obtain the *percentage* reduction in the scope of the prediction errors accomplished by using the regression line. This is the predictive ability score.

Consistent with our use of the least squares criterion, we measure the scope of the prediction errors as a sum of squared prediction errors. A predictive ability score of 0 implies that the regression line proved useless in reducing the sum of the squared prediction errors. A predictive ability score of 50 percent implies that the sum of the squared prediction errors has been reduced by 50 percent (cut in half) by using the regression line. Finally, a predictive ability score of 100 percent indicates that no prediction errors occur when the regression line is used.

1. *Scope of the prediction errors made without using the regression line.*

Lacking information with which to predict the behavior of Y other than the observations on Y itself, we would predict every observation to be at the mean value $\overline{Y}$. As we showed in Section 2.2 (page 44), based on the least squares criterion, $\overline{Y}$ is the best choice of a *single* value to predict the scores in the distribution.

Accordingly, the value for the sum of the squared prediction errors made without using the regression line is given by

$$\Sigma(Y - \overline{Y})^2$$

An abstract representation of this quantity is given in Chart 7-7(a).

2. *Scope of prediction errors made using the regression line.*

Given additional information about the independent variable X, we can base our predictions on $\hat{Y}$ instead of $\overline{Y}$. Now the minimum value for the sum of the squared prediction errors is given by

$$\Sigma(Y - \hat{Y})^2$$

In Chart 7-7(b), $\Sigma(Y - \hat{Y})^2$ is depicted as a subset of $\Sigma(Y - \overline{Y})^2$.

3. *Reduction in scope of prediction errors.*

Subtracting the second measure from the first yields

$$\Sigma(Y - \overline{Y})^2 - \Sigma(Y - \hat{Y})^2$$

This difference is represented by the blue area in Chart 7-7(b).

4. *Percentage reduction in scope of prediction errors.*

Dividing the above difference by $\Sigma(Y - \overline{Y})^2$ yields

$$\frac{\Sigma(Y - \overline{Y})^2 - \Sigma(Y - \hat{Y})^2}{\Sigma(Y - \overline{Y})^2} \quad \text{or} \quad 1 - \frac{\Sigma(Y - \hat{Y})^2}{\Sigma(Y - \overline{Y})^2} \quad (7\text{-}6)$$

The result is a fraction between 0 and 1.0, or equivalently, a percentage between 0 percent and 100 percent. In Chart 7-7(b), this result is represented by the ratio of the blue area to the total area of the circle.

Utilizing these measures, we can compute the predictive

chart 7-7

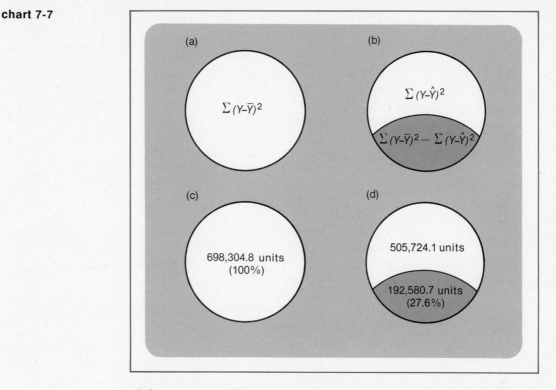

ability score of the *% low income* variable in the following four steps:

1. $\Sigma(Y - \overline{Y})^2 = 698{,}304.8$ [See Chart 7-7(c)]

In Chart 7-1, the standard deviation of *robbery rate* is reported as $s_Y = 100.6$. The formula for s_Y is

$$\sqrt{\frac{\Sigma(Y - \overline{Y})^2}{n}}$$

If we square s_Y and multiply by n, we obtain $\Sigma(Y - \overline{Y})^2$. Then

$$\Sigma(Y - \overline{Y})^2 = s_Y^2 \cdot n = (100.6)^2 \cdot 69 = 698{,}304.8$$

2. $\Sigma(Y - \hat{Y})^2 = 505{,}724.1$ [See Chart 7-7(d)]

Chart 7-6 lists the prediction errors $(Y - \hat{Y})$ for 12 precincts. The prediction errors for the remaining precincts can be found in a similar manner. Then by squaring each error and summing the squares, we obtain $\Sigma(Y - \hat{Y})^2$.

3. $\Sigma(Y - \overline{Y})^2 - \Sigma(Y - \hat{Y})^2 = 698{,}304.8 - 505{,}724.1 = 192{,}580.7$

4. $\dfrac{\Sigma(Y - \overline{Y})^2 - \Sigma(Y - \hat{Y})^2}{\Sigma(Y - \overline{Y})^2} = \dfrac{192{,}580.7}{698{,}304.8} = 0.276$, or 27.6%

In Chart 7-7(d), the blue area is 27.6 percent of the entire circle. So the regression line constructed with *% low income* as the sole predictor variable reduces the scope of the prediction errors by 27.6 percent. As such, this regression line is a relatively unsuccessful predictive device. It seems worthwhile for us to try to identify other predictors and to incorporate them into the relationship investigation.

an alternative explanation of the predictive ability score

Each of our measures of a sum of squared prediction errors is in essence a measure of *variation* — the variation of the observed values of Y about a line of predicted values $\hat{Y}$ or $\overline{Y}$. In fact, these measures are frequently labeled measures of variation. A familiarity with these labels is instructive because it will afford us an alternative explanation of the predictive ability score. The conventional labels are:

1. $\Sigma(Y - \overline{Y})^2$: *the total variation in* Y. A measure of the variation of individual Y observations about the mean $\overline{Y}$.
2. $\Sigma(Y - \hat{Y})^2$: *the unexplained variation in* Y. A measure of the variation of observed Y values about the regression line $\hat{Y}$. As such, it is considered a measure of the variation in Y that is unexplained by the regression line.
3. $\Sigma(Y - \overline{Y})^2 - \Sigma(Y - \hat{Y})^2$: *the explained variation in* Y. By subtracting the unexplained variation in Y from the total variation in Y, we obtain a measure of the amount of the total variation in Y that is explained by the regression line.
4. $\dfrac{\Sigma(Y - \overline{Y})^2 - \Sigma(Y - \hat{Y})^2}{\Sigma(Y - \overline{Y})^2}$:
   The explained variation in Y *as a percentage of the total variation in* Y, or *the percentage of the total variation in* Y *that is explained by the effect of* X.

If a predictor variable X has a score of 27.6 percent, we can state that 27.6 percent of the total variation in Y is explained by the effect of X. Therefore, 27.6 percent of the variation in *robbery rate* from precinct to precinct is explained by the effect of *% low income*.

The term *explained* has a purely statistical meaning and does not necessarily imply a cause-and-effect linkage. After all, both of the variables *robbery rate* and *% low income* may be correlated with outside factors. (The cause and effect fallacy was considered in Caveat 2 in Chapter 6, page 158, and is examined again in Caveat 4 at the end of this chapter.)

the R^2 statistic

If we were to compute the correlation coefficient $r(X, Y)$ between the variables *robbery rate* and *% low income,* we would find that

$$r(X, Y) = 0.525$$

If we then squared this value of $r(X, Y)$, we would obtain

$$[r(X, Y)]^2 = (0.525)^2 = 0.276, \text{ or } 27.6\%$$

This result is identical to the predictive ability score given to *% low income.*

> In the case of simple regression, the predictive ability score — the percentage reduction in the scope of prediction errors accomplished by the regression line —*is always equal to* the square of the correlation coefficient. In fact, the symbol for this score is R^2, and the score is often referred to as *the R^2 statistic.*[2]

Symbolically

$$R^2 = 1 - \frac{\Sigma(Y - \hat{Y})^2}{\Sigma(Y - \overline{Y})^2} = [r(X, Y)]^2 \qquad (7\text{-}7)$$

Equation (7-7) indicates that the value of the R^2 statistic — the predictive ability score — can be calculated simply by squaring the value of the correlation coefficient between X and Y.

We should think of the R^2 statistic as a version of the correlation coefficient $r(X, Y)$ that is better suited to measuring the closeness of the data points to a regression line. We would interpret the result $r(X, Y) = 0.5$ in the rather vague terms that

[2] R^2 is also called the *coefficient of determination.*

it suggests a moderate correlation between the variables. However, by squaring $r(X, Y) = 0.5$ to obtain the predictive ability score $R^2 = 25$ percent, we can state our conclusions more precisely: A 25 percent reduction in the scope of the prediction errors in Y has been accomplished by using the regression line with X as the predictor. Alternatively stated, 25 percent of the total variation in Y has been explained by the effect of the predictor X.

Moreover, this illustration suggests that a correlation

Despite a high R², this regression line is for the birds.

coefficient of 0.5 indicates that the independent variable has earned a score of only 25 percent in terms of predictive ability — not 50 percent.

Why do we concern ourselves with the correlation coefficient $r(X, Y)$ at all if its square, the R^2 statistic, allows us to make a more precise interpretation of the strength of the relationship between two variables? Think back to our investigation of the relationship between the *President's vote %* and *House support* in Chapter 6. There we were interested only in the issue of correlation — the direction and strength of the relationship — and not in the issue of regression. In such cases, $r(X, Y)$ tells us both the direction and the strength of the relationship, whereas R^2, which is always a positive quantity, gives us no information about the direction of the relationship. (Of course, in a regression investigation, we can determine the direction of the relationship by observing the sign on the slope coefficient b of the regression line.)

EXERCISES

7-12   In the simple regression $\widehat{robbery\ rate} = a + b \cdot \%\ low\ income$, the value of the R^2 statistic was 27.6 percent. One student interpreted this result as follows: "A precinct's robbery rate can be predicted perfectly by the regression equation 27.6 percent of the time." Is this a valid interpretation? (*Hint:* What does this interpretation imply about the frequency of the data points that lie right on the regression line?)

7-13   Referring to the regression equations in Exercise 7-4 on page 176, the values of the R^2 statistics are
(a) $R^2 = 54.7$ percent
(b) $R^2 = 91.9$ percent
(c) $R^2 = 83.0$ percent
(d) $R^2 = 37.7$ percent
Interpret each result, using the names of the variables involved in each case.

7-14   In Exercise 6-2 on page 151, you computed the correlation coefficient relating *food* to *income* for 1971–1976.
(a) Using the value for $r(X, Y)$ obtained there, calculate the R^2 statistic.
(b) What does the result tell us about the predictive ability of the regression line?
(c) What is the meaning of the result as a measure of explained variation?

7-15   Refer to the *cost–distance* data for the five Vermont cities given in Exercise 7-9 on page 178. The value of $r(cost, distance) = 0.974$. To verify that $R^2 = [r(X, Y)]^2$:

(a) Calculate the *total* variation in cost $\Sigma(Y - \overline{Y})^2$. Note that $\Sigma(Y - \overline{Y})^2$ is equivalent to $\Sigma Y^2 - n \cdot \overline{Y}^2$, which you can compute easily from the information given in Exhibit II.

(b) Calculate the *unexplained* variation in cost $\Sigma(Y - \hat{Y})^2$. Note that in Exercise 7-9(d), you obtained each prediction error $(Y - \hat{Y})$. If you square each value of $Y - \hat{Y}$ and sum the squares, you obtain $\Sigma(Y - \hat{Y})^2$.

(c) Calculate the *explained* variation in cost.

(d) Calculate $R^2 = \dfrac{\text{Explained variation in cost}}{\text{Total variation in cost}}$.

(e) Interpret the value you obtain for R^2.

7.4 multiple regression

Use of the predictor *% low income* succeeds in explaining 27.6 percent of the total variation in *robbery rate* among New York City precincts, leaving almost three-fourths, or 72.4 percent, of the total variation in *robbery rate* unexplained. A combination of other factors apparently plays a dominant role in determining the *robbery rate* of a precinct.

a second predictor

In a properly designed regression investigation, all of the theoretically plausible predictors should be considered. However, a list of the potential predictors of *robbery rate* would be extensive. At this time, therefore, we will introduce only one additional predictor—*density*.

Density is the number of precinct residents per square mile. *Density* reflects the congestion of living conditions in a precinct. All other factors being equal, we can expect more densely populated precincts to have larger crime rates.

How do we incorporate this new predictor into our regression investigation? We could consider performing another simple regression, using *density* in place of *% low income* as the independent variable. We could then attempt to combine the results of the two simple regression equations in some way.

As we identified more and more predictors, we could continue to amass simple regression results.

This line of reasoning should be discarded immediately.

When we feel that the dependent variable responds to two or more predictors, the appropriate form of investigation is not a series of simple regression equations but *one multiple regression equation.*

We will now examine the pitfalls of using simple regressions when we are dealing with more than one independent variable. We will then perform the appropriate multiple regression investigation.

the need for
a new approach

When we are dealing with two or more independent variables, the issues involved are an extension of the issues motivating simple regression and correlation:

1. *The predictive ability of the set of independent variables.* How well can we explain the total variation in *robbery rate* on the basis of *% low income* and *density* as dual predictors? This question asks us to compute a predictive ability score analogous to the one we found for *% low income* alone in the simple regression we performed earlier.

2. *The direction and rate of response of the dependent variable to a change in one predictor if all other predictors stay the same.* How does *robbery rate* respond to increases in *% low income* if we assume that *density* is unchanged? How does *robbery rate* respond to increases in *density* if we assume that *% low income* is unchanged?

Now we will see why neither question can be answered properly by a series of simple regressions (in this case, two). As a point of departure, a summary of the simple regression results is presented in Chart 7-8.

We already have interpreted the results of the equation $\hat{Y} = 10.7 + 6.0X_1$. The equation $\hat{Y} = 25.8 + 0.013X_2$, which relates *robbery rate* to *density*, explains 30.1 percent of the total variation in *robbery rate* and predicts that *robbery rate* will

chart 7-8

$Y =$ robbery rate	$X_1 = \%$ low income	$X_2 =$ density
$\hat{Y} = 10.7 + 6.0X_1$		$R^2 = 27.6\%$
$\hat{Y} = 25.8 + 0.013X_2$		$R^2 = 30.1\%$

respond with an increase of 0.013 robberies per 10,000 residents for every 1 unit increase in *density* (one additional person per square mile). Stated another way, this equation predicts that an additional 1.3 robberies per 10,000 residents will occur for each additional 100 residents per square mile.

The Predictive Ability of the Set of Independent Variables X_1 and X_2

The predictive ability score, R^2, of *% low income* is 27.6 percent and the R^2 for *density* is 30.1 percent. The sum of these two R^2 values is 57.7 percent. Is 57.7 percent the predictive ability score for the combination of the independent variables *% low income* and *density*? Can we add two R^2 values determined by simple regression to obtain the predictive ability score for the set of independent variables?

In general, the answer is "No, we can't." In fact, the whole typically is smaller than the sum of its parts.

The key to the problem lies in the existence of *correlation* between the predictors. The correlation coefficient of *% low income* and *density*, $r(X_1, X_2)$, is 0.485, which indicates a moderately positive correlation. As we move from precinct to precinct, *% low income* and *density* tend to increase together.

When some correlation exists between predictor variables, $r(X_1, X_2) \neq 0$, knowledge of the behavior of one variable gives us some information about the behavior of the other variable. The positive correlation between *% low income* and *density* tells us that precincts that are relatively high in *% low income* tend to be relatively high in *density*, and vice versa. Therefore, if we know the *% low income* scores for the New York City precincts, obtaining the additional data on *density* scores does not provide us with wholly *new* information. Consequently, if we were to sum the R^2 values, we would be guilty of double counting some information.

Chart 7-9 illustrates this consequence of the correlation between the two predictors *% low income* and *density*. Each

chart 7-9

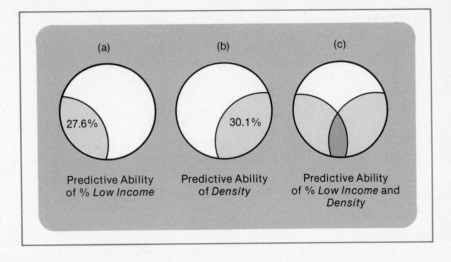

(a) (b) (c)

27.6%

30.1%

Predictive Ability
of % *Low Income*

Predictive Ability
of *Density*

Predictive Ability
of % *Low Income* and
Density

outer circle in this chart represents the total variation in *robbery rate*. The light blue area in 7-9(*a*) depicts the R^2 between *robbery rate* and *% low income*. The light blue area in 7-9(*b*) represents the R^2 between *robbery rate* and *density*. Chart 7-9(*c*) reveals that these areas overlap. The area of overlap — the dark blue area — reflects the degree of correlation between the two predictors. Due to this overlap, the portion of the total variation in *robbery rate* that is explained by the combined effects of *% low income* and *density* — the colored area in Chart 7-9(*c*) — is smaller than the sum of its two components.[3]

It is apparent that the combined predictive ability score of *% low income* and *density* will be lower than 57.7 percent — the sum of the simple R^2s. The correct score, we will see, must be determined by multiple regression.

The Direction and Rate of Response to a Change in Each Predictor

A correlation between predictors not only invalidates the addition of simple R^2s but also distorts the simple regression coefficients.

The first simple regression equation in Chart 7-8 predicts

---

[3] Theoretically, it is possible for the "whole" to exceed the sum of its parts. However, this situation occurs so infrequently that we will give it no further consideration.

an increase of 6.0 robberies per 10,000 residents for each 1 point increase in *% low income*. As *% low income* increases, however, *density* tends to increase as well. Consequently, the value of the slope coefficient 6.0 measures the response of *robbery rate* not only to the increase in *% low income* but also to the hidden changes in *density* and other variables that are correlated with *% low income*.

If we want to know how *robbery rate* responds strictly to a change in one predictor (*% low income*), we must screen out the impact of hidden changes in the other predictor (*density*). Conceptually, we might try to accomplish this goal in the following manner.

We could select a group of precincts that have identical levels of *density*. Using this group, we could then perform a simple regression relating *robbery rate* to the predictor *% low income*. Because *density* does not differ among these precincts, *density* cannot contribute to the differences in the robbery rates of these precincts. So the simple regression would reveal how *robbery rate* responds to changes in *% low income*, free from the distortion produced by hidden changes in *density*.

Similarly, we could select a group of precincts identical in terms of *% low income*. Using this group, we could then perform a simple regression predicting *robbery rate* on the basis of *density*. Because *% low income* is constant across these precincts, *% low income* cannot contribute to the differences in the robbery rates of these precincts. The simple regression therefore would indicate the influence of *density* on *robbery rate* free from the distortion of hidden changes in *% low income*.

In principle, then, the problem of correlated predictors can be eliminated by holding the value of one predictor constant while measuring the influence of the other predictor. In laboratory sciences, this goal is met by performing controlled experiments in which all factors other than the one whose effect is being studied are held constant.

In our study of crime rates, a controlled experiment is not feasible. Observations on any one predictor will be nearly identical in only a few precincts, and any attempt to devise groups of precincts with comparable *% low income* or *density* levels would fail for lack of sufficient data.

Fortunately, we do not need to resort to such a procedure.

The objective of investigating relationships under controlled conditions is achieved statistically by performing a multiple regression.

7.5   the multiple regression equation with two predictors

When we performed the simple regression relating *robbery rate* to *% low income*, we knew two facts about each precinct: the precinct's robbery rate, and the precinct's percent of low-income families. Now we know three facts—the third fact is the precinct's population density.

the regression plane

In simple regression, we can plot the observations on a two-dimensional scatter diagram and depict the regression equation as a *line* going through the center of the scatter. Now we must think in three dimensions. To plot a scatter diagram of the observations on three variables will require three axes—one axis for each variable. A regression equation relating the behavior of one variable Y to each of the other variables X_1 and X_2 must be envisioned as a *plane* going through the center of the scatter.

We can express the equation of a regression plane in general terms as

$$\hat{Y} = b_0 + b_1 X_1 + b_2 X_2 \tag{7-8}$$

where b_0, b_1, and b_2 are the regression coefficients. Given particular values of X_1 and X_2, this equation supplies a predicted value for Y.

In simple regression, we adopt the least squares criterion for selecting the regression line. In multiple regression, we do the same: We select the regression plane by finding the values for b_0, b_1, and b_2 that keep the sum of the squared prediction errors to a minimum.

The equations defining the least squares regression coefficients are complex, and presenting them here would serve little purpose other than to provide a reference for computation. Realistically, such computations are left to the computer. So we will omit the least squares equations from this discussion and concentrate on the interpretation of the computed values of the least squares regression coefficients, which are presented in Equation (7-9):

$$\hat{Y} = -12.9 + 3.9X_1 + 0.009X_2 \qquad R^2 = 39.2\% \qquad (7\text{-}9)$$

so

$b_0 = -12.9$ (intercept)
$b_1 = \quad 3.9$ (coefficient on X_1, *% low income*)
$b_2 = \quad 0.009$ (coefficient on X_2, *density*)

Chart 7-10 illustrates the equation of this regression plane. Given the observed values of *% low income*, X_1, and *density*, X_2, in each precinct, there is a point on the plane representing the predicted value of each precinct's *robbery rate*, $\hat{Y}$.

interpreting the
coefficients

Now let's see what each coefficient tells us:

$b_0 = -12.9$:   The point at the intersection of line OY (labeled b_0) denotes the vertical height of the plane directly below the origin (labeled O). The origin itself is the point at which all three variables—X_1, Y, and X_2—assume values of 0. Along line OY, variables X_1 and X_2 maintain values equal to 0. *Thus, b_0 is the value we would predict for Y when both X_1 and X_2 are equal to 0.*

Finding $b_0 = -12.9$, we would predict that -12.9 robberies per 10,000 residents would occur in a hypothetical precinct with a *% low income* score of 0 and a *density* score of 0. A negative value for *robbery rate* is impossible. The negative intercept occurred because of *extrapolation*, as explained in Caveat 1 on page 201.

Mathematically, b_0 is the Y intercept of the regression plane. If we substitute $X_1 = 0$ and $X_2 = 0$ into Equation (7-8), we obtain

$$\hat{Y} = b_0 + b_1 \cdot 0 + b_2 \cdot 0 = b_0$$

$b_1 = 3.9$:   Graphically, b_1 represents the rate of incline or decline (the former in Chart 7-10) in the height of the plane as we move from point A to point C. Line AC is parallel to the X_1 axis and perpendicular to the X_2 axis. As we proceed along line AC, the variable X_1 increases in value but the value of variable X_2 does not change. *Thus, b_1 represents the predicted rate of response of Y to an increase in X_1 while X_2 remains constant.*

Finding $b_1 = 3.9$, we would predict that, with density held constant, an increase of 3.9 robberies per 10,000 residents would occur for each 1 point increase in *% low income*.

$b_2 = 0.009$:   Graphically, b_2 represents the rate of incline or decline (the former in Chart 7-10) in the height of the plane as we move from point A to point B. Line AB is parallel to the X_2 axis and perpendicular to the X_1 axis. As we proceed along line AB, X_2 increases in value but the value of X_1 does not change. *Thus, b_2 measures the predicted rate of response of Y to an increase in X_2 while X_1 remains constant.*

Finding $b_2 = 0.009$, we would predict that if *% low income* does not change, there would be an increase of 0.009 in *robbery rate* for each additional resident per square mile. This implies that a predicted increase of slightly less than 1 robbery would occur per 10,000 residents for each additional 100 residents per square mile.

The coefficients b_1 and b_2 are called *partial regression coefficients.* Each of these coefficients measures a partial rate

chart 7-10

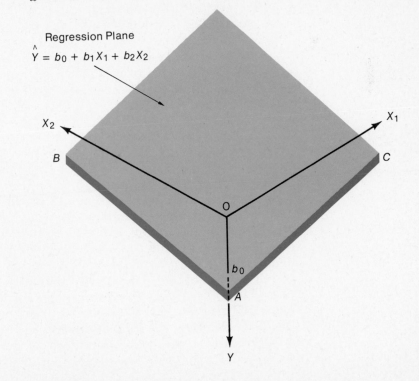

Regression Plane
$\hat{Y} = b_0 + b_1X_1 + b_2X_2$

of response—the rate of response of Y to a change in the value of one predictor while the value of the other predictor is held constant. The sign ($+$ or $-$) on each coefficient reveals the direction of the partial relationship. The positive sign on b_1 indicates that if X_2 does not change in value, then Y is positively related to X_1. The positive sign on b_2 indicates that if X_1 is held constant, then Y is positively related to X_2.

predictions

To give some perspective to these regression results, let's compare the predicted *robbery rates* for precincts A and B, assuming that

> In Precinct A: *% low income* $= 10$ and   *density* $= 10,000$
> In Precinct B: *% low income* $= 30$ and   *density* $=$ 1,000

By inserting the *% low income* (X_1) and *density* (X_2) scores into Equation (7-9), we obtain

> Precinct A: $\hat{Y}_A = -12.9 + 3.9 \cdot 10 + 0.009 \cdot 10,000 = 116.1$
> Precinct B: $\hat{Y}_B = -12.9 + 3.9 \cdot 30 + 0.009 \cdot 1,000 \; = 113.1$

Thus, we would predict that 3 more robberies per 10,000 residents would occur in Precinct A than in Precinct B.

Now subtracting the second equation from the first, we obtain

$$\hat{Y}_A - \hat{Y}_B = 0 + 3.9 \cdot (10 - 30) + 0.009 \cdot (10,000 - 1,000)$$
$$= -78 + 81 = 3$$

Because *% low income* is 10 percent in Precinct A versus 30 percent in Precinct B, we would predict that *if both precincts have the same density*, then 78 fewer robberies per 10,000 residents would occur in Precinct A. However, the *density* values are not the same: Precinct A has a much higher *density*.

Given the higher *density* in Precinct A, we would predict that *if both precincts have the same % low income*, then 81 more robberies per 10,000 residents would occur in Precinct A.

If we combine the information on *% low income* and *density*, we can see that the higher *density* in Precinct A is more than sufficient to compensate for the lower *% low income* in this precinct. Therefore, we can predict that 3 more robberies per 10,000 residents will occur in Precinct A.

a word
of caution

In general, we cannot compare the values of the partial regression coefficients in a multiple regression equation. That the coefficient on *% low income* ($b_1 = 3.9$) is larger than the coefficient on *density* ($b_2 = 0.009$) does *not* imply that changes in *% low income* have a stronger influence on *robbery rate* than do changes in *density*. First, the variables *% low income* and *density* are expressed in different units (*percent* of families whose income is below \$4,000 versus *number* of residents per square mile). Thus, a 1 unit change in *% low income* (1 percentage point) is not conceptually equivalent to a 1-unit change in *density* (1 person per square mile). Second, even if the units of the predictors X_1 and X_2 were identical, if the variability of the distributions are different, a 1 unit change will represent a more substantial change for one variable than it will for the other variable. Finally, the units used to express a variable are arbitrary. If instead of expressing *density* as the number of persons per square mile, we had expressed X_2 as the number *of thousands* of persons per square mile, then the coefficient on *density* in Equation (7-9) would have been 9.000 rather than 0.009, a higher value than the coefficient on *% low income*.

Comparisons of partial regression coefficients in the same equation are meaningful only when the predictors whose coefficients are being compared are expressed in common units and vary approximately to the same extent (have equal standard deviations). Otherwise, adjustments must be made to permit comparisons. The nature of these adjustments is discussed in textbooks that specialize in regression analysis.

the R^2 statistic

Earlier, we measured the predictive ability of a *simple* regression equation by the R^2 statistic:

$$R^2 = 1 - \frac{\Sigma(Y - \hat{Y})^2}{\Sigma(Y - \overline{Y})^2} = 1 - \frac{\text{Unexplained Variation in } Y}{\text{Total Variation in } Y}$$

$$= \frac{\text{Explained Variation in } Y}{\text{Total Variation in } Y}$$

The predictive ability score for a multiple regression equation is defined in exactly the same way. Movement from simple to multiple regression — adding a second predictor to the equation — does not change the total variation in Y but *is* expected to reduce the unexplained variation in Y, because additional information is included in the equation. (At its worst, the additional variable will have no predictive value, and the R^2 statistic will not improve.)

The R^2 statistic for Equation (7-9) — our multiple regression equation relating *robbery rate* to *% low income* and *density* — is equal to 39.2 percent. Thus, 39.2 percent of the total variation in *robbery rate* is explained by the effects of *% low income* and *density*.

This is an improvement over the predictive ability of the results of the simple regression equation ($R^2 = 27.6$ percent) that includes *% low income* as the sole independent variable. The 39.2 percent result is also an improvement over the results of the simple regression equation ($R^2 = 30.1$ percent), in which *density* is the sole predictor. But the multiple regression's R^2 of 39.2 percent is far less than the sum of the two single regression R^2s (27.6 percent + 30.1 percent = 57.7 percent). Due to the correlation between *% low income* and *density*, the sum 57.7 percent substantially exaggerates the combined predictive ability of the two independent variables.

7.6 **three or more predictors**

The principles of multiple regression represented in the two-predictor equation given in Section 7.5 extend without modification to equations incorporating three or more predictors. In the case of three predictors X_1, X_2, and X_3, we write the multiple linear regression equation

$$\hat{Y} = b_0 + b_1X_1 + b_2X_2 + b_3X_3 \qquad (7\text{-}10)$$

and compute values for the b's in a way that satisfies the least squares criterion.

The value we obtain for b_0 represents the predicted value of Y when X_1, X_2, and X_3 are all equal to 0.

The value we obtain for b_1 represents the predicted rate of response in Y to changes in X_1, if X_2 and X_3 remain constant.

Similarly, the value of b_2 gives us the predicted rate of response in Y to changes in X_2, if X_1 and X_3 remain constant. Finally, b_3 indicates the predicted rate of response in Y to changes in X_3, if X_1 and X_2 remain unchanged.

No matter how many variables (X's) are included as predictors in the multiple regression equation, each partial regression coefficient (b_1, b_2, . . .) supplies a predicted rate of response in Y to changes in one predictor, holding all other predictors in the equation constant.

EXERCISES

7-16   Consider the equation of a regression plane

$$\hat{Y} = 10 + 5X_1 - 3X_2$$

(a) What is the predicted value of Y when both X_1 and X_2 are equal to 0?
(b) What is the predicted rate of response in Y to changes in X_1 if X_2 is held constant?
(c) What is the predicted response in Y for each unit change in X_2 if X_1 does not change?
(d) What is the predicted value of Y when $X_1 = 10$ units and $X_2 = 20$ units?

7-17   Consider the following sequence of regression equations:
(a) $\hat{Y} = b_0 + b_1X_1$
(b) $\hat{Y} = b_0 + b_1X_1 + b_2X_2$
(c) $\hat{Y} = b_0 + b_1X_1 + b_2X_2 + b_3X_3$
Interpret the meaning of b_1 in each equation.

7-18   Redraw Chart 7-9(c) on page 191 to depict the multiple R^2 of a regression equation relating *robbery rate* to both % *low income* and *density,* assuming these two predictors are uncorrelated: r(% *low income, density*) $= 0.0$. What would the value of the multiple R^2 be in this event?

7-19   A political analyst used multiple regression to assess the impact of campaign expenditures on votes received. For each candidate for city council, the analyst acquired data on
 votes: The number of votes received by the candidate.
 bucks: The candidate's campaign expenditures in thousands of dollars.
 party: The number of voters from the candidate's election district who were registered members of the candidate's party.
The results were reported as

$$\widehat{votes} = 200 + 50.0\ bucks + 0.6\ party \qquad R^2 = 60\%$$

(a) What did a *buck* buy in this election?

(b) Interpret the partial regression coefficient on *party*.

(c) Why was the variable *party* included as a predictor when the analyst was interested in the relationship between *votes* and *bucks*?

(d) Interpret the value of the R^2 statistic.

(e) Predict the number of *votes* that would be received by a candidate who was not affiliated with any party and who incurred no campaign expenditures.

(f) Predict the number of *votes* that would be received by a candidate who spent \$3,000 (3 *bucks*) and whose election district contained 500 voters registered in the candidate's party.

7-20   The American Institute of Media Studies (AIMS) investigated the relationship between the television-viewing habits of U.S. families and such characteristics as family size and income level. A particular set of regression results from this study follows:

(1) $\widehat{hours} = 250 - 5.7 \cdot income$ $R^2 = 30\%$

(2) $\widehat{hours} = 75 + 34.1 \cdot size$ $R^2 = 60\%$

(3) $\widehat{hours} = 120 - 4.0 \cdot income + 28.2 \cdot size$ $R^2 = 80\%$

where

    hours: Number of TV hours watched during a test month.

    income: Family income in thousands of dollars.

      size: Number of individuals in the household.

(a) Explain why the effects of *income* and *size* on *hours* can be more accurately evaluated by using Equation (3) than by using the pair of Equations (1) and (2).

(b) The *income* of two families of the same *size* differs by 10 units (\$10,000). How many more hours of TV would you predict that the poorer family has watched?

(c) The *size* of two families with the same *income* differs by two members. How many more hours of TV would you predict that the larger family has watched?

(d) Interpret the value of the R^2 statistic for each equation.

7-21   A study of Federal Parole Board actions regarding youths sentenced under the Youthful Corrections Act reported that:

A youth is first considered for parole six months after the prison term begins.

The number of additional months the youth must spend in prison increases by an average of three months for every instance of *bad be-*

havior while in prison and by an average of five months for each unit of *severity* of crime.

If *time* denotes the months a youth spends in prison before parole, construct the implied regression equation relating the variables *time, bad behavior*, and *severity*.

Caveats on Interpreting Regression Results

Regression analysis is a powerful statistical tool. However, like all tools, it must be used thoughtfully and must not be forced into applications that are beyond its capability. We will conclude our discussion of regression analysis with several warnings against the abuse of this tool and misinterpretation of the information it provides.

CAVEAT 1:   Do not extrapolate regression results.

Extrapolation occurs when predictions of the dependent variable are made at levels of the independent variable(s) that lie outside the range of observed information. In our fuel economy illustration in Chapter 5, we tabulated the weight, X, and gas mileage, Y, of twelve automobiles with weights ranging approximately from 2,000 to 5,000 lb. A regression line relating gas mileage to automobile weight would be built on weights within this observed range. We would be *extrapolating* these results if we extended them to predict the gas mileage of automobiles that weigh less than 2,000 lb or more than 5,000 lb.

Extrapolation is inherently risky. To extrapolate a regression line is to extend it into uncharted territory. Because we lack evidence to verify the rate of response—or even the direction—of the relationship that prevails outside our range of observation on the independent variable(s), extrapolation has no foundation or support (other than good luck or extrasensory perception). The extrapolated regression line illustrated in Chart 7-11 would lead us to predict an unprecedented number

chart 7-11

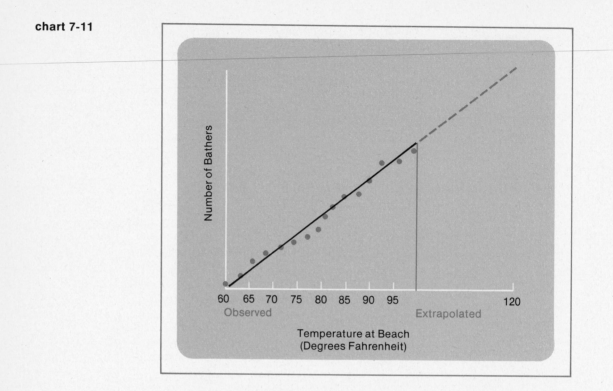

of bathers at the beach when the temperature reached 120°, whereas it is unlikely that more than a few people would venture outside their air-conditioned quarters at all at this temperature.

Not infrequently, we find that the literal interpretation of the *intercept* of a regression equation is implicitly an extrapolation. In this circumstance, the intercept plays a useful mathematical role — without an intercept, a regression line would be forced through the origin — but lacks operational meaning and should not be reported without qualification.

The problem arises when a regression equation includes one or more independent variables for which a 0 value is not observed or is theoretically inconceivable. In the equation $\hat{Y} = a + bX$ relating automobile gasoline mileage, Y, to automobile weight, X, the intercept a denotes the predicted gas mileage of a weightless car. The observation $X = 0$ (weightless car) both is outside the observed range of automobile weights and is, of course, a physical impossibility. If we report the intercept result at all, we should qualify it by saying, *"Hypothetically,* a weightless car would be predicted to travel *a* miles per gallon."

The same qualification applies to the intercepts in the *robbery rate* equations introduced in this chapter. No precinct had a *% low income* value as low as 0, although this occurrence is theoretically possible. Moreover, no precinct had a *density* value of 0 — an impossible occurrence because it implies a precinct without residents. So the negative intercept we obtained in our multiple regression investigation of *robbery rate* ($b_0 =$ -12.9) lacks operational meaning. It represents the predicted *robbery rate* for the hypothetical precinct that lacks not only low-income families but any residents at all.

CAVEAT 2:   Beware of simple regression results.

The use of simple regression when a multiple regression investigation is appropriate — or, more generally, the failure to include important independent variables in a regression investigation — is likely to generate distorted regression results. Such results do not reveal the rate of response of the dependent variable strictly to a change in the independent variable. The independent variable within the equation is apt to be correlated with the potential predictors that have been excluded from the equation, and failure to hold these other indicators constant distorts our computed regression results.

Consider the results of our *robbery rate* investigation, summarized in Chart 7-12.

For both predictors — *% low income* and *density* — the simple regression coefficients exceed the partial regression coefficients. Thus, the simple regression equation relating *robbery rate* to *% low income,* by failing to hold *density* con-

chart 7-12		PARTIAL REGRESSION COEFFICIENT (from multiple regression equation)	SIMPLE REGRESSION COEFFICIENT (from separate simple regression equations)
	% Low Income	3.9	6.0
	Density	0.009	0.013

stant, exaggerates the rate of response of *robbery rate* to changes in *% low income.*

Similarly, the simple regression equation relating *robbery rate* to *density,* by failing to hold *% low income* constant, exaggerates the rate of response of *robbery rate* to changes in *density.*

The exclusion of a correlated predictor may also cause the measured rates of response to be too *low* or even to assume the wrong sign. In the latter case, we would be led to report that the dependent variable responds positively (negatively) to changes in a predictor when the direction of response — holding all other predictors constant — is actually negative (positive).

CAVEAT 3:   Consider the possibility of nonlinear relationships.

In Caveat 1 in Chapter 6, we pointed out that the correlation coefficient $r(X, Y)$ refers only to the closeness of the *linear* relationship between two variables. The possibility of a nonlinear relationship is an issue in regression analysis as well. When we investigate the behavior of a dependent variable through a regression line or a regression plane (multiple regression), we are assuming that the response of Y to a change in any predictor is linear. Each time a predictor increases by 1 unit, the rate of response of Y is always the same.

In contrast, if a nonlinear relationship exists between two variables, then one variable's rate of response to a change in the other variable will vary under different circumstances. For example, the rate of response that results when the predictor changes from an initially low value may markedly differ from the rate of response that results when the predictor changes from an initially high value.

In the case of simple regression, we can check our assumption of linearity by plotting the data in a scatter diagram to see whether the data points appear to follow a curved (nonlinear) pattern. In the case of multiple regression, more sophisticated procedures may be necessary. We will not examine these advanced procedures here but simply point out that regression results based on linear equations should not be accepted at face value without considering the possible existence of nonlinear relationships.

CAVEAT 4:   The cause and effect fallacy revisited.

Caveat 2 in Chapter 6 outlined the dangers of assigning *cause* and *effect* labels to variables that are highly correlated. The same warning applies in regression analysis. Although we tend to think of the independent variables as *causes* and the dependent variables as *effects*, we should not ignore the possibility of mutual interaction between the dependent variable and a predictor or the possibility that the behavior of both a predictor and the dependent variable could be determined by external factors.

The robbery rate regression equations given in this chapter are a case in point. The results presumably show how precinct *robbery rate* values vary in response to changes in *% low income* and *density*. It is possible, however, that the persistence of high crime rates in an area will motivate a population exodus from that area, causing a decrease in population density. Moreover, if upper-income families move disproportionately from areas with high crime rates, the percentage of low-income families in these areas will rise. The likelihood of mutual interaction between the crime and socioeconomic variables should not only restrain us from drawing one-way cause and effect conclusions but also should motivate us to develop regression equations that permit us to measure this phenomenon explicitly.

COMING
TO
TERMS

Dependent Variable   The variable whose behavior (response) is the subject of investigation. Sometimes called the *response variable.*

Independent Variable   A variable that is believed to influence the behavior of the dependent variable. Also called a *predictor.*

Intercept of a Regression Equation   The value we would predict for the dependent variable when the independent variables all equal 0.

Least Squares Regression Line   The regression line chosen to keep the sum of the squared prediction errors to a minimum.

Partial Regression Coefficient   The predicted rate of response of the dependent variable to a change in one predictor if all other predictors in the regression equation remain unchanged.

Prediction Error   The difference between an observed value on the dependent variable Y and the value predicted using the regression equation $\hat{Y}$. Symbolically, Prediction Error $= Y - \hat{Y}$.

Predictive Ability Score   The percentage reduction in the scope of the prediction errors that is accomplished by using the regression line. Referred to as the R^2 statistic.

R^2 Statistic   A measure of the predictive ability of a regression equation. Referred to as the *predictive ability score,* because it measures the percentage reduction in the scope of the prediction errors that is accomplished by using the regression equation. Alternatively referred to as a *measure of explained variation,* because it measures the percentage of the total variation in the dependent variable that is explained by the regression equation.

Regression Equation   An equation that predicts the value of the dependent variable based on the value of one or more independent variables.

Simple Regression Equation   A regression equation that includes only one independent variable. If the relationship between the independent variable and the dependent variable is assumed to be linear, the simple regression equation is the equation of a straight line.

Multiple Regression Equation   A regression equation that includes two or more independent variables. In the case of two independent variables, the regression equation is the equation of a plane. When more than two predictors are included, no graphic representation is possible.

Slope of a Simple Regression Equation   The predicted rate of response of the dependent variable to changes in the independent variable.

Total Variation in Y The variation of individual Y observations from the mean $\overline{Y}$, measured by $\Sigma(Y - \overline{Y})^2$.

Unexplained Variation in Y The variation of individual Y observations from the regression line, measured by $\Sigma(Y - \hat{Y})^2$.

ISSUES
FOR
ANALYSIS

1. A cosmetics firm wishes to determine whether it will be more profitable to spend its advertising budget on television or newspaper advertising. A recent industry-wide study revealed that:

   (1) $\hat{Y} = 150,000 + 3.0X_1$ $R^2(X_1, Y) = 0.25$
   (2) $\hat{Y} = 250,000 + 2.0X_2$ $R^2(X_2, Y) = 0.75$

   where

   Y denotes the revenue (dollars of sales) of the cosmetics firm.

   X_1 denotes expenditures (dollars spent) on television advertising by the cosmetics firm.

X_2 denotes expenditures (dollars spent) on newspaper advertising by the cosmetics firm.

(a) Carefully explain what these results indicate about the relative merits of television versus newspaper advertising.

(b) Where would you advise the company to place its advertising dollars if it had budgeted a total of $100,000 for advertising?

2. The least squares criterion for the regression line $\hat{Y} = a + bX$ requires us to find the line that keeps the sum of the *squared* prediction errors to a minimum—that is, to minimize $\Sigma(Y - \hat{Y})^2$.

(a) Why is the least squares criterion used rather than the conceptually simpler criterion: Find the line that keeps the sum of the unsquared predictions as close to 0 as possible—that is, minimize the size of $\Sigma(Y - \hat{Y})$?

(b) What line—a line with what slope and intercept—always satisfies the criterion $\Sigma(Y - \hat{Y}) = 0$?

3. A statistician developed a regression equation to be used by National Motors to forecast future automobile sales. The equation included the variables

   sales: The number of cars sold per year.

   scrap: The number of cars sold six years ago.

   income: The U.S. income level during the year in billions of dollars.

   trend: A variable set equal to 1 in 1950 (when the company began production), to 2 in 1951, and so on, to 29 in 1978.

   The results for two equations were reported:

   (1) $\widehat{sales} = 40{,}000 + 10{,}000 \; trend$ $R^2 = 46.9\%$

   (2) $\widehat{sales} = -25{,}000 + 0.5 \; scrap + 150 \; income$ $R^2 = 63.8\%$

   (a) Use Equation (1) to report the average annual increase in the number of cars sold by National Motors between 1950 and 1978.

   (b) Use Equation (1) to forecast *sales* for 1979.

   (c) Interpret the coefficient of *scrap* in Equation (2).

   (d) Actual *sales* in 1973 were 250,000. Income in 1979 is estimated to be $2,000 billion. Based on this information, forecast 1979 sales.

   (e) Compare the 1979 forecasts based on Equation (1) and Equation (2). In which one would you have more faith?

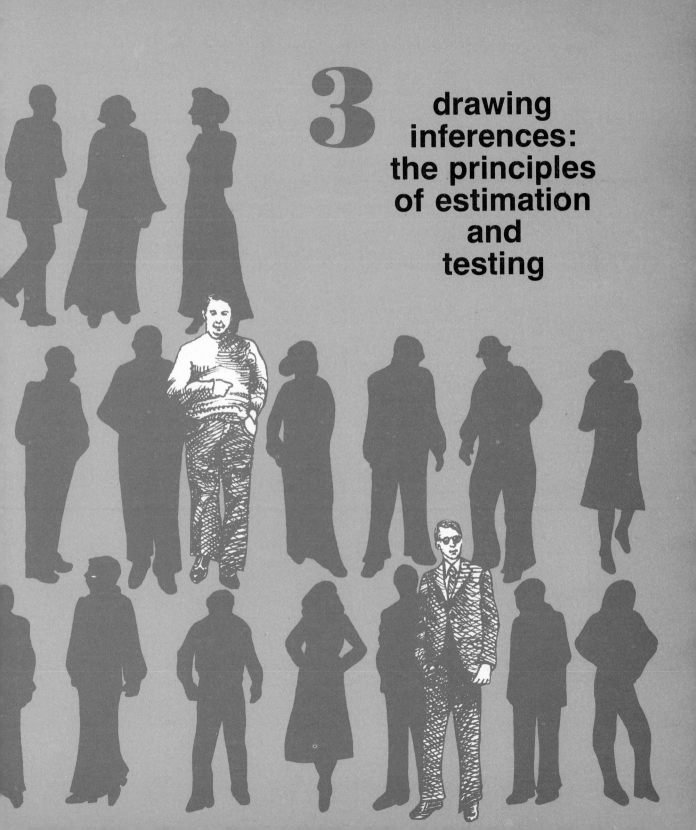

3

drawing inferences: the principles of estimation and testing

COLLEGE BULLETIN
President Releases Faculty Salary Data

In last month's *Bulletin*, we reported that faculty salaries at the college average substantially below the national mean of $18,000 for private colleges of comparable size. Our statement was based on the responses of 21 of our 100 faculty members to the questionnaire circulated by the *Bulletin*.

Troubled by the implications of this report, the president last week agreed to permit publication of faculty salaries at the college. The faculty members' names have been withheld, although their ranks and office locations have been provided.

In contrast to the results of the questionnaire, complete data show that the mean faculty salary at the college is $18,805, several hundred dollars above the comparable national average.

Faculty Salaries

Instructors: Mean Salary = $12,000

ID	BUILDING	SALARY	ID	BUILDING	SALARY	ID	BUILDING	SALARY
00	A	$10,000	04	B	$12,000	07	D	$12,000
01	A	12,000	05	C	10,250	08	D	13,250
02	A	14,000	06	C	13,750	09	E	11,500
03	B	10,750						

Assistant Professors: Mean Salary = $16,000

ID	BUILDING	SALARY	ID	BUILDING	SALARY	ID	BUILDING	SALARY
10	A	$14,250	20	B	$15,500	30	D	$15,500
11	A	14,500	21	B	18,000	31	D	15,750
12	A	15,000	22	B	17,000	32	D	16,000
13	A	15,600	23	B	17,850	33	D	16,200
14	A	16,100	24	C	14,250	34	D	17,000
15	A	16,500	25	C	15,250	35	D	17,250
16	A	17,000	26	C	16,000	36	E	15,250
17	A	17,500	27	C	16,750	37	E	15,250
18	B	14,500	28	C	17,750	38	E	16,150
19	B	15,500	29	D	14,100	39	E	16,750

Associate Professors: Mean Salary $20,000

ID	BUILDING	SALARY	ID	BUILDING	SALARY	ID	BUILDING	SALARY
40	A	$18,050	54	B	$20,700	67	D	$19,450
41	A	18,200	55	B	21,800	68	D	19,750
42	A	18,500	56	B	20,250	69	D	19,800
43	A	19,000	57	B	20,250	70	D	20,250
44	A	20,500	58	C	18,100	71	D	20,600
45	A	21,500	59	C	18,750	72	D	21,450
46	A	22,500	60	C	18,750	73	D	21,650
47	A	20,250	61	C	19,550	74	E	19,000
48	A	20,750	62	C	20,750	75	E	19,250
49	A	20,750	63	C	23,000	76	E	20,450
50	B	18,250	64	D	18,250	77	E	21,000
51	B	18,750	65	D	18,750	78	E	21,300
52	B	19,250	66	D	19,000	79	E	22,000
53	B	19,900						

Full Professors: Mean Salary = $24,000

ID	BUILDING	SALARY	ID	BUILDING	SALARY	ID	BUILDING	SALARY
80	A	$23,000	87	B	$24,000	94	D	$24,500
81	A	24,000	88	C	25,500	95	D	25,000
82	A	25,550	89	C	24,000	96	E	23,100
83	A	26,000	90	C	24,000	97	E	23,000
84	B	23,200	91	D	23,100	98	E	23,500
85	B	23,200	92	D	23,600	99	E	23,700
86	B	24,450	93	D	23,600			

All Ranks Mean Salary = $18,805

sampling

8.1   drawing an inference

Statistics, we said in Chapter 1, seeks to make the most of data, and in Parts 1 and 2 we examined various methods for describing the data at hand.

(Frequently, the data at hand result from a *sampling* process and provide us with only incomplete information about the variables of interest. To make the most of sample data, we must do more than *describe what we observe*. We also must formulate *inferences about* our variables. The quality of these inferences depends on the sampling process we use and on our understanding of the effects of chance in sampling.)

(To "draw an inference" is to reach a conclusion or to pass judgment on the basis of incomplete evidence.) It is said that the French naturalist Baron Curvier (1769–1832) could correctly describe a whole animal from the examination of a single bone—a considerable talent for his time. The analysis of fingerprints—and now even of voiceprints—permits nearly flawless inferences to be made about the identity of an individual. Some professional recruiters claim that they can "size up" a job applicant after the first 20 seconds of an interview.

We must make statistical inferences constantly in daily life. Let's consider two examples.

1. Economic policymakers in the U.S. government require timely information about many factors, including the nation's unemployment rate. The U.S. unemployment rate is the portion (fraction or percentage) of the labor force that is out of work and seeking jobs. Currently, there are approximately 100 million individuals in the U.S. labor force.

   In principle, all 100 million members could be surveyed each month to determine what fraction of the labor force is unemployed. However, time and budget constraints as well as the inaccessibility of some members to interviewers make such a complete survey of the labor force impractical. So the U.S. government interviews only a small part, or *sample*, of the labor force. From the unemployment rate observed among the members who are sampled, analysts infer the approximate rate of unemployment in the total labor force.

2. In the past decade, voluminous data have been published on the pathological effects of smoking marijuana. In a typical study, a group of test subjects is selected from among marijuana users. The physiological or psychological responses of each member in this sample are then measured and compared with similar data from a control group—subjects who have characteristics similar to those of the members in the test group but who have not smoked marijuana—in an attempt to determine whether marijuana smokers have medical problems that are not evident in the nonsmoking population.

   Obviously, not every individual who is a marijuana user can be tested. Moreover, researchers are concerned not only with the effects on current marijuana users but with the probable effects on future users. So the research results are of interest not specifically for what they reveal about the effects of the drug on the *subjects tested* but for what can be inferred from these effects about the potential risk faced by any current or future smoker.

In both of these cases, samples were selected and used to draw inferences about the characteristics of a larger group. We call the scores of this larger group the *population of inference.*)

(**POPULATION OF INFERENCE**   All the possible scores on a subject of interest. The term *universe* is sometimes used in place of *population.*)

So we speak of the employment status of all individuals in the labor force and all possible responses to marijuana use as populations of inference.

sample statistics and population parameters

(Statisticians distinguish between the characteristics of *populations,* which they call *parameters,* and the characteristics of *samples,* which they call *statistics.*)

(**PARAMETER**   A numeric measure of some feature of a population of inference, such as the population mean, the population standard deviation, or a population regression coefficient.

STATISTIC   A numeric measure of some feature of a sample, such as the sample mean, the sample standard deviation, or a sample regression coefficient.)

(By convention, statisticians use Greek letters to represent parameters and either Arabic letters or Greek letters with hats (^) as superscripts to denote statistics. Chart 8-1 lists the symbols for parameters and statistics that we will use frequently in the remainder of the book.)

chart 8-1

FEATURE	SYMBOL FOR POPULATION PARAMETER	SYMBOL FOR SAMPLE STATISTIC
Mean	μ (mu)	$\bar{X}$
Standard Deviation	σ (sigma)	s and $\hat{\sigma}$
Regression Coefficient	β (beta)	b
Proportion	π (pi)	$\hat{\pi}$

We have already learned the equations for the sample statistics $\overline{X}$ and s. The equations for the corresponding population parameters μ and σ are identical, except that n (sample size) is replaced by N (population size), and $\overline{X}$ is replaced by μ. For example, the equations

$$\mu = \frac{\Sigma X}{N} \qquad \text{and} \qquad \sigma = \sqrt{\frac{\Sigma(X - \mu)^2}{N}}$$

correspond to the equations

$$\overline{X} = \frac{\Sigma X}{n} \qquad \text{and} \qquad s = \sqrt{\frac{\Sigma(X - \overline{X})^2}{n}}$$

(In the terminology of parameters and statistics, we say that a primary objective of statistical inference is to draw inferences about the values of parameters using the appropriate statistics.)

So in the unemployment survey, we could say that the intention is to find the proportion of the sample members who are unemployed—a statistic $\hat{\pi}$—and to use this proportion to estimate the proportion of the members of the labor force who are unemployed—a parameter π.

In one marijuana study, the intention was to compare the average reaction times of marijuana users and nonusers—statistics—and from these statistics infer whether or not differences exist in the average reaction times of the two populations—parameters.

Of course, values of sample statistics cannot be expected to correspond exactly to the values of population parameters.

(Because a sample is only a portion of the population of inference—often a minute portion—the value of a sample statistic inevitably provides an imperfect estimate of the population parameter of interest. Without complete information, we cannot expect to hit the bull's eye.)

How closely we can expect to estimate a parameter from a statistic depends critically on *how* we select the sample.

EXERCISES

8-1   The following statements represent inferences about population parameters. In each case:
(a) Identify the population of inference.
(b) Indicate the appropriate symbol for the parameter on which the inference is drawn.
(c) Name the statistic on which the inference is based and state its value.

STATEMENT 1: It is estimated that the spectators in attendance at
Civic Center events during 1977 spent $1.62 on the average for refresh-
ments, according to the business manager of the Springfield Civic Center.

STATEMENT 2: The gestation period for a mare is estimated to vary
by an "average" of 4 days about the mean length of pregnancy, according
to a Department of Agriculture report.

STATEMENT 3: According to the latest sample survey, tuition has increased by an average of 38 percent in the past three years at all public
colleges and universities in the country.

8-2   The relative frequency equation for a sample mean, originally presented
as Equation (2-3), and the analagous formula for a population mean
follow:

$$\bar{X} = \Sigma[X \cdot p(X)]$$ (2-3)
$$\mu = \Sigma[X \cdot p(X)]$$

The two equations are identical except for their perspective. In the case of
the sample mean, X denotes the possible values for the scores in a sample,
whereas in the case of the population mean, X denotes the possible values
for all the scores in the population of inference. Similarly, $p(X)$ denotes the
relative frequency (percentage or fraction of times) with which each value
occurs: in one case, in the sample; in the other, in the population.

Given the following population distribution, compute μ:

NUMBER OF CARS X	RELATIVE FREQUENCY OF FAMILIES $p(X)$
0	0.23
1	0.55
2	0.20
3	0.015
4	0.005
	1.00

8-3   Analagous to Equation (3-4) — the relative frequency equation for a sample standard deviation — is the relative frequency equation for the standard deviation of a population of inference σ:

$$\sigma = \sqrt{\Sigma[X^2 \cdot p(X)] - \mu^2}$$

Compute σ for the population of the *number of cars per family* in Exercise 8-2.

8.2   random sampling

Sampling procedures range from casual to scientific, but all have a common objective: To obtain a distribution of scores that is *representative* of the population of inference. The ideal sample is a *miniature* of the population; that is, it possesses the precise features of the population, but on a smaller scale. If sampling procedures could insure such a distribution of sample scores, characteristics of that distribution — statistics — would be exact estimates of the parameters of the population of inference.

Unfortunately, we know of no mechanism that can guarantee a representative sample. But one sampling mechanism at least permits us to predict the size of the inevitable sampling errors. (When a problem can't be eliminated, perhaps the best alternative is to be able to predict its magnitude.) This mechanism is called, generically, *random sampling.*

RANDOM SAMPLING   Any procedure for selecting members from a group on the basis of chance or luck. Random sampling replaces such selection considerations as "personal judgment" and "first come, first served" with "the luck of the draw."

When we sample on the basis of personal judgment or convenience, we face subjective and therefore unpredictable risks of error. We can't always count on having good judgment, and we certainly can't place our faith in the representativeness of the most easily accessible group members. Random sampling cannot eliminate the risk of acquiring unrepresentative information (if we rely on luck, our luck can go bad). But random sampling does allow us to *measure* the risk by letting the laws of chance — the laws of probability — govern the sampling results.)

Before we examine these laws in Chapter 9, we will learn how random sampling is performed.

simple random
sampling

(**SIMPLE RANDOM SAMPLING   A procedure for
selecting members from a population that on each
drawing gives every available member an equal
chance of selection. In lay terms, a *lottery*.**)

If we want to select members from a group on the basis of the luck of the draw, the most straightforward method is a lottery. Each member of the group is given an identification number, and each number is recorded on a chip. All of the chips are placed in a bowl and stirred thoroughly. One chip is selected without looking. The number on that chip identifies the first member to be selected. The process is repeated until a sample of the desired number of members is selected.

Many counties use a lottery process to select individuals for jury duty. The population from which the selection is made typically consists of individuals who are of majority age, residents of the county, registered voters, and who have not previously been convicted of a felony. The names on the county's voter checklist are numbered, and each number is recorded on a chip. The chips are placed in a bowl and mixed thoroughly. Typically, more chips are selected than the number required to serve on a jury, since some of the individuals chosen invariably will be excused from duty.

To implement a simple random sampling procedure, it is not necessary to use a bowl filled with chips. In fact, this method of sampling is potentially dangerous. If the chips are not *thoroughly* mixed, all chips are *not* equally likely to be chosen and the result will not be a simple random sample. A simple random sample can be achieved with less fanfare and no worry about adequate mixing by using a table of random digits.

the random
digits table

Chart 8-2 is a small segment from a table of one million random digits. The table is filled with the ten digits from 0 through 9, which do not seem to appear in any systematic

chart 8-2
Random Digits Table

12651	61646	11769	75109	86996	97669	25757	32535	07122	76763
81769	74436	02630	72310	45049	18029	07469	42341	98173	79260
36737	98863	77240	76251	00654	64688	09343	70278	67331	98729
82861	54371	76610	94934	72748	44124	05610	53750	95938	01485
21325	15732	24127	37431	09723	63529	73977	95218	96074	42138
74146	47887	62463	23045	41490	07954	22597	60012	98866	90959
90759	64410	54179	66075	61051	75385	51378	08360	95946	95547
55683	98078	02238	91540	21219	17720	87817	41705	95785	12563
79686	17969	76061	83748	55920	83612	41540	86492	06447	60568
70333	00201	86201	69716	78185	62154	77930	67663	29529	75116
14042	53536	07779	04157	41172	36473	42123	43929	50533	33437
59911	08256	06596	48416	69770	68797	56080	14223	59199	30162
62368	62623	62742	14891	39247	52242	98832	69533	91174	57979
57529	97751	54976	48957	74599	08759	78494	52785	68526	64618
15469	90574	78033	66885	13936	42117	71831	22961	94225	31816
18625	23674	53850	32827	81647	80820	00420	63555	74489	80141
74626	68394	88562	70745	23701	45630	65891	58220	35442	60414
11119	16519	27384	90199	79210	76965	99546	30323	31664	22845
41101	17336	48951	53674	17880	45260	08575	49321	36191	17095
32123	91576	84221	78902	82010	30847	62329	63898	23268	74283
26091	68409	69704	82267	14751	13151	93115	01437	56945	89661
67680	79790	48462	59278	44185	29616	76531	19589	83139	28454
15184	19260	14073	07026	25264	08388	27182	22557	61501	67481
58010	45039	57181	10238	36874	28546	37444	80824	63981	39942
56425	53996	86245	32623	78858	08143	60377	42925	42815	11159
82630	84066	13592	60642	17904	99718	63432	88642	37858	25431
14927	40909	23900	48761	44860	92467	31742	87142	03607	32059
23740	22505	07489	85986	74420	21744	97711	36648	35620	97949
32990	97446	03711	63824	07953	85965	87089	11687	92414	67257
05310	24058	91946	78437	34365	82469	12430	84754	19354	72745
21839	39937	27534	88913	49055	19218	47712	67677	51889	70926
08833	42549	93981	94051	28382	83725	72643	64233	97252	17133
58336	11139	47479	00931	91560	95372	97642	33856	54825	55680
62032	91144	75478	47431	52726	30289	42411	91886	51818	78292
45171	30557	53116	04118	58301	24375	65609	85810	18620	49198
91611	62656	60128	35609	63698	78356	50682	22505	01692	36291
55472	63819	86314	49174	93582	73604	78614	78849	23096	72825
18573	09729	74091	53994	10970	86557	65661	41854	26037	53296
60866	02955	90288	82136	83644	94455	06560	78029	98768	71296
45043	55608	82767	60890	74646	79485	13619	98868	40857	19415
17831	09737	79473	75945	28394	79334	70577	38048	03607	06932
40137	03981	07585	18128	11178	32601	27994	05641	22600	86064
77776	31343	14576	97706	16039	47517	43300	59080	80392	63189
69605	44104	40103	95635	05635	81673	68657	09559	23510	95875
19916	52934	26499	09821	87331	80993	61299	36979	73599	35055
02606	58552	07678	56619	65325	30705	99582	53390	46357	13244
65183	73160	87131	35530	47946	09854	18080	02321	05809	04898
10740	98914	44916	11322	89717	88189	30143	52687	19420	60061
98642	89822	71691	51573	83666	61642	46683	33761	47542	23551
60139	25601	93663	25547	02654	94829	48672	28736	84994	13071

Source: The Rand Corporation, *A Million Random Digits with 100,000 Normal Deviates*. New York: The Free Press, 1955. Reproduced with permission of The Rand Corporation.

order. Actually, these digits are arranged in such a special pattern that it would be quite difficult for us to create a similar table without computer assistance.

First, each single digit appears with equal frequency, 1/10 of the time, so that a person whose eyes are closed is equally likely to point to 0 or 5 or 8 or any of the other single digits.

Second, each two-digit sequence (00 through 99) appears with equal frequency, 1/100 of the time, making the selection of a 21 or a 47 or a 02 or any other combination of two digits equally likely. Similarly, each three-digit sequence (000 through 999) occurs 1/1,000 of the time and is equally likely to be chosen. In fact, this is true for any sequence of k digits.

These properties make the random digits table a suitable substitute for a well-mixed bowl of chips. Selecting a number from the table is equivalent to selecting a chip from the bowl. Each number of k digits is equally likely to be selected.

Let's use the table to select a sample. Suppose that we wish to select 5 members from a group of 90 ($N = 90$, $n = 5$). With a population of 90 members (or any N between 11 and 100), we assign each member a two-digit identification number (ID). So beginning with 00, we label our group members with the IDs through 89.

We enter the table by closing our eyes and pointing a finger. Let's say it lands at the 16th row and the 11th column on the number 5. (Ignore the spaces between the rows and columns. They are strictly for viewing ease.) Reading across we obtain the two-digit number 53. The member whose ID is 53 is thereby selected.

Continuing across the row (although we could just as well move down to the next row), we read 85, then 03 (ignoring the space between 0 and 3), then 28, and then 27. Our result is a sample of five members with the IDs 53, 85, 03, 28, and 27.

If any of the two-digit numbers 90 through 99 had occurred, we would simply have discarded the number, because our population of IDs spans 00 through 89.

EXERCISES

8-4

The relative frequency distribution of the population of the *number of cars per family* appears in Exercise 8-2 on page 215. Suppose that a random sample is selected from the membership of this population.

(a) Would the relative frequency distribution of the number of cars per family sampled necessarily be identical to that given in Exercise 8-2?

(b) Would the relative frequency distribution of the number of cars per family sampled necessarily closely match that given in Exercise 8-2?

(c) If your responses to (a) and (b) were no, explain what the virtue of random sampling would be in this case, in contrast, for example, to a selection of families who are acquainted socially and professionally with your own.

8-5   In a random digits table, will the four-digit IDs 0000 and 5324 occur with the same frequency? What is the chance of selecting either of the two IDs?

8-6   Starting with the first digit in the random digits table and reading across, record the IDs that result from selecting a sample of $n = 4$ from a population of $N = 800$. (*Hint:* Let the three-digit sequence 000 through 799 be the IDs of the 800 members of the population.)

8-7   (a) Starting at the beginning of the random digits table and reading each ID from a new row (that is, reading down the columns), record the IDs that result from selecting a sample of $n = 6$ from a population of $N = 500$.

(b) How many IDs did you have to skip to obtain your sample? How could you have assigned IDs to the population membership to avoid this problem?

*8-8   (a) Starting with the 6th digit of the 20th row of the random digits table, select a simple random sample of 10 distinct two-digit ID numbers.

(b) Referring to the faculty salary data given on page 210, calculate the mean salary of the 10 faculty members whose IDs you selected in (a) and note the difference between this sample mean and the mean for the entire population of 100 faculty members.

(c) Enter the random digits table again by closing your eyes and pointing to a digit. Select another simple random sample of 10 IDs, and calculate the mean salary for the faculty members with these IDs.

(d) Did the luck of the draw grant you a more accurate estimate of the mean salary for the entire population in sampling (a) or in sampling (c)?

8.3 survey sampling

Simple random sampling, using either chips in a bowl or a random digits table, allows us to make straightforward statistical inferences. However, simple random sampling can prove difficult to implement when conducting a survey. Some of the problems that can arise are illustrated in the following case history.

a case history

An administrator of a university with approximately $N = 10,000$ matriculated students has proposed a new type of grade-recording system. Next to each grade on a student's record, a figure would be entered indicating the percentage of students in the course who received that grade or higher. (For example, Biology 19, B, 43%, would denote that 43 percent of the Biology 19 class received a grade of B or higher.) The administrator was concerned with the lack of uniformity in grading standards and wished to supply more accurate information about a student's relative performance in a course.

The student association (SA) was asked to determine student-body opinion regarding the proposal. If student opposition were sufficiently strong, the matter would be dropped.

First, SA leaders considered distributing a questionnaire to the entire student body. However, if this were done by mail, the number of nonrespondents would probably be high, and conducting personal interviews with all the students would be too time consuming. Sampling was in order. A member of the statistics faculty informed the SA that a simple random sample of 400 opinions should provide a reasonably representative picture of the population of student-body opinion.

From the registrar, SA leaders obtained a list of all matriculated students and assigned a four-digit ID (0000 through 9999) to each name on the list. Using a table of random digits, they selected 400 four-digit numbers and prepared a list of the names and addresses of the students chosen in this way. The computerized list also included each student's subject field, and this information was recorded as well. The next step was to interview each member of the sample briefly, but at this stage, SA leaders realized that they faced some difficulties.

One was that many students in the sample lived off campus. Their residences were scattered throughout the city, and some did not have telephones. Reaching these students would be difficult.

Also the SA was concerned that the sample, which had been selected by the luck of the draw, did not adequately represent all subject fields. Education majors seemed to be overrepresented, and the sample included only a few science students. Because the opinions of the students might be influenced greatly by their subject field, the SA felt that this was a major flaw in the sample.

At this point, an alternative approach was suggested. Instead of taking a random sample of students, the SA would take a random sample of *class sections* from each subject field and ask all the students in these selected sections to fill out questionnaires. With the cooperation of the professors, the questionnaires could be completed during class meetings, making it easier to contact the students and minimizing the number of nonrespondents. Finally, if the number of classes chosen from each field were proportional to the number of students in that area (and if the samples were taken from courses designed for majors), the SA could be sure that the sample adequately represented all parts of the student body.

In this way, the SA decided to use a sampling design that involved both *stratified* and *cluster* sampling—two traditional methods of survey sampling.

stratified random sampling

To stratify a group is to divide the membership into distinct subgroups, called *strata*, on the basis of one or more factors thought to influence responses (opinions) on the subject of interest.

In political polling and consumer surveys, the population is typically stratified by sex, race, and occupation. The SA leaders chose to stratify their members by subject field, because they felt this factor was most likely to influence a student's opinion on the new grade-recording proposal.

STRATIFIED RANDOM SAMPLING   The process of selecting members by chance from individual strata of members rather than from the membership of the population at large.

The use of stratified random sampling insures that some members of each stratum are selected for the sample. Frequently, the number of members taken from each stratum is in proportion to the size of that stratum in the population.

Therefore, stratified sampling is a surer way to obtain a representative sample than simple random sampling is. However, stratification can be accomplished only at a price—prior information about the population must be acquired. In this example, the price would not be high, since the subject fields of the students were readily available. Even if the price were high, it could be worth paying in return for the greater assurance of a representative sample.

cluster sampling

The SA leaders did not take a stratified random sample of the student body at large. Rather, they took a stratified random sample of groups—class sections—of students. These groups are called *clusters.*

CLUSTER SAMPLING   The process of selecting clusters of members by chance from an overall group of clusters into which the membership has been organized.

The first chip is drawn during the 1969 military draft lottery.

Examples of clusters include neighborhoods of a community and classrooms or dormitories at a school.

In 1969, the U.S. military implemented a draft lottery system based on cluster sampling. The government wanted to insure that every eligible male was equally likely to be drafted. However, randomly selecting names from a list of all eligible males would be a tedious task. Instead every potential draftee was assigned to a cluster according to his date of birth from January 1 to December 31. Each date was entered on a chip, the chips were placed in a bowl and stirred, and the first chip was drawn. The date on that chip was September 14. All eligible males born on September 14 were assigned draft position 1. The process was repeated until a complete order of draft positions was established.

Ideally, no member of the population should belong to more than one cluster. Because the SA defined each class section offered during the current semester as a cluster, students taking several classes would belong to several clusters. So the more classes a student took, the more likely it was that he or she would be included in a cluster. However, the SA decided that this was an inequity they would have to accept.

Cluster sampling is designed to make data collection more convenient, but it increases the risk of obtaining an unrepresentative sample. Geographic proximity is often correlated with uniformity of outlook. Therefore, if clusters are based on geography—and they frequently are—including all the members of a cluster may yield an unrepresentatively high number of members with one specific point of view. For this reason, cluster sampling is less hazardous when it is performed in conjunction with stratification—which is what the SA did.

follow-up

The SA defined ten strata and selected a random sample of class sections from within each stratum. The professors were cooperative and distributed the questionnaire to all the selected classes. Less than 30 percent of the students polled favored the new grade-recording system, and the proposal was scrapped.

EXERCISES
8-9

(*Continuation of Exercise 8-8*)

(a) Referring to the simple random sample of ten faculty members you selected in Exercise 8-8(c) on page 220, tabulate the number of these faculty at each faculty rank. Is your sample representative of the overall distribution of faculty ranks at the college?

(b) Stratify the faculty by rank and select a stratified random sample of ten faculty members. Make the number of faculty members you select at each rank proportional to the number of faculty at each rank in the college.

(c) Compare the salaries of the members of your stratified sample with the salaries of the members of your simple random sample. Which sample do you expect would provide a better estimate of the population mean salary? Which sample does?

8-10

(*Continuation of Exercise 8-8*) Cluster sampling was considered as a procedure for selecting a sample from the faculty population listed on page 210.

(a) How could you select a cluster sample of the salaries of the 100 faculty members? Would it be to your advantage to use this sampling procedure rather than a simple random sample? Why?

(b) What assumption would you need to make before you could safely use the results of the cluster sample to estimate the mean salary for the faculty as a whole?

8-11   Five schools entered the finals of the intercollegiate wrestling champion-ships. Each school was asked to supply 6 judges for the event, making a total of 30 judges. Three judges were to be selected by chance to officiate at each match. (Keep in mind that each match is a contest between two of the five schools.)

(a) Initially it was felt that the judges could be selected by lottery (simple random sampling); that is, 3 out of 30 chips could be selected from a bowl for each match. What is the principal shortcoming of this pro-cedure?

(b) How could a stratified random sampling procedure remedy this draw-back of the lottery?

8-12   The National Institute of Music wished to obtain data on the salary levels of musicians playing for professional orchestras throughout the country. But hundreds of professional orchestras are in existence, and each is made up of many musicians. Because the required data could be acquired only from the individual musicians, the Institute decided to sample.

(a) A *simple* random sampling procedure was ruled out on the basis of being extremely uneconomical. Why would this be so?

(b) How would the *cluster* sampling procedure save time and expense? What would be the primary disadvantage of cluster sampling in this situation?

8.4 random sampling and probability

(When we acquire data through random sampling, we are tak-ing a chance on the result. But random sampling also makes it possible to calculate the *probability* of any specific result.)

the probability
of an event

We've all used the term *probability* to express our belief in the likelihood that some event will occur. "There's a high probability that I'll Ace this course" indicates that the speaker is not entirely certain what the grade will be but feels that a grade of A is more likely to occur than the alternatives to this event.

When we sample from a population of inference, the events that interest us are related to the sample scores. To learn how to determine the probabilities of such events, we will consider a specific problem.

Town planners in a suburb of a large city wished to conduct a survey of the families in their community. Since they planned to report the results pertaining to high-, middle-, and low-income families separately, the planners were interested in knowing the likelihood of picking families from each of these categories.

When a family is selected, one of three possible results will occur:

1. The family's income will be high. This event is denoted by *high*.
2. The family's income will be in the middle-income bracket. This event is denoted by *middle*.
3. The family's income will be low. This event is denoted by *low*.

Suppose that the population of income scores is as shown in Chart 8-3.

We know that if a simple random sampling process is used to select a family, each family will have an equal chance of being selected. Since 30 percent of the families' incomes are high, the chance (probability) that the family that is selected will produce the event *high* is 30 percent. Similarly, the chances that the event *middle* and the event *low* will occur are 50 percent and 20 percent, respectively. In general, the probability that an event E will occur equals the relative frequency (fraction or percentage) of members of the population whose scores produce that event.

The probability of an event $P(E)$ must be some value from 0.0 to 1.0 (or from 0 percent to 100 percent). If there are no members in the population whose scores will produce an event E, then the event cannot occur and $P(E) = 0.0$. The probability

chart 8-3

The Population of Inference

INCOME LEVEL	RELATIVE FREQUENCY
High (More than $30,000)	0.30
Middle ($10,000 to $30,000)	0.50
Low (Less than $10,000)	0.20

of selecting a family whose income is neither *high* nor *middle,* nor *low* is equal to 0.0. At the other extreme, if the scores of *all* members of a population will produce the event E, then the event is certain and $P(E) = 1.0$. The probability that the selected family will have an income that is *high, middle,* or *low* is 1.0.

the addition rule for mutually exclusive events

In selecting a family from the population given in Chart 8-3, what chance is there that we *won't* pick a low-income family? In other words, what is the probability that either the event *high* or the event *middle* will occur?

> **ADDITION RULE FOR MUTUALLY EXCLUSIVE EVENTS** If two events E_1 and E_2 are mutually exclusive, then the probability that *either E_1 or E_2* will occur, denoted by $P(E_1 \text{ or } E_2)$, is equal to $P(E_1) + P(E_2)$.

(Two events E_1 and E_2 are said to be mutually exclusive if they cannot occur simultaneously.) Obviously, a family cannot have an income that is both *high* and *middle*. So the event *high* and the event *middle* are mutually exclusive. Similarly, *middle* and *low* are mutually exclusive events.

Therefore, from the addition rule:

1. $P(high \text{ or } middle) = P(high) + P(middle) = 0.30 + 0.50$
$$= 0.80$$

2. $P(middle \text{ or } low) = P(middle) + P(low) = 0.50 + 0.20$
$$= 0.70$$

and, by extension

3. $P(high \text{ or } middle \text{ or } low) = P(high) + P(middle) + P(low)$
$$= 0.30 + 0.50 + 0.20$$
$$= 1.0$$

the multiplication rule for independent events

Until now, we have been concerned with events that can occur when we select one family from the population. Of course, samples of only one score are rarely useful in statistical inference. So let's extend our example. What *pairs* of events— that is, what samples of two scores—can occur when we select two families from the population given in Chart 8-3, and what is the probability of each pair of events occurring?

The first part of the question can be answered simply. We want to select two families and record their income levels. The first family selected can have a high-, middle-, or low-income level, and so can the second. If we let E_1 represent the income level of the first family selected and E_2 represent the income level of the second family selected, the complete list of possible pairs of events that can result from the selection of two families is given in Chart 8-4.

We see that nine samples (pairs of events) are possible, from the one in which both families selected have high incomes to the one in which both families selected have low incomes.

Now we turn to the second part of our question: What is the probability that each of these nine possible samples will occur? At this point, we will make the assumption—a generally realistic one in sampling—that the population of inference contains a large number of families in each income bracket. Given this assumption, the selection of one family, even if we don't replace its chip in the bowl, will not noticeably change the relative frequencies of high-, middle-, and low-income families that remain in the population. Stated technically, this assumption makes the events produced by successive selections *independent*.

Two events E_1 and E_2 are said to be *independent* if the probability that the second event E_2 occurs is not altered by the occurrence or nonoccurrence of the first event E_1.

For example, the events *high* and *middle* are independent

chart 8-4
Possible Pairs of Events E_1 and E_2

high and *high*
high and *middle*
high and *low*
middle and *high*
middle and *middle*
middle and *low*
low and *high*
low and *middle*
low and *low*

if the probability of picking a middle-income family on the second selection is the same whether the first family selected has a high income or other than a high income.

(MULTIPLICATION RULE FOR INDEPENDENT EVENTS   If two events E_1 and E_2 are independent, then the probability that *both E_1 and E_2* will occur, denoted by $P(E_1$ and $E_2)$, is equal to $P(E_1) \cdot P(E_2)$. **)**

Let's use this rule to calculate the probabilities of the pairs of events given in Chart 8-4.

1. P(*high* on first selection and *high* on second selection)
   $= P$(*high* on first selection) $\cdot P$(*high* on second selection)
   $= 0.3 \cdot 0.3 = 0.09$

So the probability that both families selected will have high incomes is 9 percent.

2. P(*high* on first selection and *middle* on second selection)
   $= P$(*high* on first selection) $\cdot P$(*middle* on second selection) $= 0.3 \cdot 0.5 = 0.15$

So the probability that the first selection is a high-income family and the second selection is a middle-income family is 15 percent.

We can calculate the probabilities for the remaining pairs of events in Chart 8-4 in the same manner. These probabilities are given in Chart 8-5. Note that the probabilities sum to unity.

chart 8-5

PAIRS OF EVENTS E_1 AND E_2	$P(E_1$ AND $E_2)$
high and *high*	0.09
high and *middle*	0.15
high and *low*	0.06
middle and *high*	0.15
middle and *middle*	0.25
middle and *low*	0.10
low and *high*	0.06
low and *middle*	0.10
low and *low*	0.04
	$\Sigma = 1.00$

(If the sum were less than 1, it would indicate that we had omitted some possible samples.)

the counting rule for independent events

Chart 8-5 contains the nine possible samples resulting from the selection of two families. Why nine samples? How would we know that we had included all possible samples without calculating the probabilities to see if they sum to 100 percent? We would know by the *counting rule*.

(**COUNTING RULE** When we take a sample of *n* observations, we have *n* events representing the outcomes of the sampling. If these events are independent, then the number of possible samples is equal to V^n, where *V* denotes the number of distinct values in the population.)

The number of distinct values in the population of inference given in Chart 8-3 is three (*high, middle,* and *low*). So if $n = 2$ (two families selected), then one of 3^2, or 9, samples can arise. If we select three members of the population and list the events, we will obtain $V^n = 3^3 = 27$ possible samples of three events. A sample of 50 families is small. Yet the counting rule tells us that the number of possible samples is 3^{50}, or approximately 717,900,000,000,000,000,000,000 (717.9 septillion).

EXERCISES

8-13

According to the U.S. Bureau of the Census, the relative frequency distribution of the number of children per family in 1975 looked like this:

NUMBER OF CHILDREN	RELATIVE FREQUENCY OF FAMILIES
0	0.460
1	0.197
2	0.180
3	0.094
4 or more	0.069
	1.000

If one family is selected at random (by lottery), what is the probability that the family will have:

(a) 0 children?

(b) 4 or more children?

(c) At most 2 children?

8-14   (*Continuation of Exercise 8-13*) If two families are selected at random, what is the probability that:
(a) There will be no children in either family?
(b) Both families will have 2 children?
(c) The first family will have 1 child and the second family will have 3 children?
(d) The first family will have 3 children and the second family will have 1 child?
(e) Either (c) or (d) will occur?

8-15   (*Continuation of Exercise 8-13*) Determine the number of possible samples that can result from the selection of two families by applying the counting rule presented in this section. Treat "4 or more children" as a single value.

8-16   (*Continuation of Exercise 8-13*) Suppose that three families are selected at random.
(a) What is the probability that none of these families will have children?
(b) How many samples of three events can result from the selection of three families? (Apply the counting rule.)

8-17   Referring to the faculty salary data given on page 210, what is the probability that a faculty member selected at random from this population will be:
(a) An instructor?
(b) Either an associate or a full professor?
(c) Located in office building C?
(d) Located in office building A or E?

Caveat: Sampling Requires Considerable Skill As Well As Favorable Luck

If we are lucky, a randomly selected sample should provide a reasonably accurate representation of the population of inference. However, experience has shown that bad luck is not the only factor that can lead to a random sample that misrepresents the population of inference. If we consider bad luck to be

the source of *sampling error,* the other potential source of misrepresentation can be classified as *sampler error*—misrepresentations that result from actions taken by the sampler rather than from chance misfortune.

Government and political surveys contain ample examples of these sampler errors. Here we will examine two types of sampler error that have been committed in the past. These errors should serve as a warning that sampling requires considerable skill as well as favorable luck.

1. The scores are randomly selected from the wrong population.

In 1936 and again in 1948, certain sample surveys used to predict the outcome of the presidential election were highly inaccurate due to sampler error.

In its 1936 poll, *Literary Digest* magazine (now defunct) selected a random sample of voters from names on telephone and automobile registration lists. Based on the responses of these voters, the magazine predicted that Alf Landon would beat Franklin Roosevelt. In fact, FDR's victory was of landslide proportions.

The presidential polls of September 1948 indicated that Thomas E. Dewey would run well ahead of Harry Truman. (One newspaper headline printed for election morning actually declared Dewey the winner.) Truman's victory in November was an embarrassing surprise.

In both cases, the sampler error resulted from sampling the *wrong* population.

In the depression days of 1936, many people could not afford telephones or cars, so the list from which the *Literary Digest* selected its sample did not include these less well-to-do voters. Although the actual selection of names followed random procedures, the lists from which the names were drawn did not include the complete population of inference.

In 1948, public opinion was unstable. The election polls acquired samples from the population of voter opinion in September. The magnitude of voter-switching between September and November was not anticipated.

With luck, a random sample will give a representative picture of the population of inference—but only a picture of that population *at the time of sampling*. This is why a cautious pollster typically prefaces a prediction with the caveat, "If the elections were held at the time of the survey . . ."

2. A sizable nonresponse is ignored.

In 1977, Mayor Perk of Cleveland had city sanitation workers deliver questionnaires to more than 200,000 addresses in his city. The mayor's goal: To demonstrate community support for his planned crackdown on pornography. When the results were tabulated, the mayor announced that nearly nine out of every ten households returning the questionnaire favored a general ban on the peddling of pornographic material in Cleveland.

It is clearly beyond dispute that a great majority of the *respondents to the questionnaire* supported the mayor's anti-pornography program, but only about 13,000 responses were received—barely 5 percent of the number of questionnaires distributed. Of the households surveyed, 95 percent did not respond. So the great majority of the population of inference was actually telling the mayor, "I can't be bothered with your questionnaire." Their point was made more tellingly when in the primary election later in the year, Mayor Perk was voted out of office.

A large number of nonrespondents cannot be ignored. Generally, it cannot be assumed that the nonrespondents' scores are distributed in the same manner as the respondents' scores are. Good sampling procedure dictates that follow-up surveys be made to reduce the number of nonrespondents or that the characteristics of the nonrespondents be specifically analyzed to provide a basis for allocating their scores.

Many other types of sampler error exist. Questionnaires may be worded in a leading fashion to solicit self-serving responses. Respondents who wish to appear in a favorable light often provide the "right" (but wrong) answer. A common example is the Jack Benny syndrome, when individuals report that they are 39 years old but their ages actually range from 40 to infinity.

In the following chapters, we heroically dismiss the potential for sampler error and concern ourselves strictly with the forces of luck in random sampling.

EXERCISE

The headline story on page 210 pointed out that the mean faculty salary reported in last month's *Bulletin* was based on the responses of 21 out of 100 faculty members to a *Bulletin* questionnaire. The questionnaire mailed to each faculty member was worded:

Several department chairpersons have expressed concern that efforts to retain superior faculty as well as to attract new faculty members of a high caliber are being hampered by an inadequate salary structure. The president of the college has refused to release data on faculty salaries. Therefore, we are asking all faculty members to report their base salaries during the present academic year along with their faculty ranks. The confidentiality of your response will be preserved, and only summary figures will be published.

What factors other than sampling error could have caused the sample responses to underestimate the mean salary for the faculty?

COMING TO TERMS

Addition Rule for Mutually Exclusive Events If two events E_1 and E_2 are mutually exclusive, then the probability that either E_1 or E_2 will occur, denoted by $P(E_1 \text{ or } E_2)$, is equal to $P(E_1) + P(E_2)$.

Counting Rule When we take a sample of n observations, we have n events representing the outcomes of sampling. If these events are independent, then the number of possible samples is equal to V^n, where V denotes the number of distinct values in the population.

Independent Events Two events E_1 and E_2 are said to be independent if the probability that the second event E_2 will occur is not altered by the occurrence or nonoccurrence of the first event E_1.

Inference A conclusion or judgment based on incomplete information.

Multiplication Rule for Independent Events If two events E_1 and E_2 are independent, then the probability that both E_1 and E_2 will occur, denoted by $P(E_1 \text{ and } E_2)$, is equal to $P(E_1) \cdot P(E_2)$.

Mutually Exclusive Events Two events are said to be mutually exclusive if they cannot occur simultaneously.

Parameter A numeric measure of some feature of a population of inference. Generally, Greek letters are used to denote parameters, such as the population mean μ and the population standard deviation σ.

Population of Inference All the possible scores on a subject of interest. The term *universe* is sometimes used in place of *population*.

Probability of an Event The likelihood that an event will occur. In simple random sampling the probability of an event is equal to the relative frequency (fraction or percent) of the members of the population whose scores produce that event.

Random Sampling Any process used to select members from a group on the basis of chance or luck.

Cluster Sampling The process of selecting clusters of members by chance from an overall group of clusters into which the membership has been organized.

Simple Random Sampling A process of selecting population members that in each drawing gives every available population member an equal chance of being chosen. In lay terms, a lottery.

Stratified Random Sampling The process of selecting members by chance from each individual stratum (subgroup) of members rather than from the membership at large.

Sampler Error Error due to actions taken by the sampler rather than to misfortunes of chance.

Sampling Error Error due to chance.

Statistic A numeric measure of some feature of a sample, such as the sample mean, the sample standard deviation, or a sample regression coefficient.

1. Following the publication of a particularly controversial story, a magazine received a flood of letters to the editor. However, space was available to print only 12 letters.

Consider the following strategies for selecting the letters to be printed. From the point of view of the magazine's readers, who would like to see a representative sample of readership opinion, describe a virtue and a deficiency of each strategy. Then indicate how you would select the 12 letters if you were the magazine editor in charge.

STRATEGY 1: "First come, first serve." The first 12 letters received (or the 12 letters with the earliest postmarks) are printed.

STRATEGY 2: "Editorial judgment." The editor selects 12 letters after reading all reader responses.

STRATEGY 3: "Lottery." All of the letters are placed in a bowl and 12 are drawn.

STRATEGY 4: "Stratification." The letters are sorted initially into three strata: (1) letters generally supporting the story line; (2) letters generally opposing the story line; and (3) letters elaborating on the issue without stating an opinion about the story line. Individual letters are selected from each stratum in proportion to the number of letters in that stratum.

2. The complete list of faculty salaries on page 210 has not been publicized but you know that the distribution of faculty ranks at the college is:

RANK	NUMBER OF FACULTY
Instructor	10
Assistant Professor	30
Associate Professor	40
Full Professor	20
Total:	100

You select a simple random sample of 20 faculty and record the following observations:

RANK	NUMBER OF FACULTY	MEAN SALARY OF RESPONDENTS
Instructor	8	$11,000
Assistant Professor	8	$15,500
Associate Professor	3	$19,600
Full Professor	1	$27,300

(a) Calculate the overall mean of your sample of salary observations.
(b) Would you expect the result in (a) to be an accurate estimate of the mean of all faculty salaries at the college?
(c) Utilize the data given for the actual distribution of faculty ranks to form a more accurate estimate of the mean salary of the college faculty.

Metrorail Slowly Improves

Daily Average of Scheduled Trips Actually Operated

| | RED LINE (238 scheduled) | | BLUE LINE (218 scheduled) | |
	Actual Trips	Percentage	Actual Trips	Percentage
July	225	94.5	174	79.8
August	228	95.8	201	92.2
September	231	97.1	203	93.1
October	232	97.5	207	95.0

The Washington Post: November 1, 1977

Little by little, the reliability of Metro's subway is getting better and better.

A study of Metro's operating logs shows that, on the average, an increasing percentage of scheduled subway trains actually are completing their runs. The improvement has been marked on the Blue Line from National Airport to Stadium-Armory, which opened July 1.

In that first horrible month, only 79.8 percent of the scheduled Blue Line trains actually were dispatched on an average day. Through the first three weeks of October, that average had jumped to 95 percent.

People do not ride *average* trains, however, they ride real ones, and when the trains do not show up, people complain. . . .

Source: Douglas B. Feaver, "Metrobus Gets Complaints, Metrorail Slowly Improves; Blue Line Problems Clearing," © *The Washington Post* (November 1, 1977), page C1, C3 (*our emphasis*).

Behind the Headline

Metro commuting during the early weeks of operation was chancy. With luck, the scheduled train would arrive, although it was not always on time. By October, the fourth month of operation, the system had improved considerably. If the record established then were maintained, commuters could be about 95-percent confident that their trains would arrive.

When we randomly sample from a population of inference, we also take our chances on the results. With good luck, the sample will prove to be representative of the population of inference. With bad luck, the evidence will be misleading. The fundamental concern in statistical inference is: How confident can we be that an inference drawn from a random sample will be correct?

To express a level of confidence in an inferential statement, we must understand how chance affects the reliability of sample evidence. Fortunately, the results of random sampling are anything but chaotic; rather, they are governed by *laws of chance variation.*

In this chapter, we will introduce three laws of chance variation and apply them to the simplest problem of statistical inference —estimating a population mean.

laws of chance variation

9

"The ideal reasoner," remarked Holmes, "would, when he has once been shown a single fact in all its bearings, deduce from it not only all the chain of events which led up to it, but also all the results which would follow from it."

The Five Orange Pips

9.1 sampling distributions

The Maine Potato Packing Company (MPP) packs 5-lb sacks of Katahdin baking potatoes for distribution to New England supermarkets. The packing machines record the number of potatoes in each sack. Because the size of Katahdins varies considerably, the potato count differs from sack to sack. Chart 9-1 presents a line graph of the recorded potato counts.

We can see that each count from 12 through 18 occurs with equal frequency, 1/7 of the time, forming a *uniform height* distribution. Chart 9-1 also indicates the mean and the standard deviation of the distribution. MPP packs an average of 15.0 potatoes per sack ($\mu = 15.0$), with a standard deviation of 2.0 potatoes per sack ($\sigma = 2.0$).

Some consumers who buy sacks of baking potatoes want to know how many potatoes a sack contains. (Counting is difficult, if not impossible, without opening the sack.) So let's ask: How many potatoes can a consumer expect to get when purchasing 5-lb sacks of MPP Katahdins? This is the same as

chart 9-1
Distribution of
5-lb Sacks
of Potatoes by
Number of
Potatoes

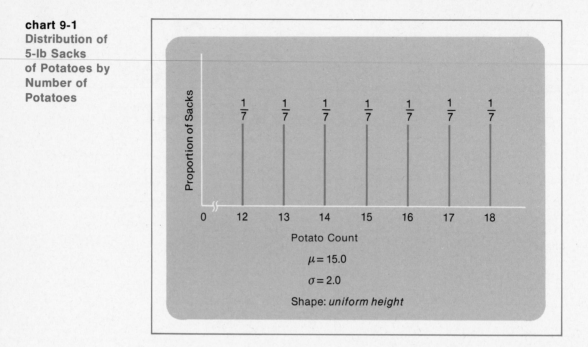

$\mu = 15.0$

$\sigma = 2.0$

Shape: *uniform height*

asking: What results can be expected when observations are selected at random from a population? In this case, the population is the potato counts recorded in Chart 9-1.

Suppose that a consumer with a large family buys two sacks of Katahdins. How many potatoes can the consumer expect to purchase; or, equivalently stated, what will the mean number of potatoes in the sample of two sacks be? (If we know the mean number in a sample of two sacks, we can determine the number of potatoes that the consumer's two sacks contain by multiplying the sample mean by 2.)

chance variation
in random
samples

A consumer who selects a random sample of two sacks must obtain one of the 49 samples (pairs of potato counts) listed in Chart 9-2.[1] Scanning the third column of Chart 9-2,

[1] The counting rule explained in Section 8.4 (page 230) tells us that exactly 49 possibilities exist. Since each sack can contain one of 7 counts ($V = 7$) and since 2 sacks are selected ($n = 2$), the number of possible samples is $V^n = 7^2 = 49$.

we see that the sample means vary considerably among samples. The mean is only 12.0 in sample 1, whereas the mean in sample 49 is 18.0 potatoes per sack. Although the population mean is $\mu = 15.0$ potatoes, the sample means vary from 12 to 18 potatoes, depending on chance selection.

The last column of Chart 9-2 lists the scores for another sample statistic—the *sample range*. We see that the sample range is as small as 0 in samples in which both sacks contain the same count and as large as 6 potatoes in some other samples. The population of potato counts has a range of 6 $(18 - 12)$, but the sample range varies from 0 to 6 depending on chance selection.

chart 9-2

POSSIBLE SAMPLE NO.	COUNT IN FIRST SACK , COUNT IN SECOND SACK	SAMPLE MEAN	SAMPLE RANGE (LARGER COUNT – SMALLER COUNT)
1	(12,12)	12.0	0
2	(12,13)	12.5	1
3	(12,14)	13.0	2
4	(12,15)	13.5	3
5	(12,16)	14.0	4
6	(12,17)	14.5	5
7	(12,18)	15.0	6
8	(13,12)	12.5	1
9	(13,13)	13.0	0
10	(13,14)	13.5	1
11	(13,15)	14.0	2
12	(13,16)	14.5	3
13	(13,17)	15.0	4
14	(13,18)	15.5	5
15	(14,12)	13.0	2
16	(14,13)	13.5	1
17	(14,14)	14.0	0
18	(14,15)	14.5	1
19	(14,16)	15.0	2
20	(14,17)	15.5	3
21	(14,18)	16.0	4
22	(15,12)	13.5	3
23	(15,13)	14.0	2
24	(15,14)	14.5	1
25	(15,15)	15.0	0

(Continued)

chart 9-2
(Continued)

POSSIBLE SAMPLE NO.	COUNT IN FIRST SACK , COUNT IN SECOND SACK	SAMPLE MEAN	SAMPLE RANGE (LARGER COUNT— SMALLER COUNT)
26	(15,16)	15.5	1
27	(15,17)	16.0	2
28	(15,18)	16.5	3
29	(16,12)	14.0	4
30	(16,13)	14.5	3
31	(16,14)	15.0	2
32	(16,15)	15.5	1
33	(16,16)	16.0	0
34	(16,17)	16.5	1
35	(16,18)	17.0	2
36	(17,12)	14.5	5
37	(17,13)	15.0	4
38	(17,14)	15.5	3
39	(17,15)	16.0	2
40	(17,16)	16.5	1
41	(17,17)	17.0	0
42	(17,18)	17.5	1
43	(18,12)	15.0	6
44	(18,13)	15.5	5
45	(18,14)	16.0	4
46	(18,15)	16.5	3
47	(18,16)	17.0	2
48	(18,17)	17.5	1
49	(18,18)	18.0	0

forming a sampling distribution

The relative frequency distributions presented in Charts 9-3(a) and (b) provide a clearer view of the pattern of chance variation for the sample mean and the sample range, respectively. To calculate the relative frequencies shown in these charts, we must recognize that each of the 49 samples listed in Chart 9-2 has the same chance of being selected, since each of the 7 counts in the population (12 through 18) occurs with equal frequency. (The relative frequency for each value of the sample mean, then, is simply the proportion of the samples whose mean equals that value.) For example, the sample mean is equal to 15 in 7 samples, so the relative frequency of the value 15 is 7/49. (The relative frequencies for the sample range are determined similarly. We denote the possible sample means as $\bar{X}_2$ ("X bar sub 2"). $\bar{X}$, of course, is the symbol for a sample mean, and the subscript 2 reminds the reader that the sample size is 2)(the consumer is buying 2 sacks of potatoes). (Similarly, sr_2 denotes the sample range.)

(The relative frequency distributions for sample statistics are referred to as *sampling distributions* to indicate that they represent frequencies that result from random sampling.)

(SAMPLING DISTRIBUTION   If we list *all* the possible random samples of a given number of observations and calculate the statistic of interest for each sample, then the relative frequency distribution of scores for the statistic is called the *sampling distribution of the statistic.*)

chart 9-3(a)

The Sampling Distribution of the Sample Mean

$\bar{X}_2$ POSSIBLE VALUES OF THE SAMPLE MEAN	$p(\bar{X}_2)$ RELATIVE FREQUENCY WITH WHICH EACH VALUE OCCURS
12.0	1/49
12.5	2/49
13.0	3/49
13.5	4/49
14.0	5/49
14.5	6/49
15.0	7/49
15.5	6/49
16.0	5/49
16.5	4/49
17.0	3/49
17.5	2/49
18.0	1/49
Any of the above	49/49

chart 9-3(b)

The Sampling Distribution of the Sample Range

sr_2 POSSIBLE VALUES OF THE SAMPLE RANGE	$p(sr_2)$ RELATIVE FREQUENCY WITH WHICH EACH VALUE OCCURS
0	7/49
1	12/49
2	10/49
3	8/49
4	6/49
5	4/49
6	2/49
Any of the above	49/49

In our example, we have listed all the possible samples of two observations (potato counts) and we have calculated two statistics for each sample: the sample mean $\overline{X}_2$ and the sample range sr_2. The results are the two sampling distributions given in Charts 9-3(a) and (b):

1. The sampling distribution of $\overline{X}_2$: The distribution of all the possible scores for the *mean* in random samples of 2 observations.
2. The sampling distribution of sr_2: The distribution of all the possible scores for the *range* in random samples of 2 observations.

The dishes represent different samples, each containing approximately the same number of seedlings. If we calculate the mean length of the seedlings within each dish, the resulting scores would form a distribution of sample means—that is, a sampling distribution of the statistic $\overline{X}$.

From the sampling distribution of $\overline{X}_2$ we learn, for example, that the value $\overline{X}_2 = 12.0$ occurs in only 1 of the 49 samples. From this, we can conclude that the probability of selecting a sample of 2 potato sacks whose mean count is 12.0 is only 1/49. Similarly, the probability that 2 sacks will have a mean count of 16.0 is 5/49. Because possible values of $\overline{X}_2$ are mutually exclusive events, their probabilities can be added together. For example, the probability that $\overline{X}_2$ will assume a value between 16.5 and 18.0 is $(4 + 3 + 2 + 1)/49 = 10/49$.

The sampling distribution of the sample range can be analyzed in the same way. In 7 of the 49 samples, both sacks contain the same potato count, resulting in the value $sr_2 = 0$. So the probability that the range of a sample of 2 sacks will equal 0 is 7/49. At the other extreme, the probability that the sample range will equal 6, making the sample range as large as the population range, is only 2/49.

EXERCISES

*9-1   In a certain city, households can receive home delivery of one or both of the city's daily newspapers: a morning paper and an evening paper. Let the variable X denote the number of newspapers delivered to each household (the newspaper count), and assume that the distribution of newspaper counts for all households in the city is

NEWSPAPER COUNT X	PROPORTION OF HOUSEHOLDS $p(X)$
0	1/3
1	1/3
2	1/3

(a) Calculate the mean μ and the standard deviation σ of this population distribution. Use the relative frequency equations:

(1) $\mu = \Sigma[X \cdot p(X)]$

(2) $\sigma = \sqrt{\Sigma[X^2 \cdot p(X)] - \mu^2}$

Express your result for σ to two decimal places and interpret the value you obtain.

(b) Two households are to be selected at random and the newspaper count for each household recorded. Following the format of Chart 9-2 on page 241, list the possible pairs of newspaper counts. Then for each of these samples, record the sample mean $\overline{X}_2$ and the sample range sr_2. Use the counting rule given in Section 8.4 (page 230) to verify that you have listed the correct number of samples (pairs of counts).

(c) Following the format used in Chart 9-3(a) on page 243, group the resulting scores for $\overline{X}_2$ into the sampling distribution of $\overline{X}_2$. Following the format used in Chart 9-3(b), group the resulting scores for sr_2 into the sampling distribution of sr_2.

(d) Referring to the sampling distribution of $\overline{X}_2$ you obtained in (c), what is the probability that a sample of two newspaper counts will have a mean $\overline{X}_2$ equal to:

(1) 0?

(2) 2?

(3) 0 or 2?

(4) 1 or more?

(5) μ as calculated in (a)?

(e) Referring to the sampling distribution of sr_2 you obtained in (c), what is the probability that a sample of two newspaper counts will have a range sr_2 equal to:

(1) 0?

(2) 1?

(3) 2?

(4) 1 or less?

(5) The range of the population of inference?

9-2   The following chart presents the distribution of the number of types of pets per household for all households in a certain county:

PET COUNT X	PROPORTION OF HOUSEHOLDS $p(X)$
0	1/4
1	1/2
2	1/4

(a) Calculate the mean μ of this distribution.

(b) Fill in the missing entry in the following sampling distribution of $\overline{X}_2$, where $\overline{X}_2$ denotes the mean of a sample of two pet counts:

POSSIBLE VALUES OF $\overline{X}_2$	RELATIVE FREQUENCY $p(\overline{X}_2)$
0	1/16
0.5	4/16
1.0	——
1.5	4/16
2.0	1/16
Any of the above	16/16

(c) Referring to the sampling distribution of $\overline{X}_2$ given in (b), what is the probability that a sample of two pet counts will have a mean $\overline{X}_2$ that is:

(1) $= \mu$?

(2) $> \mu$?

(3) $< \mu$?

(4) $\geq \mu$?

(5) $\leq \mu$?

9-3*   Referring to the distribution of potatoes per 5-lb sack given in Chart 9-1 on page 240, suppose that a consumer selected three sacks (instead of two sacks) at random and computed the sample mean $\overline{X}_3$. The result would be one of those listed in the following sampling distribution of $\overline{X}_3$:

POSSIBLE VALUES OF $\overline{X}_3$	RELATIVE FREQUENCY $p(\overline{X}_3)$
12	1/343
$12\frac{1}{3}$	3/343
$12\frac{2}{3}$	6/343
13	10/343
$13\frac{1}{3}$	15/343
$13\frac{2}{3}$	21/343
14	28/343
$14\frac{1}{3}$	33/343
$14\frac{2}{3}$	36/343
15	37/343
$15\frac{1}{3}$	36/343
$15\frac{2}{3}$	33/343
16	28/343
$16\frac{1}{3}$	21/343
$16\frac{2}{3}$	15/343
17	10/343
$17\frac{1}{3}$	6/343
$17\frac{2}{3}$	3/343
18	1/343
	343/343

(a) State a definition of the sampling distribution of $\overline{X}_3$.

(b) What is the probability that the mean $\overline{X}_3$ of a sample of three potato counts will be:

(1) $= 12$?

(2) $= 12$ or $= 18$?

(3) ≤ 15?

(4) ≥ 16?

(5) $\neq 15$?

(c) Show that the resulting number of samples of three potato counts (343) follows from the counting rule given in Section 8.4 (page 230).

(d) By comparing the sampling distribution of $\overline{X}_3$ in (a) with the sampling distribution of $\overline{X}_2$ given in Chart 9-3(a) on page 243 (both of which are derived from a population with a mean $\mu = 15.0$ potatoes per sack), determine the effect of increasing sample size from $n = 2$ to $n = 3$ on:

(1) The probability that $\overline{X}_n$ lies no more than 1 potato from μ (that is, that $\overline{X}_n$ equals from 14 to 16).

(2) The probability that $\overline{X}_n$ lies more than 2 potatoes from μ (that is, that $\overline{X}_n < 13$ or that $\overline{X}_n > 17$).

9.2 summarizing a sampling distribution of means

When we describe a frequency distribution, we typically report the average(s) and a measure of the variation of the distribution and also comment on the shape of the distribution. A sampling distribution is a frequency distribution and can be summarized in the same way. Such a summary will point out how the sampling distribution of a statistic relates to the population we are sampling and eventually will lead us to the laws of chance variation that we apply in statistical inference.

To summarize the sampling distribution of $\overline{X}_2$, we begin by determining the mean of the distribution.

expected value of the sampling distribution of $\overline{X}_2$

The mean of the sampling distribution of a statistic frequently is called the *expected value* of the distribution, because it represents an anticipated or expected result. The expected value of the sampling distribution of $\overline{X}_2$, denoted by $\mu(\overline{X}_2)$ ("mu of X bar sub 2"), is found by averaging all the possible scores for the sample mean in random samples of two potato sacks. We compute $\mu(\overline{X}_2)$ by using the relative frequency equation for a mean.

$$
\begin{aligned}
\mu(\overline{X}_2) = \Sigma[\overline{X}_2 \cdot p(\overline{X}_2)] \qquad \qquad (9\text{-}1)\\
= \quad 12.0 \cdot 1/49 \\
+ 12.5 \cdot 2/49 \\
+ 13.0 \cdot 3/49 \\
+ 13.5 \cdot 4/49 \\
+ 14.0 \cdot 5/49 \\
+ 14.5 \cdot 6/49 \\
+ 15.0 \cdot 7/49 \\
+ 15.5 \cdot 6/49 \\
+ 16.0 \cdot 5/49
\end{aligned}
$$

$$+ 16.5 \cdot 4/49$$
$$+ 17.0 \cdot 3/49$$
$$+ 17.5 \cdot 2/49$$
$$+ 18.0 \cdot 1/49$$
$$= 15.0 \text{ potatoes per sack}$$

The expected value of the sampling distribution of $\overline{X}_2$ is 15.0 because the possible scores for $\overline{X}_2$ have an average of 15.0 potatoes per sack.

the standard deviation of the sampling distribution of $\overline{X}_2$

(The standard deviation of the sampling distribution of $\overline{X}_2$, like the standard deviation of any frequency distribution, measures the "average" spread of the observations about the mean. In this case, the observations we're interested in are the means of samples of size 2, $\overline{X}_2$, and the standard deviation measures their "average" spread about $\mu(\overline{X}_2)$, the expected value of $\overline{X}_2$.)

(We denote the standard deviation of a sampling distribution of $\overline{X}_2$ by $\sigma(\overline{X}_2)$ ("sigma of $\overline{X}_2$"). As with $\mu(\overline{X}_2)$, we use a relative frequency equation to compute the value of $\sigma(\overline{X}_2)$:)

$$\left(\sigma(\overline{X}_2) = \sqrt{\Sigma[\overline{X}_2^2 \cdot p(\overline{X}_2)] - [\mu(\overline{X}_2)]^2} \right) \qquad (9\text{-}2)$$
$$= \textit{the square root of:}$$
$$(12.0)^2 \cdot 1/49$$
$$+ (12.5)^2 \cdot 2/49$$
$$+ (13.0)^2 \cdot 3/49$$
$$+ (13.5)^2 \cdot 4/49$$
$$+ (14.0)^2 \cdot 5/49$$
$$+ (14.5)^2 \cdot 6/49$$
$$+ (15.0)^2 \cdot 7/49$$
$$+ (15.5)^2 \cdot 6/49$$
$$+ (16.0)^2 \cdot 5/49$$
$$+ (16.5)^2 \cdot 4/49$$
$$+ (17.0)^2 \cdot 3/49$$
$$+ (17.5)^2 \cdot 2/49$$
$$+ (18.0)^2 \cdot 1/49$$

$$- (15.0)^2$$

$$= \sqrt{2} = 1.414 \text{ potatoes per sack}$$

So the means of samples of 2 sacks vary by an "average" of 1.414 potatoes per sack about the expected value of 15.0.

the shape of
the sampling
distribution of $\overline{X}_2$

We now have obtained measures of the average and the variation of the sampling distribution of $\overline{X}_2$, but we must still determine its shape. Based on the data in Chart 9-3(a), Chart 9-4 presents a line graph of the sampling distribution of $\overline{X}_2$. From the graph, we can see that the sampling distribution is symmetric about the value 15.0. The triangular shape formed by the line graph reveals that mean counts close to 15.0 are more likely to occur in random sampling than mean counts well below or well above 15.0.

The characteristics of the sampling distribution of $\overline{X}_2$ are summarized in the first column of Chart 9-5. The analogous characteristic of the population of potato counts appears in the second column. The relationships between the two are described in the third column. These relationships are of interest because they exist no matter what the sample size or the shape

chart 9-4
**The Sampling
Distribution of $\overline{X}_2$**

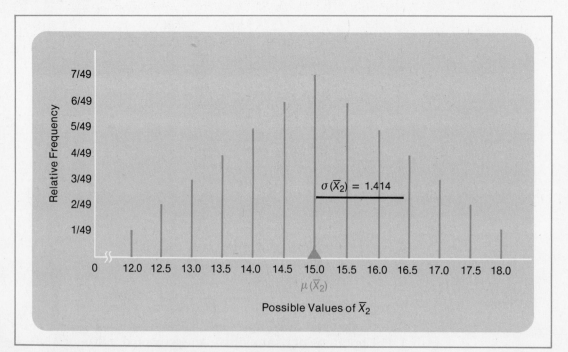

	SAMPLING DISTRIBUTION OF MEAN POTATO COUNTS $\overline{X}_2$ (Source: Chart 9-4)	POPULATION OF POTATO COUNTS (Source: Chart 9-1)	RELATIONSHIP
Mean	$\mu(\overline{X}_2) = 15.0$	$\mu = 15.0$	$\mu(\overline{X}_2) = \mu$
Standard Deviation	$\sigma(\overline{X}_2) = 1.414$	$\sigma = 2.0$	$\sigma(\overline{X}_2) < \sigma$
Shape	*Triangular*	*Uniform height*	Sampling distribution of $\overline{X}_2$ is more *normal* in shape.

chart 9-5

of the population is. From the relationships between the sampling distribution of a statistic and the distribution of scores in the original population, the *general laws of chance variation* are deduced. These laws enable us to draw inferences about population parameters when the only information we have about the population of inference is from the data we have acquired from a random sample.)

EXERCISES

*9-4   (*Continuation of Exercise 9-1, page 245*) Referring to the sampling distribution of $\overline{X}_2$ you developed in Exercise 9-1(b):
(a) Calculate $\mu(\overline{X}_2)$, the expected value of the sampling distribution of $\overline{X}_2$. Interpret the value you obtain.
(b) Calculate $\sigma(\overline{X}_2)$, the standard deviation of the sampling distribution of $\overline{X}_2$. Interpret the value you obtain.
(c) Prepare a line graph of the sampling distribution of $\overline{X}_2$.
(d) Following the format used in Chart 9-5, compare the expected value, the standard deviation, and the shape of the sampling distribution of $\overline{X}_2$ with the analogous features of the population of newspaper counts.

*9-5   (*Continuation of Exercise 9-3, page 247*) Referring to the sampling distribution of $\overline{X}_3$, where $\overline{X}_3$ denotes the mean of a sample of three potato counts:
(a) Calculate $\mu(\overline{X}_3)$, the expected value of the sampling distribution of $\overline{X}_3$. Interpret the value you obtain.
(b) Calculate $\sigma(\overline{X}_3)$, the standard deviation of the sampling distribution of $\overline{X}_3$. Interpret the value you obtain.
(c) Prepare a line graph of the sampling distribution of $\overline{X}_3$.
(d) What is the relationship among:
(1) $\mu(\overline{X}_3)$, $\mu(\overline{X}_2)$, and μ?
(2) $\sigma(\overline{X}_3)$, $\sigma(\overline{X}_2)$, and σ?
(3) The shape of the sampling distribution of $\overline{X}_3$ and the shape of the sampling distribution of $\overline{X}_2$?

9.3   three laws of chance variation

We have used the symbol $\overline{X}_2$ to denote the mean value of a random sample of two observations. We often use a subscript with $\overline{X}$ to indicate the size of the sample for which $\overline{X}$ is the mean value. For example

$\overline{X}_4$ is used to denote the mean of a random sample of four observations:

$$\overline{X}_4 = \frac{\Sigma X}{4}$$

$\overline{X}_{10}$ is used to denote the mean of a random sample of ten observations:

$$\overline{X}_{10} = \frac{\Sigma X}{10}$$

and in general
$\overline{X}_n$ is used to denote the mean of a random sample of n observations:

$$\overline{X}_n = \frac{\Sigma X}{n}$$

We can specify a particular sampling distribution by using the same notation.

SAMPLING DISTRIBUTION OF $\overline{X}_n$ The relative frequency distribution of all the possible scores for the mean in random samples of *n* observations.

We can now use this notation to state three laws of chance variation in $\overline{X}_n$.

the expected value law

In Chart 9-5 we see that the expected value of the mean potato count in random samples of two sacks is equal to the mean of the population of potato counts; that is $\mu(\overline{X}_2) = \mu$. The *Expected Value Law for $\overline{X}_n$* is a generalization of this relationship.

EXPECTED VALUE LAW FOR $\overline{X}_n$ The expected value of a sampling distribution of sample means is equal to the mean of the population of inference. This law holds for all sample sizes n and for all populations with a mean value.

Equation (9-3) states the Expected Value Law for $\overline{X}_n$ in algebraic notation:

$$\mu(\overline{X}_n) = \mu \qquad \text{for all } n \tag{9-3}$$

The Expected Value Law for $\overline{X}_n$ does *not* imply that the mean of any random sample selected from a population will be equal to the population mean. Sample means vary by chance from sample to sample, providing no assurance that the mean of any one sample will equal μ. The Expected Value Law for $\overline{X}_n$ *does* imply that the potential that $\overline{X}_n$ underestimates (is smaller than) μ and the potential that $\overline{X}_n$ overestimates (is larger than) μ balance one another. If we use $\overline{X}_n$ to estimate μ, we will not always hit the bull's-eye, but we will be aiming in the right direction.

When the expected value of a statistic is equal to the parameter the statistic is supposed to estimate, we have not biased our results in the direction of overestimation or underestimation. So we say that such a statistic provides an *unbiased estimate*.

UNBIASED ESTIMATE A statistic provides an unbiased estimate if its expected value is equal to the population parameter that it is being used to estimate.

Estimating the range of the population of inference on the basis of the range of a random sample, sr, is quite different from estimating the population mean on the basis of the sample mean. Chart 9-6 is a line graph of the sampling distribution of sr_2 that we tabulated in Chart 9-3(b). The expected value of the sampling distribution of sr_2 is $\mu(sr_2) = 2.3$, but the population range is 6. Thus sr_2 is a severely biased estimate: $\mu(sr_2)$ is less than half of the population range. Unfortunately, unless the shape of the population is known, an unbiased estimate of the population range cannot be obtained.

chart 9-6
The Sampling
Distribution of sr_2

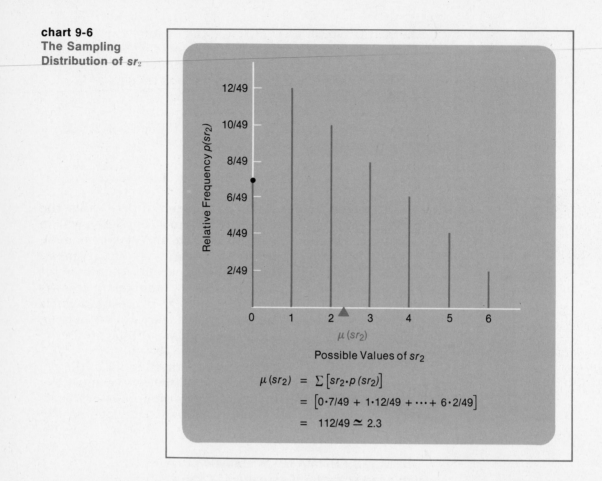

$$\mu(sr_2) = \sum \left[sr_2 \cdot p(sr_2) \right]$$
$$= \left[0 \cdot 7/49 + 1 \cdot 12/49 + \cdots + 6 \cdot 2/49 \right]$$
$$= 112/49 \simeq 2.3$$

**the standard
error law**

(Whereas the Expected Value Law for $\overline{X}_n$ applies to the *average* of all the possible mean scores in a random sample of n observations, the Standard Error Law pertains to the extent of *variation* in $\overline{X}_n$ from sample to sample. The Standard Error Law answers the question: How far from μ is the mean value of a random sample likely to be?)

(The standard deviation of a sampling distribution of $\overline{X}_n$, denoted by $\sigma(\overline{X}_n)$, measures the "average" spread in the scores for $\overline{X}_n$ about $\mu(\overline{X}_n)$. We know from the Expected Value Law for $\overline{X}_n$ that $\mu(\overline{X}_n) = \mu$. So, equivalently, $\sigma(\overline{X}_n)$ measures the "average" spread in the possible values of $\overline{X}_n$ about the population mean μ. Consequently, we can interpret $\sigma(\overline{X}_n)$ as the "average" error we would make if we used the sample mean to estimate the population mean. For this reason, we call the standard deviation of a sampling distribution a *standard error* and we call $\sigma(\overline{X}_n)$ the *standard error of the mean*.)

Referring to Chart 9-5, we see that the sampling distribution of $\overline{X}_2$ has a standard error (standard deviation) $\sigma(\overline{X}_2) = 1.414$ potatoes per sack. When random samples of two potato sacks are selected, the sample mean will differ from μ by an "average" of 1.414 potatoes per sack. So we make an "average" error of 1.414 potatoes per sack if we use $\overline{X}_2$ to estimate μ.

Chart 9-5 also shows that $\sigma(\overline{X}_2) < \sigma$: The standard error of $\overline{X}_2$ is smaller than the standard deviation of the population. This means that the "average" error that results from choosing the mean count in two potato sacks to estimate μ is less than the "average" error that results from using the potato count in a single sack to estimate μ.

(So $\sigma(\overline{X}_n)$ measures the "average" size of the error we can expect to make in our estimate. If $\sigma(\overline{X}_n)$ is very small, we know that the mean of any sample selected by chance will not differ much from the population mean. Conversely, if $\sigma(\overline{X}_n)$ is large, the luck of the draw becomes important because we might choose a sample whose mean differs greatly from μ.)

All we know about the value of $\sigma(\overline{X}_n)$ up to this point is that when $n = 2$, $\sigma(\overline{X}_2) < \sigma$.(However, a formal mathematical relationship exists between $\sigma(\overline{X}_n)$ and σ, and we call this relationship the *Standard Error Law*.)

(**THE STANDARD ERROR LAW FOR $\overline{X}_n$ The standard error of $\overline{X}_n$ is equal to the standard deviation of the population of inference divided by the square root of the sample size n.**)

(Symbolically the law is expressed

$$\sigma(\overline{X}_n) = \frac{\sigma}{\sqrt{n}} \quad) \tag{9-4}$$

We can verify the Standard Error Law for the sampling distribution of $\overline{X}_2$ in our example. From Chart 9-1, we know that $\sigma = 2.0$. So

$$\sigma(\overline{X}_2) = \frac{\sigma}{\sqrt{n}} = \frac{2.0}{\sqrt{2}} = 1.414$$

LIFE'S LAWS

MURPHY'S LAW: If anything can go wrong it will.

O'TOOLE'S COMMENTARY ON MURPHY'S LAW: Murphy was an optimist.

*Law of SELECTIVE GRAVITY
An object will fall so as
to do the most damage.*

*JENNING'S COROLLARY
The chance of bread falling
with the buttered side down
is in direct proportion to the
cost of the carpet.*

*MAER'S Law
If the facts do not
conform to the theory
they must be disposed of.*

BOREN'S LAW: When in doubt, mumble!

Source: Text excerpts used by permission of Price/Stern/Sloan Publishers, Inc., Los Angeles, Calif. from *Murphy's Law and Other Reasons Why Things Go Wrong* by Arthur Bloch. Copyright © 1977. Original drawings by Evanell Towne.

This is precisely the result we obtained earlier using the relative frequency equation, Equation (9-2), on page 249.

(The Standard Error Law indicates that the size of $\sigma(\overline{X}_n)$ depends on two factors: the standard deviation of the population of inference σ and the sample size n:

1. σ: For any sample size n, $\sigma(\overline{X}_n)$ increases as σ increases. The greater the spread is among observations in the population of inference, the greater the spread will be among the means of random samples of n observations.) For example, if σ had been equal to 4 instead of 2 potatoes per sack in our illustration, then $\sigma(\overline{X}_2)$ would

have been equal to $4/\sqrt{2} = 2.818$ potatoes—a value twice as large as the one in our example.

$\big($2. n: For a given value of σ, $\sigma(X_n)$ decreases as sample size n increases. So the mean of a large sample is *likely to be* closer to μ than the mean of a small sample is. Or stated another way, increasing sample size reduces the consequences of bad luck in selecting the sample.$\big)$

$\big($How much do you gain by increasing sample size? Since the denominator of Equation (9-4) contains the *square root of* n, a fourfold increase in sample size will be required to reduce the standard error to half its former value.$\big)$If we had tabulated the sampling distribution of $\overline{X}_8$ instead of $\overline{X}_2$, we would have found that

$$\sigma(\overline{X}_8) = \frac{\sigma}{\sqrt{8}} = \frac{2}{\sqrt{8}} = 0.707 \text{ potatoes}$$

The "average" error that results from using $\overline{X}_8$ to estimate μ is half the "average" error that results from using $\overline{X}_2$ to estimate μ.

the law of normality

$\big($We have seen how $\sigma(\overline{X}_n)$ decreases as sample size n increases, implying that the sampling distribution of $\overline{X}_n$ becomes more concentrated about μ. The sampling distribution of $\overline{X}_n$ also changes *shape* as n increases.$\big)$

Chart 9-7 illustrates the shapes of sampling distributions of $\overline{X}_n$ from our population of potato counts for various values of n. The population of inference—the same *uniform height* distribution depicted in Chart 9-1—appears in (a). The sampling distribution of $\overline{X}_2$, which we have been analyzing in detail, is represented in (b). The sampling distribution of $\overline{X}_4$, the relative frequency distribution of the possible scores for the mean of a sample of four observations is illustrated in (c). Finally, the sampling distributions of $\overline{X}_{10}$ and $\overline{X}_{25}$ are presented in (d) and (e) respectively.

Notice how the shape evolves from the *uniform height* distribution in (a) to the virtually *normal* curves in (d) and (e). $\big($In Chapter 4, we indicated that a distribution of mean values tends to be *normal* in shape. We can now formalize that statement as the *Law of Normality*.$\big)$

chart 9-7

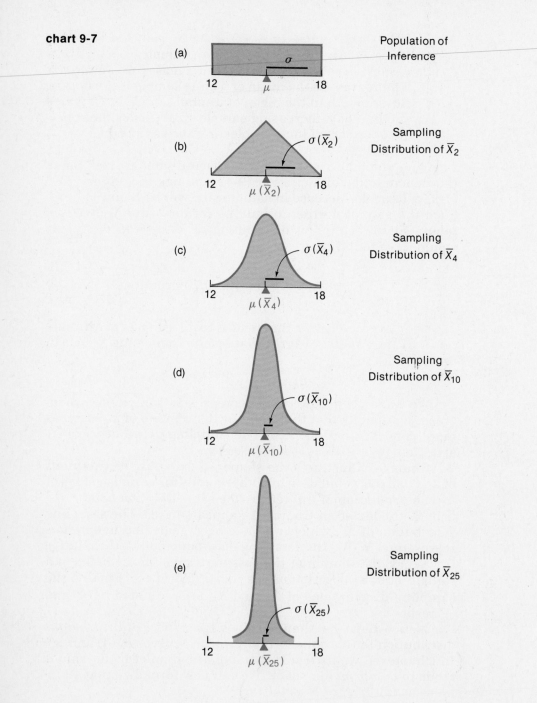

(a) Population of Inference

(b) Sampling Distribution of $\overline{X}_2$

(c) Sampling Distribution of $\overline{X}_4$

(d) Sampling Distribution of $\overline{X}_{10}$

(e) Sampling Distribution of $\overline{X}_{25}$

chart 9-8

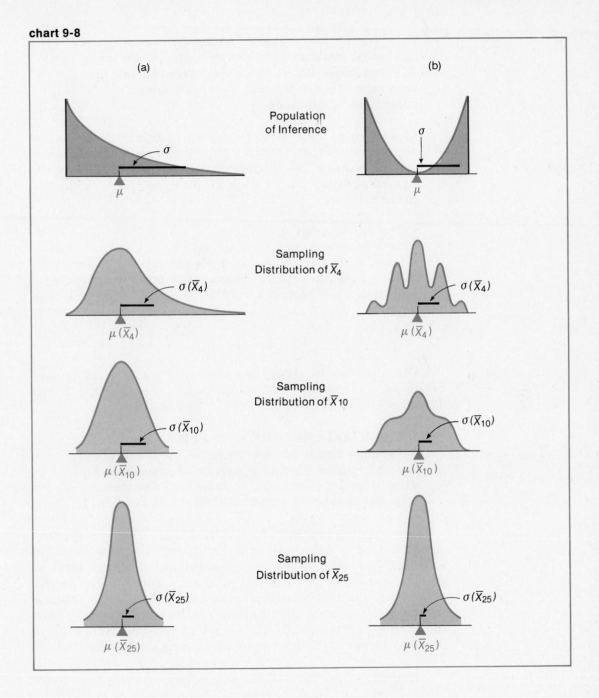

(a)   (b)

Population
of Inference

σ

μ

σ

μ

Sampling
Distribution of $\overline{X}_4$

$\sigma(\overline{X}_4)$

$\mu(\overline{X}_4)$

$\sigma(\overline{X}_4)$

$\mu(\overline{X}_4)$

Sampling
Distribution of $\overline{X}_{10}$

$\sigma(\overline{X}_{10})$

$\mu(\overline{X}_{10})$

$\sigma(\overline{X}_{10})$

$\mu(\overline{X}_{10})$

Sampling
Distribution of $\overline{X}_{25}$

$\sigma(\overline{X}_{25})$

$\mu(\overline{X}_{25})$

$\sigma(\overline{X}_{25})$

$\mu(\overline{X}_{25})$

THE LAW OF NORMALITY   As sample size n increases, the shape of the sampling distribution of $\overline{X}_n$ approaches that of a *normal* curve. This law prevails regardless of the shape of the original population of inference.

Chart 9-8(a) depicts a population of inference that is skewed to the right, and Chart 9-8(b) depicts a U-shaped population. The value of the population mean is indicated by the balance point symbol ▲ on each graph.

In both charts:

- For any value of n, $\mu(\overline{X}_n) = \mu$.
- As n increases, $\sigma(\overline{X}_n)$ decreases.
- As n increases, the shape of the sampling distribution of $\overline{X}_n$ approaches that of the *normal* curve, evolving into a virtually bell-shaped curve by the time sample size n reaches 25.

the central limit theorem

Often we see the three laws of chance variation in $\overline{X}_n$ combined in a composite theorem, called the *Central Limit Theorem* (CLT).

THE CENTRAL LIMIT THEOREM FOR $\overline{X}_n$ Provided the sample size is sufficiently large, the mean of any one random sample $\overline{X}_n$ can be considered to be a score from a *normal* distribution that is centered on μ and that has a standard error equal to $\sigma/\sqrt{n}$.

The building blocks of the Central Limit Theorem are presented in Charts 9-9(a), (b), and (c). Chart 9-9(a) illustrates the Law of Normality. If we select a sufficiently large random sample, we can safely assume that the sample mean is one point from a *normal* distribution of scores for $\overline{X}_n$. The specific sample size n that is *sufficiently large* depends on the shape of the population from which the sample is selected. Here are two rules of thumb for determining adequate sample size:

1. Unless the shape of the population is markedly skewed, as it is in Chart 9-8(a), or U-shaped, as it is in Chart

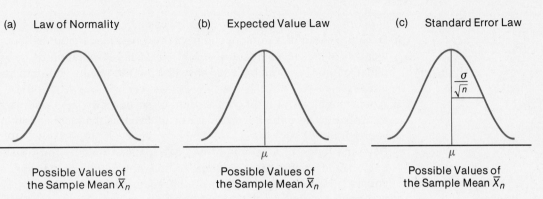

(a)   Law of Normality   (b)   Expected Value Law   (c)   Standard Error Law

Possible Values of   Possible Values of   Possible Values of
the Sample Mean $\overline{X}_n$ the Sample Mean $\overline{X}_n$ the Sample Mean $\overline{X}_n$

chart 9-9

9-8(b), a sample size of $n = 15$, will be sufficient to satisfy the Central Limit Theorem.
2. Regardless of the shape of the population, a sampling distribution of $\overline{X}_n$ will form a virtually *normal* curve when $n \geq 25$.

(Chart 9-9(b) illustrates the Expected Value Law. We have positioned the population mean μ in the center of the *normal* distribution of sample means, since $\mu(\overline{X}_n) = \mu$.

Finally, Chart 9-9(c) illustrates the Standard Error Law. We have indicated that a distance of 1 standard deviation in the *normal* distribution of sample means is equal to $\sigma/\sqrt{n}$, since $\sigma(\overline{X}_n) = \sigma/\sqrt{n}$.

The Central Limit Theorem provides the theoretical foundation for many of the statistical procedures that are used to draw inferences about the values of population averages. In the next section, we will learn how to apply the CLT to estimate a population mean.)

EXERCISES

9-6   Explain what each of the following symbols denotes:
(a) μ
(b) σ
(c) $\overline{X}_n$
(d) $\mu(\overline{X}_n)$
(e) $\sigma(\overline{X}_n)$

*9-7   (*Continuation of Exercises 9-1 and 9-4, pages 245 and 251*)

(a) Does the mean of a random sample of two newspaper counts $\overline{X}_2$ provide an unbiased estimate of the mean of the population of newspaper counts μ? Support your answer from the results you obtained in Exercise 9-4.

(b) Would the mean of a random sample of seven newspaper counts $\overline{X}_7$ also provide an unbiased estimate of μ? What is the justification for your answer?

(c) Does the range of a sample of two newspaper counts sr_2 provide an unbiased estimate of the range of the population of newspaper counts? (*Hint:* Calculate $\mu(sr_2)$ by the equation $\mu(sr_2) = \Sigma[sr_2 \cdot p(sr_2)]$ and compare your result with the value of the population range.)

(d) Show that the value you calculated for $\sigma(\overline{X}_2)$ in Exercise 9-4 can be determined from the Standard Error Law.

9-8   A population of inference is known to have a mean $\mu = 100$ units and a standard deviation $\sigma = 20$ units. A random sample of n scores is to be selected from that population and the mean value $\overline{X}_n$ is to be computed.

(a) What two factors determine the size of the "average" error you would make if you used $\overline{X}_n$ to estimate μ?

(b) Determine $\sigma(\overline{X}_n)$, the standard error of $\overline{X}_n$, if:
    (1) $n = 4$
    (2) $n = 16$
    (3) $n = 64$

(c) How large a sample is required to bring $\sigma(\overline{X}_n)$ down to 1 unit?

(d) Prepare a graph showing the relationship between sample size n on the horizontal axis and $\sigma(\overline{X}_n)$ on the vertical axis. Characterize the response of $\sigma(\overline{X}_n)$ to increases in n.

9-9   (a) In which of the following cases will $\sigma(\overline{X}_n)$ be the smaller weight?
    (1) Case A: $\sigma = 60$ lb, $n = 25$
    (2) Case B: $\sigma = 240$ lb, $n = 100$

(b) Explain why the effect on $\sigma(\overline{X}_n)$ of quadrupling the standard deviation from $\sigma = 60$ to $\sigma = 240$ is not offset by the effect of quadrupling sample size from $n = 25$ to $n = 100$.

9-10   (*Continuation of Exercise 9-5, page 251*) Referring to your answers to Exercise 9-5(d):

(a) Verify that the Expected Value Law holds.

(b) Confirm that the value of $\sigma(\overline{X}_3)$ can be derived from the Standard Error Law.

(c) Do your findings also support the Law of Normality?

9-11

(a) Sketch a frequency curve for a sampling distribution of $\overline{X}_{100}$. Label the horizontal axis.

(b) In preparing the graph in (a), why is it unnecessary for you to know the shape of the population of inference?

(c) Assuming that all the possible scores for $\overline{X}_{100}$ were derived by sampling from a population of inference with $\mu = 40$ and $\sigma = 40$ place the *numeric values* of $\mu(\overline{X}_{100})$ and $\sigma(\overline{X}_{100})$ in their appropriate locations on the frequency curve you sketched in (a).

9-12

True or False? (Explain each answer.)

(a) As sample size n increases, $\mu(\overline{X}_n)$, the expected value of $\overline{X}_n$, becomes closer to μ.

(b) As sample size n increases, $\sigma(\overline{X}_n)$, the standard error of $\overline{X}_n$, becomes closer to σ.

(c) A sampling distribution of $\overline{X}_n$ will always be *normal* in shape.

(d) A sampling distribution of $\overline{X}_n$ will only be *normal* in shape if the population of inference has a *normal* distribution.

9.4 estimating a population mean

A few years ago, the Social Services Agency of a state government drew criticism for its method of making payments to licensed child day care centers. The agency had been paying centers $0.90 for each hour of child care authorized. Critics from the state legislature argued that centers should be paid for each hour that children were actually in attendance.

The costs of the two payment methods could differ substantially. For example, the agency would pay $180 monthly ($0.90·200 hours) for each child authorized to have 200 hours of day care, regardless of how many hours of care the child received. If that child actually received only 100 hours of care, the agency would pay only $90 if the payment system changed.

To determine how much money could be saved by amending the payment system, the legislature asked the agency how many hours of child care clients actually received in a month. Since it would be too expensive to answer this question for all children in day care centers, the agency was faced with a sampling and estimation problem.

The agency planned to estimate the mean (and total) hours the children spent at day care centers on the basis of the data from a sample of these children. We now summarize the agency's approach.

The Population of Inference: The number of actual hours that each child attended a day care center during a designated month.

Parameter to Be Estimated: The population mean μ—the true mean number of hours of attendance that month. (The estimated value of μ multiplied by the number of children in day care centers will provide an estimate of the total number of actual hours of attendance.)

Sample to Be Obtained: A simple random sample of 36 records ($n = 36$) from the agency files for children authorized to receive day care support. For each child selected, a report of the actual hours of attendance during the designated month is to be requested from the day care centers.

Statistic to Be Calculated: The sample mean $\overline{X}_{36}$ of the 36 counts of hours of attendance. This sample mean is used to estimate the population mean. The estimate will allow a 5-hour *margin for error*, since the mean of a random sample can vary by chance from the value of μ.

The data collected by the agency are shown in Chart 9-10.

chart 9-10
Actual Hours of Day Care Provided to a Sample of 36 Children

124	111	149	151	102	117
132	140	94	91	156	122
120	138	158	117	126	150
115	143	107	130	106	98
133	107	148	113	145	142
112	102	100	138	136	127

The sample mean $\overline{X}_{36} = 125$ hours. Consequently, the agency drafted the following statement for the legislature.

> From a random sample of 36 cases, we have estimated that the agency children averaged between 120 and 130 hours of actual attendance in day care centers during the month. In contrast, the agency authorized an average of 160 hours of day care per child for this period.

The agency's statement can be expressed this way: "The population mean μ lies within an interval of 125 ± 5 hours = 120 to 130 hours." Technically, such a statement is called an *interval estimate of μ* and has two components:

1. 125 hours: (The value of the sample mean — a value referred to as a *point estimate of μ.*)
2. 5 hours: (The *margin for error* — the margin allowed for anticipated sampling error.)

Let's raise and answer three basic questions about the agency's estimate:

1. How certain can the agency be that its estimate is correct — that the value of μ actually lies within the interval of 120 to 130 hours? In other words, what *level of confidence* can the agency attach to its estimate?
2. Why did the agency allow a margin for error of 5 hours? How are the size of the margin for error and the level of confidence in the estimate related?
3. Was the agency's choice of sample size $n = 36$ reasonable? How would a change in sample size affect the estimate that can be made?

computing the
level of
confidence

To determine the level of confidence, we apply the Central Limit Theorem for sample means.

The theorem tells us that the mean of the agency's random sample $\overline{X}_{36} = 125$ hours is a point from a *normal* distribution of scores for $\overline{X}_{36}$. It also tells us that this sampling distribution of $\overline{X}_{36}$ is centered on μ and has a standard error of $\sigma(\overline{X}_{36}) = \sigma/\sqrt{36} = \sigma/6$, where σ denotes the standard deviation of the population of inference. Chart 9-11 illustrates these facts.

chart 9-11

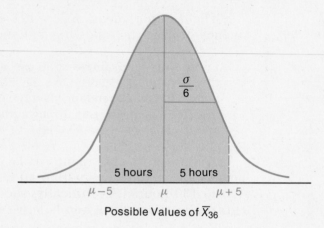

$$\frac{\sigma}{6}$$

5 hours | 5 hours

$\mu - 5$ μ $\mu + 5$

Possible Values of $\overline{X}_{36}$

The agency has allowed a margin for error of 5 hours. If the sample mean $\overline{X}_{36}$ is within 5 hours of μ (within the interval from $\mu - 5$ to $\mu + 5$), the agency's interval estimate will be correct. Otherwise, the estimate will be incorrect. The probability that the agency's estimate is correct, then, is simply the probability that $\overline{X}_{36}$ will lie within the interval from $\mu - 5$ to $\mu + 5$. (This probability is represented by the portion of the area under the *normal* curve shaded in blue in Chart 9-11. We call this probability the *level of confidence.*)

(**LEVEL OF CONFIDENCE**   The probability that a sample mean lies within the designated margin for error about the population mean. Alternatively stated, the probability that the difference between the values for $\overline{X}_n$ and μ is no greater than the margin for error.)

To calculate the level of confidence in the agency's estimate, therefore, we must determine what portion of the area under the *normal* curve in Chart 9-11 lies within the interval from $\mu - 5$ hours to $\mu + 5$ hours.

(At this point, we should recall from Chapter 4 the method for calculating the area under a *normal* curve within an interval centered on the mean:

1. Calculate the Z score of the upper bound of the interval; that is, determine the number of standard deviations

between the center and the upper bound of the interval.
2. Find the Z score in the margin of the *normal* table. The corresponding entry in the body of the table indicates the portion of the area under a *normal* curve between the center and the upper bound.
3. Double the value of the entry to obtain the portion of the area under a *normal* curve between the lower bound and the upper bound of the centered interval.)

To express the location of the upper bound of the interval as a Z score, $Z_{\mu+5}$, we need to know the size of one standard deviation under the *normal* curve in Chart 9-11. As we noted earlier, the standard deviation is the standard error of $\overline{X}_{36}$, $\sigma(\overline{X}_{36}) = \sigma/6$.

The upper bound of the interval is a distance of 5 hours from μ, since a 5-hour margin for error was allowed. The number of standard deviations between μ and the upper bound — its Z score — is therefore

$$Z_{\mu+5} = \frac{\text{Margin for error}}{\sigma(\overline{X}_{36})} = \frac{5}{\sigma/6}$$

To complete the calculation of the Z score, we need to determine the value for σ, the standard deviation of the population. (σ measures the "average" spread in the population of actual hours of attendance of agency children.) Although σ could be estimated from the sample data (in Chapter 10, we will examine the estimation of σ in detail), the agency chose instead to assume that the standard deviation of the actual hours of attendance is equal to the standard deviation of the hours authorized, which is known to be $\sigma = 18$.[2] So

$$Z_{\mu+5} = \frac{5}{\sigma/6} = \frac{5}{18/6} = \frac{5}{3} = 1.6$$

The upper bound of the interval, $\mu +$ Margin for error, lies 1.6 standard errors above the center of the *normal* curve in Chart 9-11.

[2] This assumption was made to simplify the estimation procedure, but this kind of assumption should be made only after careful consideration. If the actual value of σ differs greatly from the assumed value, the stated level of confidence will be incorrect.

Turning to the *normal* table (Appendix Table B, page 495), we locate the Z score of 1.60 and read the corresponding table entry .4452, or 44.52 percent. This result tells us that 44.52 percent of the possible values for $\overline{X}_{36}$ in random sampling will lie within the interval from μ to $\mu + 5$ hours. Doubling the entry gives us $.4452 \cdot 2 = .8904$, or 89.04 percent, which tells us that approximately 89 percent of the scores for $\overline{X}_{36}$ lie within the interval from $\mu - 5$ hours to $\mu + 5$ hours.

The level of confidence for an interval estimate is defined as the probability that a sample mean lies within the designated margin for error about the population mean. Since 89 percent of the $\overline{X}_{36}$ scores lie within 5 hours of μ, the interval $\overline{X}_{36} \pm 5$ hours has a level of confidence of 89 percent and we say that $\overline{X}_{36} \pm 5$ hours is an 89-percent *confidence interval for μ.*

In conclusion, the agency could be 89-percent confident that the true mean of the distribution of actual hours of attendance μ has a value between 120 and 130 hours.

Let's summarize the steps taken to compute the level of confidence.

1. Determine a sample size n:

$$n = 36$$

2. Establish a margin for error:

$$\text{Margin for error} = 5 \text{ hours}$$

3. Compute the standard error of $\overline{X}_n$, $\sigma(\overline{X}_n)$:
   Given $\sigma = 18$ hours

$$\sigma(\overline{X}_n) = \frac{\sigma}{\sqrt{n}} = \frac{18}{\sqrt{36}} = \frac{18}{6} = 3 \text{ hours}$$

4. Calculate the Z score of the upper bound of the interval, $\mu + \text{Margin for error}$:

$$Z_{\mu+5} = \frac{\text{Margin for error}}{\sigma(\overline{X}_{36})} = \frac{5}{3} = 1.6$$

(5. Use the *normal* table to obtain the area within the interval $\mu \pm$ Margin for error:)

$$\text{Area} = 0.89$$

(6. Express the result as a confidence interval for μ, $\overline{X}_n \pm$ Margin for error.)

We can be 89 percent confident that μ lies within the interval

$$\overline{X}_n \pm \text{Margin for error}$$
$$= 125 \pm 5$$
$$= \{120 \text{ to } 130\} \text{ hours}$$

varying the confidence level

The legislators were unhappy with the agency's estimate because it carried only an 89-percent confidence level, indicating that there was an 11-percent risk that the estimate was incorrect ($100\% - 89\% = 11\%$): 11 percent of the scores for $\overline{X}_{36}$ differ from μ by more than 5 hours—the margin for error. The legislators asked the agency to estimate μ so that there would be only a 5-percent risk of a misstatement. So the agency proceeded to form a 95-percent confidence interval for μ.

Logically, a 95-percent confidence interval for μ must allow a wider margin for error than the 89-percent confidence interval does, because the margin for error must include 95 percent of all the possible scores for $\overline{X}_{36}$. The agency has to determine how wide this margin for error must be. In Chart 9-12, the required margin for error is shown as the distance on either side of μ needed to encompass 95 percent of the area under the *normal* curve. The lower and upper bounds of this interval are denoted L and U, respectively. The agency must compute the numeric values of L and U.

First the Z scores of the points L and U, denoted by Z_L and Z_U, respectively, must be determined. The Z score for the entry .4750 (half of 0.95) is 1.96. Since $Z_U = +1.96$ and the interval from L to U is a centered interval, $Z_L = -1.96$. The points L and U lie 1.96 standard deviations on either side of the center of the *normal* curve.

chart 9-12

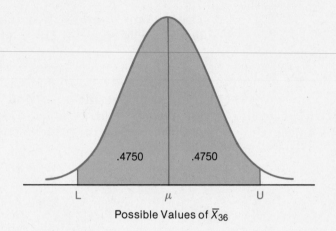

.4750   .4750

L μ U

Possible Values of $\overline{X}_{36}$

The standard deviation of the *normal* curve in Chart 9-12 is the standard error $\sigma(\overline{X}_{36})$ previously found to be equal to 3 hours. (Under the assumption that $\sigma = 18$ hours, $\sigma(\overline{X}_{36}) = \dfrac{18}{\sqrt{36}} = 3$ hours.) So,

$$L = \mu - 1.96 \cdot \sigma(\overline{X}_{36}) = \mu - 1.96 \cdot 3 = \mu - 5.88 \text{ hours}$$
$$U = \mu + 1.96 \cdot \sigma(\overline{X}_{36}) = \mu + 1.96 \cdot 3 = \mu + 5.88 \text{ hours}$$

A 95-percent confidence interval for μ requires a margin for error of 5.88 hours. Accordingly, the agency can be 95-percent confident that its sample mean $\overline{X}_{36}$ and the population mean μ differ by no more than 5.88 hours. Having found that $\overline{X}_{36} = 125$ hours, the 95-percent confidence interval for μ is 125 ± 5.88 hours = {119.22 to 130.88} hours. The agency presented its new estimate to the legislature:

At the 95-percent confidence level, we estimate that the average number of hours of child care our clients received was between 119 and 131 hours.

Let's review the steps the agency took to form a confidence interval for μ at the 95-percent level of confidence.

$\Big($1. Determine the sample size n:$\Big)$

$$n = 36$$

$\Big($2. Find the Z score bounds for the required confidence level:$\Big)$

The probability that Z is between -1.96 and $+1.96$ is 0.95, so ± 1.96 are the Z score bounds for a 95-percent confidence level.

$\Big($3. Compute the standard error $\sigma(\overline{X}_n)$:$\Big)$

$$\sigma(\overline{X}_{36}) = 3 \text{ hours}$$

$\Big($4. Calculate the margin for error (Z score $\cdot$ Standard error):$\Big)$

$$\text{Margin for error} = 1.96 \cdot 3 = 5.88 \text{ hours}$$

$\Big($5. Express the result as a confidence interval for μ:$\Big)$

The agency can be 95 percent confident that μ lies within the interval

$$\overline{X}_n \pm 5.88$$

$$= 125 \pm 5.88$$

$$= \{119.22 \text{ to } 130.88\} \text{ hours}$$

$\Big($Statisticians typically report confidence intervals to which at least a 90-percent confidence level can be attached. The most common confidence levels are 90 percent, 95 percent, and 99 percent. For this reason, it is useful to remember the tabulated Z scores associated with 90-percent, 95-percent, and 99-percent confidence levels. These values are presented in Chart 9-13.$\Big)$

chart 9-13

DESIRED LEVEL OF CONFIDENCE	Z SCORE	MARGIN FOR ERROR
90%	1.64	$1.64 \cdot \sigma(\overline{X}_n)$
95%	1.96	$1.96 \cdot \sigma(\overline{X}_n)$
99%	2.58	$2.58 \cdot \sigma(\overline{X}_n)$

the confidence–
precision
trade-off

(We have seen that a confidence interval for μ has three components:

1. The value of $\overline{X}_n$ calculated from the random sample; the point estimate of μ.
2. The margin for error.
3. The level of confidence.

(The size of the margin for error can be expressed symbolically as

$$\text{Margin for error} = Z \cdot \sigma(\overline{X}_n)$$

$$= Z \cdot \frac{\sigma}{\sqrt{n}} \quad) \qquad (9\text{-}5)$$

(where the value of the Z score depends on the level of confidence, as shown in Chart 9-13.)

(The size of the margin for error determines the *precision* of our estimate of μ. A precise estimate has a *small* margin for error.)

(We would like the estimate of μ not only to be precise but also to carry a high level of confidence. Chart 9-13 shows that there is a trade-off between precision and confidence. We see that as the level of confidence increases from 90 percent to 99 percent the size of the margin for error increases from 1.64 to 2.58 standard errors.)

The general nature of the trade-off between confidence and precision is illustrated by the following problem. A coroner has been asked to estimate the time of death of a homicide victim. The coroner offers two estimates:

1. I can place the time of death between 11 P.M. and 11:30 P.M. with 60-percent confidence.
2. I can place the time of death between 8 P.M. and 2 A.M. with 99-percent confidence.

The first estimate is very precise, providing a margin for error of 15 minutes before or after 11:15 P.M., but the low level of confidence would make it unreliable evidence at the trial.

The second estimate is very trustworthy, providing a confidence level of 99 percent, but it is much too imprecise to implicate the principal suspect.

$\Big($ Once we have selected a sample of size n, we can achieve a more precise estimate of μ only by accepting a lower level of confidence. Conversely, we can achieve a higher level of confidence only by accepting a less precise estimate (by permitting a larger margin for error). $\Big)$

$\Big($ A careful look at Equation (9-5), however, reveals that the precision of an estimate can be improved without reducing the level of confidence. Holding the confidence level represented by the Z score constant, the margin for error will decline as sample size n increases. So we can buy our way out of the trade-off between precision and confidence by selecting a larger sample. $\Big)$

the effect of
sample size

Suppose that two requirements are imposed on the agency's estimate of μ, the average number of actual hours of child care:

1. *Precision Requirement:* A margin for error no larger than 2 hours.
2. *Confidence Requirement:* A confidence level of 95 percent.

These requirements are illustrated in Chart 9-14. The problem is to find the sample size n that simultaneously satisfies both the precision and the confidence requirements.

chart 9-14

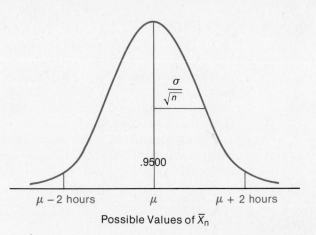

Possible Values of $\overline{X}_n$

We begin with Equation (9-5), which determines the margin for error:

$$\text{Margin for error} = Z \cdot \frac{\sigma}{\sqrt{n}}$$

Rearranging terms to single out sample size on the left side of the equation, we obtain

$$\sqrt{n} = \frac{Z \cdot \sigma}{\text{Margin for error}}$$

and

$$n = \left(\frac{Z \cdot \sigma}{\text{Margin for error}}\right)^2 \qquad (9\text{-}6)$$

For a 95-percent confidence level, $Z = 1.96$ (see Chart 9-13), which we will round to $Z = 2$ for convenience. To fulfill the precision requirement, the margin for error must be 2 hours. The value of σ — the standard deviation of the population — is assumed to be 18 hours, as before.

Substituting $Z = 2$, Margin for error $= 2$, and $\sigma = 18$ into Equation (9-6) gives us

$$n = \left(\frac{Z \cdot \sigma}{\text{Margin for error}}\right)^2$$

$$= \left(\frac{2 \cdot 18}{2}\right)^2$$

$$= 324$$

To estimate μ with a 95-percent confidence level and a margin for error of 2 hours, the agency must acquire a simple random sample of 324 scores — a considerably larger sample than the 36 scores selected.

Let's summarize the steps for calculating the sample size n that will fulfill the confidence–precision requirements of an estimate.

1. Set precision and confidence level requirements:

Margin for error: 2 hours
Confidence level: 95 percent

$\Bigg($ 2. Determine the Z score for the required confidence level: $\Bigg)$

For 95-percent confidence
$Z = 1.96$, or approximately 2

$\Big($ 3. Solve Equation (9-6) for n:

$$n = \left(\frac{Z \cdot \sigma}{\text{Margin for error}}\right)^2 \Bigg)$$

$$= \left(\frac{2 \cdot 18}{2}\right)^2$$

$$= 324$$

EXERCISES

9-13

Suppose that the only fact you know about a certain population of inference is that its standard deviation $\sigma = 10$ units. Now for the purposes of estimating the mean value of the population μ, you select a random sample of 25 scores and plan to compute the sample mean $\overline{X}_{25}$.

(a) Determine the probability that $\overline{X}_{25}$ will lie within a margin for error of:
   (1) 1.0 unit of μ (that is, within the interval $\mu - 1.0$ to $\mu + 1.0$)
   (2) 2.5 units of μ
   (3) 5.0 units of μ
(b) You find $\overline{X}_{25} = 50$ units. Form and interpret a confidence interval for μ that allows a margin for error of:
   (1) 1.0 unit
   (2) 2.5 units
   (3) 5.0 units
(c) Rank the confidence intervals you obtained in (b) from (1) the *most precise* to (3) the *least precise*. Now rank the confidence intervals from (1) the interval with the *highest confidence level* to (3) the interval with the *lowest confidence level*. What is the relationship between the *precision* of a confidence interval and the *level of confidence* it provides?

*9-14

How large is the size of a hamster's litter on the average? In 25 cases recently observed, female hamsters gave birth to an average $\overline{X}_{25}$ of 7.0 hamsters. The standard deviation was reported to be 2.5 hamsters per litter. Compute and interpret 90-percent, 95-percent, and 99-percent confidence intervals for μ, the expected size of a hamster's litter.

*9-15   (*Continuation of Exercise 9-14*) Now apply the same question to the rabbit (the animal, not the car). In 25 rabbit deliveries observed, the average number of rabbits per litter $\overline{X}_{25}$ proved to be 7.0, with a standard deviation of 3.0.
(a) Compute a 95-percent confidence interval for μ, the expected size of a rabbit's litter.
(b) Compare the margin for error of the 95-percent confidence interval for the rabbit mean with the margin for error of the 95-percent confidence interval for the hamster mean. Which animal's average litter size can be estimated more precisely and why?

9-16   During the past year, you took the same flight from City A to City B on 49 occasions and calculated an average flight time of $\overline{X}_{49} = 96.7$ minutes. The standard deviation of flight times was reported to be 14 minutes. You wish to estimate μ, the average flight time to be expected on this route.
(a) State a point estimate of μ.
(b) State an interval estimate of μ that allows a margin for error of 5 minutes.
(c) How confident can you be that your interval estimate for μ is correct?

9-17   In sampling to estimate the value of a parameter, a major decision is the size of the sample. If data acquisition is costly, you want to keep sample size n to the minimum number that will meet the requirements of confidence and precision. Indicate whether each of the following changes will increase or decrease required sample size:
(a) The standard deviation of the population of inference is smaller than previously thought.
(b) The desired confidence level is increased from 90 percent to 95 percent.
(c) The margin for error is increased.

9-18   Determine the sample size necessary to meet each of the following confidence–precision requirements in estimating μ if $\sigma = 15$:
(a) 90-percent confidence; margin for error = 5 units.
(b) 99-percent confidence; margin for error = 10 units.
(c) 99-percent confidence; margin for error = 5 units.

9-19   To project demands on its sewage treatment system, a municipal board wishes to estimate the average daily water usage of households within its jurisdiction. The estimate is to be based on a random sample of households, but how many households should be selected? Assume that the standard deviation σ is 3.0 gallons per day; that is, that daily water usage varies on the "average" by 3 gallons from the daily household mean. How large a sample is required if the municipal board wants to be 95-percent confident that its sample mean $\overline{X}_n$ will not differ from the true mean μ by more than 1.0 gallon per day?

Caveats on the Laws of Chance Variation

In this chapter we have examined three laws of chance variation for sample means and have learned how to apply the laws to estimate a population mean. We have stressed the implications of these laws in estimating μ, but we also should understand what the laws do *not* imply. The laws of chance variation govern *distributions of sample means*, not distributions of observations in a single sample — a fact that is too readily forgotten. Failure to realize this distinction leads to four frequent misstatements.

> *Misstatement 1.* The mean of a random sample of n scores is equal to the mean of the population from which the sample was selected.

Misstatement 1 indicates confusion about the meaning of the Expected Value Law for $\overline{X}_n$, which tells us that $\mu(\overline{X}_n)$, the expected value of a distribution of sample means, is equal to μ. The Expected Value Law for $\overline{X}_n$ does *not* imply that the mean of any one sample of n observations must equal μ. The mean of a single sample is just one score from a distribution of sample means — few if any of which will equal μ.

So we replace Misstatement 1 with Correct Statement 1:

> *Correct Statement 1.* The mean of any random sample of n scores $\overline{X}_n$ provides an unbiased point estimate of

the population mean. $\overline{X}_n$ is not biased in the direction of underestimation or overestimation of μ.

Misstatement 2. The mean of a larger sample will always be closer to μ than the mean of a smaller sample.

Misstatement 2 indicates confusion about the meaning of the Standard Error Law for $\overline{X}_n$. This law tells us that $\sigma(\overline{X}_n)$ decreases as sample size increases, implying that the distribution of means from larger samples will be more concentrated than the distribution of means from smaller samples. Alternatively stated, the Standard Error Law for $\overline{X}_n$ indicates that the "average" error that results when the mean of a larger sample is used to estimate μ will be less than the "average" error that results when the mean of a smaller sample is used to estimate μ.

But the Standard Error Law does *not guarantee* that the mean of any one sample of 100 observations, for example, will necessarily be closer to μ than the mean of any one sample of 25 observations. As illustrated in Chart 9-15, the mean of a sample of 100 observations will be a score from the sampling distribution of $\overline{X}_{100}$, whereas the mean of a sample of 25 scores will be a score from the sampling distribution of $\overline{X}_{25}$. By the luck of the draw in random sampling, we can obtain a score from the sampling distribution of $\overline{X}_{25}$ with a value of, say, point A, which is closer to μ than the many possible values for $\overline{X}_{100}$

chart 9-15

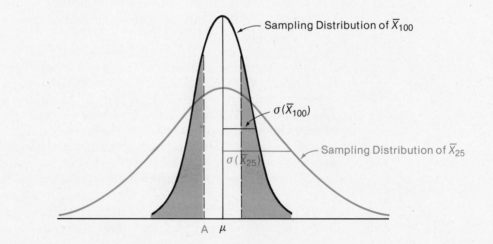

Sampling Distribution of $\overline{X}_{100}$

$\sigma(\overline{X}_{100})$

Sampling Distribution of $\overline{X}_{25}$

$\sigma(\overline{X}_{25})$

A μ

depicted by the gray area under the curve of the sampling distribution of $\overline{X}_{100}$.

> *Correct Statement 2.* The mean of a larger sample is *more likely* to be close to μ than the mean of a smaller sample. For a specified confidence level, a confidence interval for μ based on the mean of a larger sample will be more precise (will have a smaller margin for error) than a confidence interval for μ based on the mean of a smaller sample.

> *Misstatement 3.* As sample size increases, the shape of the distribution of observations in the sample will approach the bell shape of a *normal* curve.

Misstatement 3 indicates confusion about the meaning of the Law of Normality for $\overline{X}_n$. This law tells us that as n increases, the distribution of the means of random samples of size n will approach the shape of a *normal* curve.

The Law of Normality tells us *nothing* about the shape of the distribution of n observations in any one sample. As sample size increases, the distribution of sample observations will more closely reflect the shape of the population from which the sample is selected, and that population's shape may differ markedly from the *normal* curve.

> *Correct Statement 3.* The distribution of scores in a random sample will more or less reflect the shape of the population, but the distribution of mean values in random samples of n scores will always approach the bell shape of a *normal* curve as sample size increases.

To present our final misstatement, we'll return to the illustrative example on confidence intervals for μ. The Social Services Agency estimated the mean μ of the distribution of hours of child care its clients received during one month. The point estimate of μ found from the mean of a random sample of 36 children was $\overline{X}_{36} = 125$ hours. Application of the Central Limit Theorem (the composite statement of the three laws of chance variation for $\overline{X}_n$) produced the 95-percent confidence interval for μ:

$$\overline{X}_n \pm \text{Margin for error}$$
$$= 125 \pm 5.88 \text{ hours}$$
$$= \{119 \text{ to } 131\} \text{ hours}$$

> *Misstatement 4.* We can be 95-percent confident that each child spent between 119 and 131 hours in a day care center during the month studied.

Misstatement 4 indicates confusion about the meaning of a confidence interval for μ. A confidence interval for μ is a statement about the approximate location of the *mean value* of a population, *not* a statement about the range of scores in the population.

Sample means, we've learned, form a more homogeneous group of scores than the observations in the population (or in any sample from that population). The Central Limit Theorem enables us to say that 95 percent of the possible means in samples of 36 observations will lie within a margin for error of 5.88 hours of μ. The theorem does *not* imply that 95 percent of the population of actual hours of attendance will lie within 5.88 hours of μ. In fact, in the agency's sample of 36 children (Chart 9-10, p. 264), only 6 of the 36 children (17 percent) attended between 119 and 131 hours during the month studied.

EXERCISE

Two samples are selected randomly from the *same* population of inference, but one sample contains more observations than the other. The sample means are denoted, respectively, by $\overline{X}_L$ (mean of the larger sample) and by $\overline{X}_S$ (mean of the smaller sample). The population mean is denoted by μ.

(a) Does it follow from the Expected Value Law that the value of $\overline{X}_L$ must equal the value of $\overline{X}_S$?

(b) Does it follow from the Expected Value Law that $\mu(\overline{X}_L) = \mu(\overline{X}_S)$?

(c) Does it follow from the Standard Error Law that the difference between the value of X_L and μ must be smaller than the difference between the value of $\overline{X}_S$ and μ?

(d) Does it follow from the Standard Error Law that $\sigma(\overline{X}_L) < \sigma(\overline{X}_S)$?

(e) Does it follow from the Law of Normality that the distribution of scores in the larger sample will be more *normal* in shape than the distribution of scores in the smaller sample?

(f) Does it follow from the Law of Normality that if the sampling distribution of $\overline{X}_S$ is *normal* in shape, then the sampling distribution of $\overline{X}_L$ must be *normal* in shape?

(g) Does it follow from the Law of Normality that if the sampling distribution of $\overline{X}_L$ is *normal* in shape, then the sampling distribution of $\overline{X}_S$ must be *normal* in shape?

COMING
TO
TERMS

Central Limit Theorem for $\overline{X}_n$ If the sample size is sufficiently large, the mean of any one random sample $\overline{X}_n$ can be considered to be a score from a *normal* distribution that is centered on μ and that has a standard error $\sigma/\sqrt{n}$.

Confidence Interval   An interval estimate of a parameter that is centered on a point estimate and allows a margin for error due to chance variation.

Expected Value Law for $\overline{X}_n$ The expected value of a sampling distribution of sample means is equal to the mean of the population of inference. Symbolically, $\mu(\overline{X}_n) = \mu$. This law prevails for any sample size n and for all populations with a mean value.

Expected Value of a Statistic   The mean of the sampling distribution of a statistic, denoted by μ(statistic). For example, $\mu(\overline{X}_n)$ represents the expected value of a sampling distribution of $\overline{X}_n$.

Law of Normality   As sample size n increases, the shape of the sampling distribution of $\overline{X}_n$ approaches a *normal* curve. This law prevails regardless of the shape of the original population of inference.

Level of Confidence   The probability that a sample mean lies within the designated margin for error about the population mean.

Margin for Error   The distance that is included on either side of a point estimate in a confidence interval to allow for chance variation.

Precise Estimate   An interval estimate that has a small margin for error.

Sampling Distribution of a Statistic   If *all* possible random samples of a given number of observations are listed and the statistic of interest is calculated for each sample, then the relative frequency distribution of the scores for that statistic is the sampling distribution of the statistic.

Standard Error Law for $\overline{X}_n$ The standard error of $\overline{X}_n$ is equal to the standard deviation of the population of inference divided by the square root of the sample size n.

Standard Error of a Statistic   The standard deviation of the sampling distribution of that statistic.

Unbiased Estimate   An unbiased estimate results if a statistic's expected value is equal to the population parameter it is being used to estimate.

ISSUES FOR ANALYSIS

1. A group of students who are not familiar with the Standard Error Law for sample means, $\sigma(\overline{X}_n) = \sigma/\sqrt{n}$, wish to determine how the size of the standard error is affected by the standard deviation of the population σ. The students perform an experiment (called a *simulation*) in which:
   (1) They have a computer generate sampling distributions of $\overline{X}_{25}$ for various values of σ.
   (2) For each sampling distribution, the standard error $\sigma(\overline{X}_{25})$ is obtained.
   (3) Finally, designating $\sigma(\overline{X}_{25})$ as the dependent variable and σ as the predictor, they define the regression equation

$$\sigma(\overline{X}_{25}) = b_0 + b_1 \cdot \sigma$$

   What values should they obtain for b_1 and b_0, respectively? (*Hint:* Consider what the Standard Error Law tells you in the case of $n = 25$.)

2. Let C denote the central value of a *normally* distributed population. So C is the population mean and the median (as well as the mode and the midrange). Let $\overline{M}_n$ denote the median of a random sample of n observations from that population. The laws of chance variation for $\overline{M}_n$ tell us that:
   (1) $\mu(\overline{M}_n) = C$

   (2) $\sigma(\overline{M}_n) = \dfrac{1.25\sigma}{\sqrt{n}}$

   (3) If the population is *normal* in shape, the sampling distribution of $\overline{M}_n$ (as well as the sampling distribution of $\overline{X}_n$) will be *normal in shape* for any n.
   (a) On a single set of axes, sketch:
      (1) The sampling distribution of $\overline{M}_n$.
      (2) The sampling distribution of $\overline{X}_n$.
      Label the location of C and label the distances represented by $\sigma(\overline{M}_n)$ and $\sigma(\overline{X}_n)$.

(b) Would you use $\overline{M}_n$ or $\overline{X}_n$ to compute an estimate of C? Explain your answer, taking into consideration both the issues of bias and precision.

(c) Assuming $\sigma = 20$ units, how large a sample size n is required to estimate C:
 (1) from $\overline{M}_n$
 (2) from $\overline{X}_n$
 given that in both cases the confidence level requirement is 95 percent and the precision requirement is a 3-unit margin for error?

Long after this advertisement first appeared in the press, the metropolitan news editor of a well-known newspaper assigned an investigative reporter to look into the methods and accomplishments of dietary clinics. Were the clients successfully achieving weight reduction, and if so, was it being sustained? The reporter's investigation begins on page 288, in the section entitled "A Practical Problem."

the
t distribution
and
the *t* test

10

In Chapter 9, we studied the Central Limit Theorem (CLT) for sample means and learned to apply the CLT to form a confidence interval for a population mean μ. That application, however, contained an unrealistic assumption: the assumption that independent information was available about the approximate value of the population standard deviation σ. Remember that the Social Services Agency was able to assign an approximate value to σ ($\sigma = 18$ *actual* hours), based on the known standard deviation of the distribution of *authorized* hours.

Usually, we will not have independent information that enables us to assign an approximate value to σ. Rather, our only source of information will be the sample of scores we have selected.

So we have a problem. To draw inferences about μ, we need to determine a value for the standard error $\sigma(\overline{X}_n)$. The *Standard Error Law* for $\overline{X}_n$ tells us that $\sigma(\overline{X}_n) = \sigma/\sqrt{n}$. If we do not have independent information on σ, then we must estimate the value of σ from our sample data.

10.1   a sample estimate of σ

It seems logical to choose the standard deviation of a random sample, s, to provide an estimate of the standard deviation of the population of inference, σ. Whereas σ measures the "average" spread of the observations in the population about the population mean, s measures the "average" spread of the scores sampled about the sample mean. Recall that s is defined by the equation

$$s = \sqrt{\frac{\Sigma(X - \overline{X})^2}{n}} \qquad (n \text{ denotes sample size})$$

whereas

$$\sigma = \sqrt{\frac{\Sigma(X - \mu)^2}{N}} \qquad (N \text{ denotes population size})$$

Since s is a sample statistic, it varies in value from sample to sample, depending on the luck of the draw. With good luck, we will obtain a sample with a standard deviation that is nearly equal to σ; with bad luck, we will acquire a sample with an "average" spread that is much greater or smaller than the spread of the population.

We know that the sample mean $\overline{X}_n$ provides an unbiased estimate of the population mean μ: The expected value of $\overline{X}_n$ is equal to μ. Unfortunately, the sample standard deviation s does not share this virtue. In fact, s tends to underestimate σ.

If we were to compute the standard deviation of each sample s from the table listing all the samples of two potato sacks (Chart 9-2 on page 241) and group the results into a sampling distribution for s, we would find $s < 2$ in 37 of the 49 cases. Moreover, the expected value of s would be equal to 1.14, whereas, in fact, σ is equal to 2.

So the sample standard deviation tends to underestimate the population standard deviation. Ideally, we would like to define a statistic that provides us with an unbiased estimate of σ. The ideal, unfortunately, is impossible to achieve; no sample statistic always provides an unbiased estimate of σ.

However, we can define a statistic that always provides an unbiased estimate of σ^2, the population variance. This statistic is denoted by $\hat{\sigma}^2$ ("sigma-hat squared") and is defined by Equation (10-1):

$$\hat{\sigma}^2 = \frac{\Sigma(X - \overline{X})^2}{n - 1}$$ (10-1)

Note that $\hat{\sigma}^2$ is simply the sample variance s^2 computed with $n - 1$ in the denominator in place of n. Because $n - 1$ is less than n, $\hat{\sigma}^2$ is always larger than s^2. Similarly, the square root of $\hat{\sigma}^2$, $\sqrt{\hat{\sigma}^2} = \hat{\sigma}$, is always larger than the sample standard deviation s:

$$\hat{\sigma} = \sqrt{\frac{\Sigma(X - \overline{X})^2}{n - 1}} \qquad \text{whereas} \qquad s = \sqrt{\frac{\Sigma(X - \overline{X})^2}{n}}$$ (10-2)

Since the value of $\hat{\sigma}$ necessarily exceeds the value of s, $\hat{\sigma}$ is less likely to understate σ than s is. Therefore, $\hat{\sigma}$ provides us with a less biased estimate of σ.

The quantity $n - 1$ in the denominator of the equations for $\hat{\sigma}^2$ and $\hat{\sigma}$ is called the *number of degrees of freedom*, abbreviated *df*. The number of degrees of freedom *df* is used in lieu of sample size n to obtain an unbiased estimate of a population variance and a less biased estimate of a population standard deviation.

In our listing of random samples of two potato sacks, the possible scores for $\hat{\sigma}$ average to 1.6 potatoes per sack, which is considerably closer to $\sigma = 2$ than the expected value of s, which is 1.14. Moreover, $\hat{\sigma}$ is smaller than σ in 29 of the 49 possible samples of size 2, whereas s is smaller than σ in 37 of these 49 samples.

So $\hat{\sigma}$, not s, will serve as our estimate of σ. We then can apply the Standard Error Law to compute an estimate of $\sigma(\overline{X}_n)$. Denoting *the estimated standard error of* $\overline{X}_n$ by $\hat{\sigma}(\overline{X}_n)$, we write

$$\hat{\sigma}(\overline{X}_n) = \frac{\hat{\sigma}}{\sqrt{n}}$$ (10-3)

Equation (10-3) gives us a method for estimating the standard error of $\overline{X}_n$ when σ is unknown. It tells us to compute $\hat{\sigma}$ from our sample observations and divide $\hat{\sigma}$ by the square root of the sample size. If $\hat{\sigma}$ must be calculated by hand, we can use the computational equations in Section 3.2 (pages 23–26), if we recognize that $\hat{\sigma} = s \cdot \sqrt{n/(n-1)}$.)

(Sometimes the estimated standard error is written $s/\sqrt{n}$, because many textbooks do not differentiate between $\hat{\sigma}$ and s. They use $n-1$ and Equation (10-1) to define s, making s and $\hat{\sigma}$ one and the same. In our discussion of inference, we will follow this tradition and refer to $\hat{\sigma}$ as the sample standard deviation.)

a practical
problem

A reporter assigned to write a feature story on weight-reduction clinics wishes to estimate the mean one-year weight loss for people who remain in such a program to give prospective clients some idea of how much weight they might expect to lose.[1] The reporter selects a random sample of 16 clients from those who have been enrolled at the local branch of Slim and Trim, Inc., for at least one year. From the clinic's files, she obtains data on the weight lost during one year's participation in the program. Her findings appear in Chart 10-1.

chart 10-1
One-Year
Weight Loss

CLIENT NO.	WEIGHT LOSS* (lb)
1	8.2
2	2.0
3	13.7
4	−3.5
5	−1.5
6	13.2
7	14.3
8	−4.1
9	12.8
10	−2.6
11	−4.2
12	14.4
13	14.0
14	−3.8
15	13.1
16	−2.9
$\bar{X}_{16} =$	5.2
$\hat{\sigma} =$	8.3

*A negative entry indicates a weight gain.

[1] How much the *average* weight loss tells the prospective client about the effectiveness of the program will depend on the shape of the distribution of weight losses and on the amount of variation, but at least it's a starting point.

The sample mean weight loss $\overline{X}_{16}$ is 5.2 lb. Since the reporter needs to estimate the standard error of $\overline{X}_{16}$ to draw an inference about μ, she calculates $\hat{\sigma}$ (rather than s) and finds $\hat{\sigma} = 8.3$. Then she uses Equation (10-3) to make her estimate:

$$\hat{\sigma}(\overline{X}_{16}) = \frac{\hat{\sigma}}{\sqrt{16}} = \frac{8.3}{4} = 2.1 \text{ lb}$$

This result indicates that the scores for $\overline{X}_{16}$ vary by an estimated "average" of 2.1 lb about μ. So the "average" error that will be made if $\overline{X}_{16}$ is used to draw an inference about μ is estimated to be 2.1 lb.

EXERCISES
10-1

(*Continuation of Exercise 9-1, page 245*) In Exercise 9-1(b), you listed each possible sample of two scores.

(a) Now for *each* of these nine samples, compute the two statistics

$$s = \sqrt{\frac{\Sigma(X - \overline{X})^2}{n}} \quad \text{and} \quad \hat{\sigma} = \sqrt{\frac{\Sigma(X - \overline{X})^2}{n - 1}}$$

Note that 18 separate calculations are *not* required. The respective values of s and $\hat{\sigma}$ are identical across several of the samples, and both values are equal to 0 in several samples.

(b) Group the results you obtained in (a) into sampling distributions of s and $\hat{\sigma}$ by completing the following chart:

POSSIBLE VALUES OF s	POSSIBLE VALUES OF $\hat{\sigma}$	RELATIVE FREQUENCY $p(s)$ OR $p(\hat{\sigma})$

(c) Calculate $\mu(s)$, the expected value of the sampling distribution of s, and $\mu(\hat{\sigma})$, the expected value of the sampling distribution of $\hat{\sigma}$.

NOTE: $\mu(s) = \Sigma[s \cdot p(s)]$ and $\mu(\hat{\sigma}) = \Sigma[\hat{\sigma} \cdot p(\hat{\sigma})]$

(d) In Exercise 9-1, you found that $\sigma = 0.816$. Based on the result you obtained in (c), why is it desirable to compute an estimate of σ based on $\hat{\sigma}$ rather than on s?

(e) Does your answer to (d) imply that in any single sample, $\hat{\sigma}$ will be closer to σ than s is? Consult the chart you completed in (b).

10-2   From a randomly selected sample of 25 scores, we find $\hat{\sigma} = 10$ units. Using this result, calculate the estimated standard error $\hat{\sigma}(\overline{X}_{25})$ and interpret its value.

10-3   From a randomly selected sample of 16 scores, we find that the quantity $\Sigma(X - \overline{X})^2 = 60$ units. Calculate $\hat{\sigma}$ and then calculate the standard error $\hat{\sigma}(\overline{X}_{16})$. Interpret both results.

10.2   the t distribution

Having estimated $\sigma(\overline{X}_n)$ by $\hat{\sigma}(\overline{X}_n)$, can the reporter still employ the *normal* table, as we did in Chapter 9, to compute a confidence interval for μ? The answer, unfortunately, is no. Instead of the *normal* distribution the reporter will have to use the so-called *t distribution*.

the t distribution versus the *normal* distribution

Before explaining the t distribution, let's define a new term. A deviation between $\overline{X}_n$ and μ in units of *estimated* standard errors is called a *t score* and is defined by

$$t_{\overline{X}_n} = \frac{\overline{X}_n - \mu}{\hat{\sigma}/\sqrt{n}}$$

In contrast, the Z score, which measures a deviation between $\overline{X}_n$ and μ in units of *true* standard errors, is defined by

$$Z_{\overline{X}_n} = \frac{\overline{X}_n - \mu}{\sigma/\sqrt{n}}$$

The two equations differ in their denominators: σ appears in the denominator of the Z score and $\hat{\sigma}$ appears in the denominator of the t score. A Z score of 1.5 indicates that the distance between $\overline{X}_n$ and μ is 1.5 standard errors, whereas a t score of 1.5 indicates that the distance between $\overline{X}_n$ and μ is *estimated* to be 1.5 standard errors.

The Central Limit Theorem tells us that the distribution of Z scores is *normal* and centered at 0, so that, for example, 95

chart 10-2
Centered Intervals
Required to
Include 95 Percent
of the Possible
Scores for $t_{\bar{X}_n}$

-2.571	when $df = 5$	$+2.571$
-2.228	when $df = 10$	$+2.228$
-2.131	when $df = 15$	$+2.131$
-2.086	when $df = 20$	$+2.086$
-2.060	when $df = 25$	$+2.060$
-2.042	when $df = 30$	$+2.042$
-1.96	when $df = \infty$	$+1.96$

percent of the possible values for $\overline{X}_n$ will have Z scores between -1.96 and $+1.96$. Accordingly, a 95-percent confidence interval for μ requires a margin for error of 1.96 $\sigma/\sqrt{n}$.

(The distribution of t scores is also centered at 0. However, as we've just learned, $\hat{\sigma}$ tends to understate σ, and therefore $\hat{\sigma}/\sqrt{n}$ tends to be smaller than $\sigma/\sqrt{n}$.)So although 95 percent of the possible Z scores will be between -1.96 and $+1.96$, a wider interval will be required to include 95 percent of the possible t scores.How much wider depends on sample size or, more technically, on the number of degrees of freedom ($df = n - 1$). Chart 10-2 shows, for selected numbers of degrees of freedom, the width of an interval centered at 0 that is required to include 95 percent of the possible t scores $t_{\bar{X}_n}$.)

For random samples with $df = 5$ (or samples of $n = 6$), the required interval is from -2.571 to $+2.571$.As df increases, the required interval diminishes, until at $df = 30$ it is nearly as narrow as the Z score interval.)

(The intervals displayed in Chart 10-2 are based on the *t distribution*, which was developed by the Irish statistician William S. Gossett (1876–1936). Writing under the pseudonym "Student," Gossett showed that

If the population of inference has a *normal* shape, the t scores $t_{\bar{X}_n} = \dfrac{\overline{X} - \mu}{\hat{\sigma}/\sqrt{n}}$ will form a t distribution with $df = n - 1$.)

(Chart 10-3 depicts the t distribution curves for $df = 3$ and $df = 10$ and the *normal* curve. t distribution curves, like the *normal* curve, are symmetric and bell-shaped, but they are flatter and more elongated than the *normal* curve. The extra width indicates that t scores tend to lie farther from 0 than Z scores do, because estimated standard errors tend to be smaller

chart 10-3

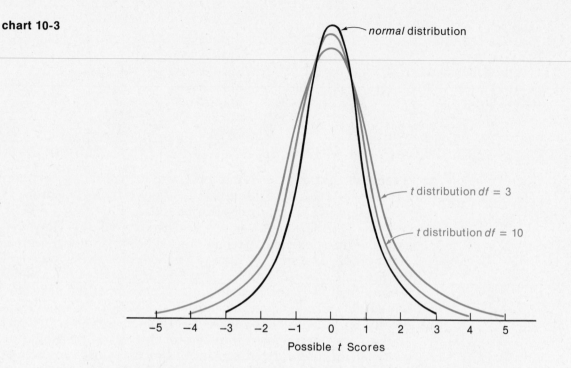

than true standard errors. As sample size *n* increases (that is, as *df* increases), the corresponding *t* distribution curve begins to look more and more like the *normal* curve, because $\hat{\sigma}$ is less likely to underestimate σ as *df* increases.)

(Gossett's finding, however, was based on the assumption that the population of inference itself is *normally* distributed. This imposes a severe restriction on the use of the *t* distribution. What if the population is not *normal* in shape or what if we don't know the shape of the population? Then how can we draw inferences about μ from our sample evidence?)

the *t* distribution law for means

(Experiments based on *t* scores subsequent to Gossett's pioneering work permit us to state the following guidelines, which we will call the *t Distribution Law for Means:*

THE *t* DISTRIBUTION LAW FOR MEANS   Provided that the shape of the population of inference is not markedly skewed and that *df* $\geq$ 15, a *t* distribution with $df = n - 1$ will closely approximate the distribution of possible t scores, $t_{\bar{X}_n} = \dfrac{\bar{X}_n - \mu}{\hat{\sigma}/\sqrt{n}}$. As df increases beyond 15, the t distribution approximation of the sampling behavior of $t_{\bar{X}_n}$ will continue to improve, no matter what the shape of the population of inference. $\big)$

$\Big($ So for small samples in which $df < 15$, the t distribution should *not* be used unless we have a theoretical reason to believe that the population of inference is nearly *normal* in shape. For mildly skewed distributions, the t distribution can be used to draw inferences about the mean when $df \geq 15$. For more highly skewed population distributions, considerably greater sample sizes may be required. In some cases, the t distribution may never be the appropriate choice (see Caveat 1 at the end of this chapter). $\Big)$

10.3   a confidence interval for μ

Assuming that the distribution of weight reductions is fairly symmetric, the t Distribution Law for Means provides the reporter with the necessary tool to form a confidence interval for μ, the true mean weight loss at one year.[2]

The reporter decides $\big($ to form a 95-percent confidence interval. If σ were known, the interval would be determined by

$$\bar{X}_n \pm Z \cdot \frac{\sigma}{\sqrt{n}}$$

where $Z \cdot \sigma/\sqrt{n}$ provides the margin for error. But σ is unknown and must be estimated by $\hat{\sigma}$, so she restates the expression of a confidence interval for μ as

$$\bar{X}_n \pm t \cdot \frac{\hat{\sigma}}{\sqrt{n}}$$

and $t \cdot \hat{\sigma}/\sqrt{n}$ becomes the margin for error. $\big)$

---

[2] The assumption of approximate symmetry is valid as long as the clinic does not cater to extremely overweight people. A 100-lb weight loss is unlikely to be matched by a 100-lb weight gain, so in that circumstance the distribution would be asymmetric.

With σ known, a 95-percent confidence interval for μ requires a margin for error of 1.96 $\sigma/\sqrt{n}$ ($Z = 1.96$). However, the reporter realizes that when σ is estimated by $\hat{\sigma}$, a 95-percent confidence interval must allow a wider margin for error than 1.96 $\hat{\sigma}/\sqrt{n}$ (t must be > 1.96). To find the appropriate value for t, she consults the t table.

Turning to the t table (Appendix Table D reproduced as Chart 10-4), the reporter finds that it is structured differently from the *normal* table. (Each column of the t table has two headings: one to be used when the probability (area) in one tail of a t distribution is needed, and another to be used when the probability (area) in the two tails of a t distribution is needed.) The column labeled "One-Tailed P Value = 0.05," for example, is the one the reporter turns to if she wants to place 5 percent of the area under *one* tail. She also turns to this column if she wants to place a total of 10 percent of the area in the two tails (or 90 percent of the area between the tails). The column labeled "Two-Tailed P Value = 0.05" places a total of 5 percent of the area in the two tails, leaving 95 percent of the area between the tails. This is the appropriate column to use for a 95-percent confidence interval.

(The rows are labeled "Degrees of Freedom.") Since the reporter interviewed 16 clients, $n = 16$ and $df = n - 1 = 16 - 1 = 15$. So she seeks the entry in the t table at the intersection of the column "Two-Tailed P Value = 0.05" and the row where $df = 15$, finding $t = 2.131$. She can now complete the computation for a 95-percent confidence interval for μ.

(The confidence interval for μ is

$$\overline{X}_n \pm t \cdot \frac{\hat{\sigma}}{\sqrt{n}} \;)$$

$$= 5.2 \pm 2.131 \cdot \frac{8.3}{\sqrt{16}}$$

$$= 5.2 \pm 4.422$$

$$= \{0.778 \text{ to } 9.622\} \text{ lb}$$

"We can be 95 percent confident," the reporter concludes, "that the average one-year weight loss of Slim and Trim clients will be between 0.8 and 9.6 lb."

Let's review the steps the reporter took to find a 95-percent confidence interval for μ.

chart 10-4
The *t* Table

DEGREES OF FREEDOM	ONE-TAILED *P* VALUE					
	0.25	0.10	0.05	0.025	0.01	0.005
	TWO-TAILED *P* VALUE					
	0.50	0.20	0.10	0.05	0.02	0.01
1	1.000	3.078	6.314	12.706	31.821	63.657
2	0.816	1.886	2.920	4.303	6.965	9.925
3	.765	1.638	2.353	3.182	4.541	5.841
4	.741	1.533	2.132	2.776	3.747	4.604
5	.727	1.476	2.015	2.571	3.365	4.032
6	.718	1.440	1.943	2.447	3.143	3.707
7	.711	1.415	1.895	2.365	2.998	3.499
8	.706	1.397	1.860	2.306	2.896	3.355
9	.703	1.383	1.833	2.262	2.821	3.250
10	.700	1.372	1.812	2.228	2.764	3.169
11	.697	1.363	1.796	2.201	2.718	3.106
12	.695	1.356	1.782	2.179	2.681	3.055
13	.694	1.350	1.771	2.160	2.650	3.012
14	.692	1.345	1.761	2.145	2.624	2.977
15	.691	1.341	1.753	2.131	2.602	2.947
16	.690	1.337	1.746	2.120	2.583	2.921
17	.689	1.333	1.740	2.110	2.567	2.898
18	.688	1.330	1.734	2.101	2.552	2.878
19	.688	1.328	1.729	2.093	2.539	2.861
20	.687	1.325	1.725	2.086	2.528	2.845
21	.686	1.323	1.721	2.080	2.518	2.831
22	.686	1.321	1.717	2.074	2.508	2.819
23	.685	1.319	1.714	2.069	2.500	2.807
24	.685	1.318	1.711	2.064	2.492	2.797
25	.684	1.316	1.708	2.060	2.485	2.787
26	.684	1.315	1.706	2.056	2.479	2.779
27	.684	1.314	1.703	2.052	2.473	2.771
28	.683	1.313	1.701	2.048	2.467	2.763
29	.683	1.311	1.699	2.045	2.462	2.756
30	.683	1.310	1.697	2.042	2.457	2.750
35	.682	1.306	1.690	2.030	2.438	2.724
40	.681	1.303	1.684	2.021	2.423	2.704
45	.680	1.301	1.680	2.014	2.412	2.690
50	.680	1.299	1.676	2.008	2.403	2.678
55	.679	1.297	1.673	2.004	2.396	2.669
60	.679	1.296	1.671	2.000	2.390	2.660
70	.678	1.294	1.667	1.994	2.381	2.648
80	.678	1.293	1.665	1.989	2.374	2.638
90	.678	1.291	1.662	1.986	2.368	2.631
100	.677	1.290	1.661	1.982	2.364	2.625
120	.677	1.289	1.658	1.980	2.358	2.617
∞	.674	1.282	1.645	1.960	2.326	2.576

Source: Adapted from *Scientific Tables*, published by Ciba-Geigy Limited, 1962, pages 32–34.

1. Determine the degrees of freedom:

$$df = 16 - 1 = 15$$

2. Locate the entry in the *t* table for two-tailed *P* value = 0.05:

$$t = 2.131$$

3. Compute the margin for error:

$$\text{Margin for error} = 2.131 \cdot \frac{\hat{\sigma}}{\sqrt{n}} = 4.422$$

4. Form the confidence interval as $\bar{X}_n \pm$ Margin for error:

$$5.2 \pm 4.422 = \{0.8 \text{ to } 9.6\} \text{ lb}$$

EXERCISES

10-4
A population has a mean $\mu = 40$ units and a standard deviation $\sigma = 6.5$ units. A random sample of 25 observations selected from this population has a mean $\bar{X}_{25} = 38$ units and a standard deviation $\hat{\sigma} = 5$ units.
(a) Calculate $\sigma(\bar{X}_{25})$ and $\hat{\sigma}(\bar{X}_{25})$. Interpret each result.
(b) Calculate $Z_{\bar{x}_{25}}$ and $t_{\bar{x}_{25}}$. Interpret each result.

10-5
Referring to the faculty data set given on page 210, a simple random sample of 4 faculty members revealed a mean salary of $\bar{X}_4 = \$21,500$ with a standard deviation $\hat{\sigma} = \$1,600$. Calculate the t score of the sample mean of $\$21,500$ and interpret your result.

10-6
Following the format of Chart 10-2 on page 291, prepare a chart showing the centered intervals that are required to include 90 percent of the possible values for t. Make an interpretive statement based on the interval for $df = 15$.

10-7
Determine the entry in the t table required for:
(a) A 99-percent confidence interval for μ; $n = 22$.
(b) A 99-percent confidence interval for μ; $n = 16$.
(c) A 95-percent confidence interval for μ; $n = 20$.
(d) A 95-percent confidence interval for μ; $n = 30$.
(e) A 90-percent confidence interval for μ; $n = 41$.
(f) A 90-percent confidence interval for μ; $n = 53$.
(Use the table entry closest to the actual number of degrees of freedom.)

10-8
Referring to Exercise 9-14 on page 275 and assuming that the value of the standard deviation represents the estimate $\hat{\sigma}$, recompute the 90-percent confidence interval for μ (the expected size of a hamster's litter).

10-9
Referring to Exercise 9-15 on page 276 and treating the value of the standard deviation as the estimate $\hat{\sigma}$, recompute a 95-percent confidence interval for μ (the expected size of a rabbit's litter). Why is this interval estimate slightly less precise than the one you obtained in Exercise 9-15?

10-10
A recent survey by the American Hospital Association of a random sample of 16 community hospitals revealed that the mean cost per patient day was $\$196.40$ with a standard deviation $\hat{\sigma}$ of $\$30.20$. Assuming that the shape of the distribution of costs per patient day among all community hospitals is not markedly skewed, calculate and interpret a 90-percent confidence interval for μ, the mean of the population of inference.

10-11   The following data were reported to the personnel director of a large company:

	SICK-LEAVE ESTIMATES FOR PAST YEAR	
	Men	Women
Mean $\bar{X}_n$	5.6 days	4.8 days
Standard Deviation $\hat{\sigma}$	2.0 days	3.0 days
Sample Size n	25	16

(a) Calculate a 95-percent confidence interval for μ_{male}, the true mean of the distribution of the number of days of sick leave taken last year by male employees of the company.

(b) Calculate a 95-percent confidence interval for μ_{female}, the true mean days of sick leave taken last year by female employees of the company.

(c) Is the interval estimate for μ_{male} or μ_{female} more precise? Why? (Discuss the implication of differences in both the sample sizes n and the standard deviations $\hat{\sigma}$.)

10.4   introduction to hypothesis testing

$\Big($Populations are generally sampled
* to estimate the value of an unknown parameter
                          and/or
* to evaluate a claim or hypothesis concerning a parameter value. $\Big)$
The difference in these two approaches is illustrated in the following cases.

CASE 1   A prospective advertiser may want to estimate the average annual income of a magazine's readers *or* may want to evaluate the magazine executive's claim that readers of the magazine earn more than $20,000 a year on the average.

CASE 2   A city engineer may want to estimate the average number of cars that pass through a busy intersection daily *or* may want to test the hypothesis that this traffic flow is no different from last year's mean number of 35,000 cars per day.

CASE 3   Instead of estimating the mean weight loss after one year, the reporter in our example may want to test the hypothesis that on the average a participant in the program would weigh no less after one year than he or she weighed on entering the clinic. The reporter may hope that such a test will reveal whether or not the Slim and Trim program helps the average participant at all.

**the null
and alternative
hypotheses**

(Every hypothesis test begins with a statement of the *null* and *alternative* hypotheses. *Null* means "of no consequence." A null hypothesis is a claim that something—some event, some program, some level of a variable—is of no consequence. The alternative hypothesis is just what its name implies: the alternative to the null hypothesis. Included in the alternative hypothesis are all the possibilities of consequence to those performing the test.)

CASE 1   The Prospective Advertiser
If μ denotes the average annual income of the magazine's readers, then

$$\text{Null hypothesis: } \mu \leq \$20,000$$
$$\text{Alternative hypothesis: } \mu > \$20,000$$

The prospective advertiser will not give the company's account to the magazine without the assurance that the ads will reach higher income readers. From the advertiser's point of view, an average income level not above $20,000 is of no consequence to the company.

CASE 2   The Traffic Study
If μ denotes the average number of cars passing through the busy intersection per day, then

$$\text{Null hypothesis: } \mu = 35,000$$
$$\text{Alternative hypothesis: } \mu \neq 35,000$$

The city engineer considers a traffic flow at last year's mean of 35,000 cars per day to be of no consequence, because it would not require a change in city street programs. A substantial increase or decrease from last year's traffic flow would be of consequence, because it would require rethinking current transportation programs.

CASE 3   The Weight-Reduction Clinic
In this case, μ denotes the mean weight *loss* of program participants, so that

$$\text{Null hypothesis: } \mu = 0$$
$$\text{Alternative hypothesis: } \mu > 0$$

The null hypothesis represents the claim that the clinic program does not promote sustained weight reduction on the average, whereas the alternative hypothesis asserts that the program is effective on the average (is of some consequence in reducing weight).

the logic of a hypothesis test

The reporter's null hypothesis is $\mu = 0$. In her sample, however, she found $\overline{X}_{16} = 5.2$ lb. Obviously, the observed evidence, $\overline{X}_{16} = 5.2$ lb, and the claim, $\mu = 0$, differ. The discrepancy can be explained in one of two ways:

1. The null hypothesis is true: Mean weight reduction after one year *is* equal to 0, and the deviation of the sample mean from 0 is *due to chance*. By the luck of the draw, the reporter has obtained a sample with a positive mean, even though the population mean is truly equal to 0.
2. The alternative hypothesis is true: Mean weight reduction is positive, and the positive value for the sample mean arises because the population mean is positive.

(A hypothesis test is a logical procedure used to determine whether or not the chance explanation for the discrepancy can be ruled out beyond a reasonable doubt. The test leads to one of two findings:

1. The discrepancy is statistically significant; it is too great to be attributed to chance. On finding this, we can reject the null hypothesis; that is, we can reject the chance explanation for the discrepancy as farfetched.
2. The discrepancy is not statistically significant; it is small enough to reasonably be due to chance. On finding this, we cannot reject the null hypothesis based on the evidence at hand. This amounts to saying that we must reserve judgment as to the truth of the null hypothesis—*not* that we have proved the null hypothesis to be true.)

(Accordingly, to test a hypothesis, we compute a probability based on the chance explanation. This probability is referred to as a *P value*.

> **P VALUE** A measure of the plausibility of the chance explanation for the discrepancy between the evidence and the claim (the null hypothesis). A very low *P* value implies that the chance explanation is implausible, permitting us to reject the null hypothesis. A high *P* value implies that the chance explanation is plausible and should dissuade us from rejecting the null hypothesis.)

(Hypotheses are rarely rejected unless the *P* value is < 0.10; usually they are not rejected unless the *P* value is < 0.05.)

one-tailed versus two-tailed tests

(To compute a *P* value, we have to pay attention to the form of the alternative hypothesis. When the null hypothesis is in the form $\mu = 0$, the alternative hypothesis will be one of the following statements:

1. $\mu > 0$: μ is greater than 0
2. $\mu < 0$: μ is less than 0
3. $\mu \neq 0$: μ is unequal to 0)

(CASE 1   Null: $\mu = 0$ Alternative: $\mu > 0$

These are the hypotheses that the reporter is considering. The sample evidence is summarized by obtaining a value for $\overline{X}_n$. So the discrepancy between this evidence and the null hypothesis is simply the difference between the value found for $\overline{X}_n$ and 0.

Because the alternative hypothesis is $\mu > 0$, only a *positive* discrepancy (a positive value for $\overline{X}_n$) can possibly lead us to reject the null hypothesis. Only an average weight loss *greater than 0* in the sample can support the alternative hypothesis that the population mean weight loss is greater than 0.

Accordingly, we will reject the null hypothesis only if chance cannot explain why $\overline{X}_n$ is *at least as positive* as the value observed for $\overline{X}_n$ in the sample.)

(The probability that chance will give us a sample
whose mean is at least as positive as our observed
sample mean is called a *one-tailed P value for* $\overline{X}_n$.
The name *one-tailed P value* is used because
only one tail (the positive tail) of the distribution
of possible values for $\overline{X}_n$ supports the alternative
hypothesis.)

(CASE 2   Null: $\mu = 0$ Alternative: $\mu < 0$

If the alternative hypothesis is $\mu < 0$, then only a *negative*
discrepancy (a negative value for $\overline{X}_n$) supports the alternative
hypothesis. We will reject the null hypothesis only if chance
cannot explain why $\overline{X}_n$ is *at least as negative* as the value observed for $\overline{X}_n$ in the sample. /

(The probability that chance will give us a sample
whose mean is at least as negative as our observed
sample mean is also called a *one-tailed P value for*
$\overline{X}_n$. But this time only the negative tail of the
distribution of possible values for $\overline{X}_n$ supports the
alternative hypothesis. /

(CASE 3   Null: $\mu = 0$ Alternative: $\mu \neq 0$

When the alternative hypothesis is in the form $\mu \neq 0$, *any*
discrepancy (any positive or negative value for $\overline{X}_n$) will support
the alternative hypothesis.) For example, in an experiment to
determine whether a drug modifies reaction times on the average, the null hypothesis of no modification on the average,
($\mu = 0$, will be rejected if we find that the sample exhibits too
large a positive or a negative change) in mean reaction time to
be due to chance.

(The probability that chance will give us a sample
whose mean is at least as far as our observed sample
mean is from the μ value claimed under the null
hypothesis is called a *two-tailed P value for* $\overline{X}_n$.
Values in both tails of the distribution of possible
values for $\overline{X}_n$ support the alternative hypothesis.)

A two-tailed test

EXERCISES

Each of the following exercises presents an issue to be defined in terms of hypotheses for statistical testing. In each case:
(a) State the null hypothesis symbolically and explain what it asserts.
(b) State the alternative hypothesis symbolically and explain what it asserts.
(c) Indicate whether the alternative hypothesis implies a one-tailed or a two-tailed test.
(d) Verbally describe the meaning of the P value that must be determined to reach a decision about the null hypothesis.

*10-12   Does the newly patented gasoline additive Surge improve fuel economy on the average? The sample data are measurements of the change in miles per gallon for each automobile after adding Surge.

*10-13   Is time-in-rank (time until promotion) shorter on the average for U.S. Army majors with PhDs than the current average of 5.7 years for all Army majors? The sample data are measurements of the time served in the rank of major by PhDs promoted to lieutenant colonel last year.

*10-14   Would the reappraisal of houses in a community have an impact on the average property-tax bills for houses that are at least 20 years old? The sample data are measurements of the change in the property-tax bill following the reappraisal for houses built at least 20 years ago.

*10-15   Do male offspring tend to grow taller than their fathers on the average? The sample data are measurements of the difference between the mature height of a son and the height of his father.

10.5   the one-sample *t* test

The reporter wishes to test the null hypothesis of no weight reduction on the average, $\mu = 0$, against the alternative hypothesis of a positive mean weight reduction, $\mu = 0$.

Her random sample of $n = 16$ clients lost an average of $\overline{X}_{16} = 5.2$ lb, and she notes that the discrepancy between this result and the null hypothesis appears to support the alternative hypothesis. But the reporter wants to be sure that the chance explanation for this discrepancy can be rejected beyond a reasonable doubt.

(Accordingly, she must compute a P value. To do so, she applies the t Distribution Law for Means.) Assuming that the population of weight-reduction scores is not markedly skewed, the law tells her that the t score of her sample mean $t_{\overline{X}_{16}}$ will be a point from a t distribution with $df = 16 - 1 = 15$.

Under the null hypothesis that $\mu = 0$

$$t_{\overline{X}_{16}} = \frac{\overline{X}_{16} - \mu}{\hat{\sigma}/\sqrt{n}} = \frac{5.2 - 0.0}{8.3/\sqrt{16}} = 2.51$$

If $\mu = 0$, the t score of the sample mean is equal to 2.51; that is, the sample mean lies an estimated 2.51 standard errors to the right of 0. This result is depicted in Chart 10-5.

The reporter needs to know the one-tailed P value that is depicted by the blue area—the area to the right of $t = 2.51$—in the chart. This area represents the probability that a positive

chart 10-5
t distribution
with *df* = 15

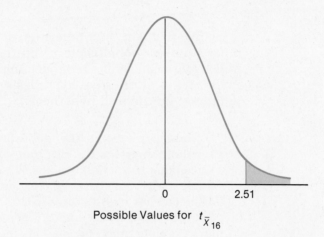

Possible Values for $t_{\bar{X}_{16}}$

discrepancy at least as large as 2.51 estimated standard errors will arise by chance.

determining the
P value

(To compute the *P* value, the reporter turns to the *t* table, locates the row where *df* = 15, and looks for an entry that is equal to 2.51. She doesn't find an entry that is exactly equal to 2.51, so she selects the entries that are just smaller and just larger than 2.51. We call these entries the *t score bounds:*

The *lower bound* (the entry just below 2.51) is 2.131.
The *upper bound* (the entry just above 2.51) is 2.602.)

The lower bound 2.131 is listed in the column headed "One-Tailed *P* Value = 0.025," which tells the reporter that 2.5 percent of the area under the curve lies to the right of *t* = 2.131. Since her *t* score of 2.51 lies farther out in the tail, the blue area is less than 2.5 percent of the area under the curve.

The upper bound 2.602 is listed in the column headed "One-Tailed *P* Value = 0.01," which tells the reporter that 1.0 percent of the area under the curve lies to the right of *t* = 2.602. Since her *t* score of 2.51 is closer to 0 than 2.602, the blue area is greater than 1.0 percent of the area under the curve. So the blue area is less than 2.5 percent but more than 1 percent of the area under the curve.

(*Conclusion:* The chance explanation for the discrepancy has a *P* value between 0.01 and 0.025.)

Now that she has determined the *P* value, the reporter must state her conclusion or inference about the truth of the null hypothesis. Will she reject the null hypothesis $\mu = 0$, in favor of the alternative hypothesis $\mu > 0$, or will she reserve judgment on the null hypothesis?

If the reporter rejects the null hypothesis, the computed *P* value represents the risk that she will be rejecting a true claim. If the null hypothesis is true—if μ is equal to 0—then a *P* value between 0.01 and 0.025 indicates that discrepancies with *t* scores ≥ 2.51 will occur by chance between 1.0 percent and 2.5 percent of the time. Her own sample may be one of these unrepresentative samples from the population of inference.

So the reporter's decision as to whether or not to reject the null hypothesis depends on what she considers to be an acceptable risk. Generally, in statistical research, the acceptable risk level chosen is the 1-percent level, the 5-percent level, or the 10-percent level. (A risk greater than 10 percent is almost always considered to be excessive, whereas a risk smaller than 1 percent is characteristically dismissed as negligible.) The selection of an acceptable risk level depends on how cautious the researcher wishes to be, which, in turn, depends on the cost of rejecting the null hypothesis when it is true.

If the reporter's criterion of acceptable risk were the 10-percent or the 5-percent level, finding a *P* value less than 0.025 would lead her to reject the null hypothesis, because in doing so she would face less than a 5-percent risk of having rejected the truth.

Before collecting her data, the reporter had decided that she would not reject the null hypothesis that the average weight loss was 0 unless she could do so with less than a 1-percent risk. Although her *P* value is less than 0.025, it is greater than 0.01. So she reports that her evidence does *not* refute the null hypothesis that $\mu = 0$.

The evidence that weight loss results from a one-year membership in the clinic is not statistically significant at the 1-percent level of significance.

(The expression *statistically significant* is used commonly in research reports. To state that the evidence is statistically significant is an alternative way of asserting that the discrepancy between the sample evidence and the null hypothesis is too large to be attributed to chance. The related expression *level of significance* refers to the level of risk involved. To say that the evidence is statistically significant at the 5-percent level, for example, means that the null hypothesis (or the chance explanation) can be rejected with less than a 5-percent risk.) Similarly, by stating that her evidence on the effectiveness of the weight-reduction clinic was *not* statistically significant at the 1-percent level, the reporter is saying that she *cannot* reject the null hypothesis without facing more than a 1-percent risk of rejecting a true claim.

(If the acceptable level of significance is chosen in advance, a researcher can determine whether or not to reject the null hypothesis simply by comparing the observed *t* score with the *t* table entry for that significance level. An observed *t* score larger than the table entry means that the *P* value is below the acceptable risk level, permitting us to reject the null hypothesis. An observed *t* score smaller than the table entry means that the *P* value is above the acceptable risk level, and we cannot reject the null hypothesis.) In this case, the reporter chose 1 percent as an acceptable level of significance (level of acceptable risk). The *t* table entry for a one-tailed *P* value of 0.01 is 2.602. Since her observed *t* score is 2.51 (a smaller value), she cannot reject the null hypothesis.

(Generally, the actual *P* value should be reported, rather than simply the presence or the lack of statistical significance. Then each of us can decide whether or not this risk level is acceptable and whether or not to reject the null hypothesis.)

summarizing the
procedure

(Let's review the steps the reporter followed in performing the *t* test for μ:

1. State the null and alternative hypotheses:

Null: $\mu = 0$ Alternative: $\mu > 0$
(The weight-reduction   (The weight-reduction
program is ineffective on   program is effective on
the average.)   the average.)

2. Select a random sample and compute the mean $\overline{X}_n$ and the standard deviation $\hat{\sigma}$ of the sample:

$$\overline{X}_{16} = 5.2 \text{ lb}$$
$$\hat{\sigma} = 8.3 \text{ lb}$$

3. If the discrepancy between the sample mean and the value of the population mean given by the null hypothesis supports the alternative hypothesis and if the conditions for the *t* Distribution Law for Means apply, compute the *t* score of the sample mean:

$$t_{\overline{X}_{16}} = \frac{\overline{X}_{16} - \mu}{\hat{\sigma}/\sqrt{n}} = \frac{5.2 - 0.0}{8.3/\sqrt{16}} = 2.51$$

4. Obtain the appropriate *P* value from the *t* table:
   At $df = n - 1 = 16 - 1 = 15$, the one-tailed *P* value is between 0.01 and 0.025.

5. Report the test results.
   If the reporter rejects the null hypothesis, she faces less than a 2.5-percent but more than a 1-percent risk of rejecting a true hypothesis. (Or, alternatively stated, the sample evidence is statistically significant at the 2.5-percent level but not at the 1-percent level.) Since the reporter chose 1 percent as an acceptable level of significance, she cannot reject the null hypothesis but must declare that the data do not demonstrate that the weight-reduction program is effective on the average.)

procedure for a
two-tailed test

In the case of the reporter's evaluation of Slim and Trim, Inc., the appropriate hypothesis test is one tailed. The hypothesis testing procedure changes in only one respect when a two-tailed procedure is required. Let's conduct a *t* test for μ for Case 2 on page 299 to see how a two-tailed test is performed.

(1. State the null and alternative hypotheses.
   The question the city engineer wishes to answer is whether the traffic level at a busy intersection differs from last year's mean level of 35,000 cars per day. If μ denotes the average number of cars passing through the

intersection per day, the engineer can state the hypotheses as

Null: $\mu = 35{,}000$ cars   Alternative: $\mu \neq 35{,}000$ cars
(The average number of   (The average number of
cars has not changed from   cars has changed from
the preceding year.)   the preceding year.)

2. Select a random sample and compute the mean $\overline{X}_n$ and the standard deviation $\hat{\sigma}$ of the sample.

The engineer records the number of cars at the intersection on each of 20 randomly selected days and obtains the following summary statistics:

$$\overline{X}_{20} = 38{,}000 \text{ cars}$$
$$\hat{\sigma} = 6{,}800 \text{ cars}$$

3. If the conditions for the use of the t distribution hold, compute the t score of the sample mean.

$$t_{\overline{X}_{20}} = \frac{\overline{X}_{20} - \mu}{\hat{\sigma}/\sqrt{n}} = \frac{38{,}000 - 35{,}000}{6{,}800/\sqrt{20}} = 1.973$$

We do not need to check whether the discrepancy between the sample mean and the value claimed for the population mean by the null hypothesis supports the alternative hypothesis—a necessary step in the one-tailed test. (In a two-tailed test, discrepancies from the null hypothesis in either direction favor the alternative hypothesis.)

4. Obtain the appropriate P value from the t table.

Referring to the t table, at $df = n - 1 = 20 - 1 = 19$, the engineer finds that the t score bounds are 1.729 and 2.093—the t scores associated with the two-tailed P values of 0.10 and 0.05, respectively. So the two-tailed P value of 1.973 is less than 0.10 but greater than 0.05.

5. Report the test results.

If the engineer rejects the null hypothesis, he faces between a 5-percent and a 10-percent risk of rejecting a true hypothesis. (The sample evidence is statistically significant at the 10-percent level but not at the 5-percent level).

The engineer gave his report to the city manager. The manager felt that a 10-percent risk was reasonable and recommended that a traffic light be installed at the intersection.

EXERCISES

10-16   To test the null hypothesis $\mu = 0$, researchers randomly selected a sample of 16 observations and found $\overline{X}_{16} = 5$ units and $\hat{\sigma} = 12$ units.

(a) Calculate $\hat{\sigma}(\overline{X}_{16})$. Interpret this value.

(b) Calculate $t_{\overline{X}_{16}}$, assuming that the null hypothesis is true. Interpret this value.

(c) If the alternative hypothesis is $\mu > 0$, obtain the P value for the result you obtained in (b). Can you reject the null hypothesis without facing more than a 10-percent risk of rejecting a true claim?

(d) If the alternative hypothesis is $\mu \neq 0$, obtain the P value for the result you obtained in (b). Now can you reject the null hypothesis without facing more than a 10-percent risk of rejecting a true claim?

(e) Explain the reason for the difference in the test decisions made in (c) and (d).

(f) If the alternative hypothesis were $\mu < 0$, would you bother to determine a P value for the result you obtained in (b)? Explain your answer.

*10-17   The Educational Testing Service (ETS) claims that special preparation for its Scholastic Aptitude Test (SAT) does not improve test scores on the average. Dubious about the claim, a research team selected a random sample of 25 high-school seniors and gave the students a brief course in SAT testing strategy. The team administered prototype verbal aptitude examinations to these students before and after this course. The results showed a mean improvement $\overline{X}_{25} = 4.8$ points with a standard deviation $\hat{\sigma} = 20$ points. The shape of the distribution of improvement scores was reasonably symmetric.

(a) State the ETS claim as the null hypothesis, and then state the alternative hypothesis.

(b) Does the test procedure call for a one-tailed or a two-tailed test?

(c) Compute the P value of the sample evidence on this issue.

(d) Would you reject the ETS claim based on the evidence acquired? If you do reject the claim, what risk do you face of rejecting a true claim?

10-18   (*Continuation of Exercise 10-12, page 303*) Given the sample evidence, $n = 25$, $\overline{X}_{25} = 1.2$ mpg, and $\hat{\sigma} = 1.5$ mpg:

(a) Is the chance explanation for the discrepancy between the evidence and the null hypothesis plausible? Report the appropriate P value and interpret this value.

(b) Does the result you obtained in (a) imply Surge will improve the fuel economy of all cars in which it is used? Support your answer.

10-19 (*Continuation of Exercise 10-13, page 304*) Given the sample evidence, $n = 64$, $\overline{X}_{64} = 5.4$ years, and $\hat{\sigma} = 1.9$ years:

(a) Can you reject the null hypothesis without facing more than a 10-percent risk of rejecting a true claim?

(b) Does the result you obtained in (a) imply that *no* Army major will receive a more rapid promotion because he or she holds a PhD degree?

10-20 (*Continuation of Exercise 10-14, page 304*) Given the sample evidence, $n = 36$ houses, $\overline{X}_{36} = \$50$, and $\hat{\sigma} = \$120$:

(a) Is the discrepancy between the sample evidence and the null hypothesis statistically significant at the 5-percent level?

(b) Can you reject the null hypothesis without facing more than a 1-percent risk of rejecting a true claim?

10-21 (*Continuation of Exercise 10-15, page 304*) Given the sample evidence, $n = 9$, $\overline{X}_9 = 1.3$ in., and $\hat{\sigma} = 0.9$ in.:

(a) Is the discrepancy between the sample evidence and the null hypothesis statistically significant at the 1-percent level?

(b) Can you justify using the t distribution in (a)?

10.6 the two types of error

An irate letter to the editor followed the publication of the reporter's study on the effectiveness of weight-reduction clinics. In it, the president of Slim and Trim, Inc., denounced the reporter's story as a slanted evaluation of the company's program and suggested that she and her editor learn more about Type II error.

The president's complaint is legitimate. (Rejecting a true null hypothesis, sometimes called a *Type I error*, is not the only error we can make in hypothesis testing. We can *fail to reject* a *false* null hypothesis, which is called a *Type II error*.) The reporter was concerned about declaring the clinic program to be effective when it was not—a Type I error. The president's concern is that the reporter has committed a Type II error by not

Chart 10-6

IF IN REALITY →	The null hypothesis is true	The null hypothesis is false
AND WE DECIDE ↓		
Not to reject the null hypothesis	A correct decision	A Type II error
To reject the null hypothesis	A Type I error	A correct decision

THEN
THE CONSEQUENCE OF OUR DECISION IS ⌐

declaring the clinic program to be effective when, in fact, it is helping its clients.

TYPE I ERROR Rejecting a true null hypothesis.

TYPE II ERROR Failing to reject a false null hypothesis.

The possible outcomes of a hypothesis test are summarized in Chart 10-6.

A Type I error will harm clients who spend time and money on the weight-reduction program in the mistaken belief that it will help them lose weight. A Type II error will clearly be detrimental to the interests of Slim and Trim, Inc., but will also discourage potential clients from taking advantage of a beneficial program.

The same kinds of conflicting concerns occur in the traffic study in Case 2. If the city manager decides that changes need to be made when in fact they do not (a Type I error), time and money will be spent unnecessarily. On the other hand, if the null hypothesis is not rejected and no changes are made when changes are needed (a Type II error), two possible types of consequences can occur: If traffic has increased, traffic tie-ups and accidents may increase; if traffic has decreased, inaction means resources will be used unnecessarily.

P values and proof of the null hypothesis

In reporting test results that are not statistically significant, people often conclude with the statement: The null hypothesis must be accepted.

(This statement should be avoided. It implies that the only alternative to rejecting the null hypothesis as false is accepting the null hypothesis as true. In fact, the decision being made is to reserve judgment on the null hypothesis.)

(We can never prove that the null hypothesis is true. A high P value suggests only that the chance explanation for the discrepancy between the sample evidence and the null hypothesis is plausible. A high P value does not prove that the null hypothesis is true, because the possibility of committing a Type II error always exists.)

An analogy is often made between hypothesis testing and a jury trial. In a jury trial, the defendant is assumed innocent until proven guilty beyond a reasonable doubt. (In the terminology of hypothesis testing, we say that the null hypothesis of innocence is not rejected unless the discrepancy between the evidence presented and the claim of innocence has a very low P value.)

The jury's verdict is either *guilty* (the null hypothesis of innocence is rejected) or *not guilty* (the claim of innocence cannot be rejected because a reasonable doubt exists). (Not guilty does not imply proof of innocence — only insufficient evidence of guilt. In any *not guilty* decision, there is some risk of freeing a guilty party — a Type II error.)

trade-off between
Type I and
Type II errors

Jurors frequently find it difficult to define what constitutes reasonable doubt. What is the acceptable risk of convicting an innocent party? Ideally, the acceptable risk of committing a Type I error should be very close to 0, making it nearly impossible to convict an innocent party. However, as the risk of convicting an innocent party is lowered, a greater amount of evidence is required to gain a conviction, thereby increasing the likelihood that the jury will fail to convict a guilty party.

(In hypothesis testing, we are faced with a trade-off between two types of risk. As we lower the acceptable risk level of committing a Type I error (rejecting a true null hypothesis), we increase the risk of committing a Type II error (failing to reject a false null hypothesis).)

The president of Slim and Trim, Inc., complained that by refusing to declare the diet clinic program effective unless the evidence of weight loss were statistically significant at the 1-percent level, the reporter was taking an excessively large risk of committing a Type II error—the error of failing to declare the diet clinic program effective when, in truth, it is. If the acceptable level of risk for committing a Type I error had been set at 5 percent or even at 2.5 percent, instead of at 1 percent, the evidence of weight loss would have been declared statistically significant and the null hypothesis would have been rejected.

In contrast, a citizens' committee concerned with excessive taxation objected to the city manager's decision to install a traffic light at the intersection on the basis of results that were *not* statistically significant at the 5-percent level but only at the 10-percent level. The committee felt that a 10-percent risk of committing a Type I error was too high, giving insufficient weight to the risk of spending tax revenues unnecessarily. If the acceptable risk of committing a Type I error had been set at 5 percent, the null hypothesis would not have been rejected and the money for a new traffic light would not have been spent.

(In our examination of confidence intervals in Chapter 9, we learned that by increasing sample size, we could buy our way out of the trade-off between the level of confidence and precision (the width of the confidence interval). The same principle works in hypothesis testing as well: Increasing sample size reduces the combined risk of making Type I and Type II errors.)

(By increasing sample size, we reduce the risk of committing a Type II error without increasing the risk of committing a Type I error. A Type II error can be made only if we *fail to reject* the null hypothesis. Having established an acceptable level of risk for committing a Type I error, *we will correctly reject the null hypothesis more often as sample size increases*. This is because as sample size increases, the possible scores for $\overline{X}_n$ concentrate more closely about μ, thereby requiring smaller discrepancies between the observed value of $\overline{X}_n$ and the null value of μ for rejection of the null hypothesis.)

EXERCISES
10-22   (*Continuation of Exercise 10-17, page 310*)

(a) Suppose that the ETS claim (the null hypothesis) is true and your conclusion is to reject the null hypothesis. What type of error would you be committing and what effect would it have on students who believed your conclusion?

(b) Suppose that the ETS claim is false and your conclusion is that the null hypothesis cannot be rejected. What type of error would you be committing and what effect would it have on students who believed your conclusion?

(c) What can you do to reduce the combined risk of making a Type I and a Type II error?

10-23   In testing a null hypothesis $\mu = 0$, you find that your sample evidence is not statistically significant at the 10-percent level (that is, you find that $\bar{X}_n$ does not differ significantly from 0 at the 10-percent level). Why should you avoid stating, "Accordingly, we must *accept* the null hypothesis"? Instead how should you phrase your conclusion?

10-24   A medical researcher claims to have developed a diagnostic procedure to detect the presence of precancerous cells in the digestive tract.

(a) The diagnostic procedure is not foolproof. Describe two types of faulty diagnoses that could result and explain the potential harm that could be done in each case.

(b) Let the null hypothesis be that an individual patient has no precancerous cells in his or her digestive tract. Following the format of Chart 10-6, display the consequences of the diagnoses that could result.

10-25   A study based on a random sample of 15 cigarette smokers who had switched from high tar to low tar brands reported a 35-percent increase in the number of cigarettes smoked per day on the average ($\bar{X}_{15} = 35$ percent). The result, however, was not statistically significant at the 1-percent level.

(a) From the data, can you conclude that switching from high tar to low tar brands does not increase the number of cigarettes smoked per day? Explain your answer.

(b) What steps could be taken to reduce the risk of a Type II error?

Caveats on the *t* Distribution and Hypothesis Testing

CAVEAT 1: Limitation of the *t* distribution.

In this chapter, we have discussed the *t* distribution and its role in drawing inferences about a population mean. But we must also know when to avoid the use of the *t* distribution.

When we presented the *t* Distribution Law for Means, we indicated that the law cannot be used to find *exact* probabilities (*P* values) unless the population of inference has a *normal* shape. To the extent that the shape of the population departs from the bell-shaped curve, the column headings in the *t* table will only approximate the true probabilities. Fortunately, except for markedly skewed populations, these approximations will be close when $df \geq 15$.

When the population is markedly skewed, the *t* distribution approximation can be relied on only for large samples. However, even if the sample size is large, we may not want to use the *t* distribution in these cases. If the population is markedly skewed, the median is a better measure of the center of the distribution than the mean is, and we may want to use our sample data to make inferences about a population median rather than a population mean. Procedures giving exact probabilities (*P* values), even for very small samples, are available for drawing inferences about a population median.[3]

CAVEAT 2: Statistical versus practical significance.

The president of Slim and Trim argued with the reporter about whether or not her sample evidence on weight loss was statistically significant. The reporter said no at the 1-percent level of significance; the president said yes at the 2.5-percent level of significance. But we should also ask about *practical* significance.

[3] These are called nonparametric procedures and are examined in such books as E.L. Lehmann, *Nonparametrics: Statistical Methods Based on Ranks* (San Francisco: Holden-Day, Inc., 1975).

Statistical significance does not imply practical significance. When we say that the evidence is statistically significant, we mean only that it is unlikely that the discrepancy between the evidence and the null hypothesis resulted from chance variation in random sampling. If our sample size is extremely large, even a minute discrepancy between the evidence and the null hypothesis can be *statistically* significant.

Consider this example. Suppose that the reporter tests the null hypothesis $\mu = 0$ against the alternative hypothesis $\mu > 0$, using a sample of $n = 400$. The reporter will declare her results to be statistically significant at a significance level of 1 percent if the observed t score is greater than 2.326 (the t table entry for a one-tailed P value = 0.01). If $\hat{\sigma} = 8$ lb, then any positive value of $\overline{X}_{400}$ in excess of 1 lb will be statistically significant at the 1-percent level. We can see this by substituting $\overline{X}_{400} = 0.93$ lb into the equation for a t score.

$$t_{\overline{X}_{400}} = \frac{\overline{X} - \mu}{\hat{\sigma}/\sqrt{n}}$$

$$= \frac{0.93 - 0}{8/\sqrt{20}} = 2.326$$

Even if the reporter had found a sample mean weight loss of 1 lb, she would have had to conclude that the evidence of a 1-lb weight loss on the average is statistically significant at the 1-percent level. Yet a 1-lb weight loss is clearly not of *practical* significance.

EXERCISES

1. Refer to the data on the household electricity bills for a sample of 21 households in a congressional district given in Chart 2-5 (page 32).
   (a) Assuming that the households were randomly selected and given that $\overline{X}_{21} = \$27.76$ and $\hat{\sigma} = \$15.10$, compute and interpret a 99-percent confidence interval for μ.
   (b) Referring to the histogram of the distribution of sample scores that appears in Chart 2-6 (page 33), should you be concerned that the results you obtained in (a) may be invalid? Why or why not? Would you have the same concern if the score for household 21 were $\$29.25$ instead of $\$79.25$ (assuming the appropriate adjustments were made for $\overline{X}_{25}$ and $\hat{\sigma}$)?

2. (*Continuation of Exercise 10-17, page 310*) To test the ETS claim, a random sample of 400 students were selected as test subjects and,

again, that the standard deviation was $\hat{\sigma} = 20$ points. How large an improvement in test scores on the average (that is, how large a value for $\overline{X}_{400}$) would be required for you to conclude that the evidence of improvement is significant at the 1-percent level? Would this finding necessarily imply that the course in SAT test-taking strategy is worthwhile, keeping in mind that scores on the verbal aptitude test ranged from 200 to 800?

COMING
TO
TERMS

Alternative Hypothesis The alternative to the null hypothesis. This hypothesis includes all the possibilities of consequence to those performing the hypothesis test.

Degrees of Freedom *df* The number used in lieu of sample size to obtain an unbiased estimate of a population variance. The number of degrees of freedom associated with a *t* score determines which *t* distribution curve is used to represent the sampling distribution of *t* scores.

Estimated Standard Error of $\overline{X}_n$, $\hat{\sigma}(\overline{X}_n)$ The estimate of the standard error of $\overline{X}_n$, defined as $\hat{\sigma}(\overline{X}_n) = \hat{\sigma}/\sqrt{n}$. When it is clear by context that the true standard error is unknown, the term *standard error* is frequently used in place of *estimated standard error*.

Level of Significance Another term for *P value*, or the risk of committing a Type I error. Saying that the evidence is statistically significant at the 5-percent level, for example, means that the null hypothesis can be rejected with less than a 5-percent risk of committing a Type I error.

Null Hypothesis In general, the claim that something is of no consequence.

One-Tailed *t* **Test** A hypothesis test in which the results in only one tail of the distribution of *t* scores support the alternative hypothesis.

P **Value** A measure of the plausibility of chance as an explanation for the discrepancy between the results observed and the results expected under the null hypothesis. A very low *P* value implies that the chance explanation is implausible, permitting us to reject the null hypothesis. A high *P* value implies that the chance explanation is plausible and should dissuade us from rejecting the null hypothesis. The *P* value is also called the *level of significance*, or the risk of committing a Type I error.

t **Distributions** A class of distributions, similar in shape to the *normal* distribution curves, that represents the sampling distributions of *t* scores.

t **Score** The distance of an individual score from the mean in units of estimated standard deviations. Used as an alternative to the *Z* score when the population standard deviation σ is unknown.

Two-Tailed *t* Test   A hypothesis test in which the results in either tail of the distribution of *t* scores support the alternative hypothesis.

Type I Error   Declaring the null hypothesis to be false when it is true.

Type II Error   Declaring the null hypothesis to be true when it is false.

ISSUES FOR ANALYSIS

1. After studying the effects of an experimental nutrient on the weights of $n = 30$ cows, a dairy researcher reported that a 90-percent confidence interval for the expected weight increase μ was {10.2 to 25.8} pounds.
   (a) Can we be more than 90 percent confident that $\mu > 0$? Explain your answer.
   (b) How confident can we be that $\mu > 10.2$? Explain your answer.
   (c) Can we reject the null hypothesis that $\mu = 0$ in favor of the alternative hypothesis that $\mu > 0$ without facing more than a 1-percent risk of making a Type I error? (*Hint:* The information you need to conduct a hypothesis test can be found in the components of the 90-percent confidence interval.)

2. A family counseling service wishes to estimate the mean duration μ of U.S. marriages that ended in divorce during the past year. The desired confidence level is 99 percent, and the margin for error is limited to no more than 1.5 years. The sample size required to achieve these confidence–precision requirements must be determined.

   If the value of σ (the standard deviation of the population of the duration of U.S. marriages) were known, then the required sample size could be determined from Equation (9-6):

   $$n = \left(\frac{Z \cdot \sigma}{\text{Margin for error}}\right)^2$$

   But σ is not known.
   (a) To obtain an estimate of σ, the counseling service selects a sample of 16 cases and calculates $\hat{\sigma} = 5.7$ years. Determine whether a sample size of $n = 16$ is sufficient to meet the confidence–precision requirements for the estimate of μ. (*Hint:* Calculate the margin for error as $t \cdot \hat{\sigma}/\sqrt{n}$.)
   (b) Assuming that σ is equal to the value of $\hat{\sigma}$ you obtained from the sample in (a), determine the sample size that will meet the confidence–precision requirements for the estimate of μ. (*Hint:* For any $n \geq 30$, the margin for error can be approximated closely as $Z \cdot \sigma/\sqrt{n}$.)
   (c) Verify the result you obtained in (b) by calculating the margin for error of a 99-percent confidence interval, based on the sample size you determined in (b).

Does Air Pollution Shorten Lives?

In their study entitled "Does Air Pollution Shorten Lives?," Lave and Seskin used a multiple regression to estimate the effects of air pollution indicators on mortality rates. A mortality rate MR is the number of deaths recorded per 10,000 population. Using data collected from 117 Standard Metropolitan Statistical Areas (SMSAs) during 1960. Lave and Seskin reported the following results:

In Equation (1), the total mortality rate is "explained" by:

Mean P: Mean of the biweekly suspended particulate readings in $\mu g/m^3$ [micrograms per cubic meter].

Min S: Smallest of the biweekly sulfate readings in $\mu g/m^3 \times 10$.

P/M^2: Population density in the SMSA in people per square mile.

%NW: Percentage of the SMSA population who are nonwhite $\times$ 10; that is, the number per thousand.

% $\geq$ 65: Percentage of the SMSA population 65 and older $\times$ 10.

$$MR = 19.607 + 0.041\ Mean\ P + 0.071\ Min\ S + 0.001\ \frac{P}{M^2} + 0.041\ \%NW + 0.687\ \% \geq 65 \qquad (1)$$
$$\quad\ \ (2.53) \qquad\qquad (3.18) \qquad\qquad (1.67) \qquad (5.81) \qquad\quad (18.94)$$
$$R^2 = 0.827$$

These independent variables explain the total mortality rate across 117 SMSAs extremely well, since 82.7 percent of the variation is explained ($R^2 = 0.827$). The effect of each variable on mortality is estimated by the [regression] coefficient; each is statistically significant (as shown by the t statistics below the coefficients). Note that both measures of air pollution are significant explanatory factors of the mortality rate across cities.

Source: Lester B. Lave and Eugene P. Seskin, "Does Air Pollution Shorten Lives?" from John W. Pratt (ed.), *Statistical and Mathematical Aspects of Pollution Problems* (New York: Marcel Dekker, Inc., 1974), pages 223–44.

Behind the Headline

Until this point, our discussion of statistical inference has been limited to issues that are concerned with the population *mean*. But when we described distributions in Part 1, we examined other types of averages as well as measures of variation. In Part 2, we learned how correlation and regression coefficients can be used to describe relationships between variables.

Statistical procedures can be employed to draw inferences from sample data about all of these measures. Some of these procedures are simple; others are complex. We cannot describe them all in this book. However, we will examine how the principles of estimation and hypothesis testing can be extended to draw inferences from regression results.

regression analysis 11

"You may not be aware that the deduction of a man's age from his writing is one which has been brought to considerable accuracy by experts. In normal cases, one can place a man in his true decade with tolerable confidence. I say normal cases, because ill health and physical weakness reproduce the signs of old age, even when the invalid is a youth."

Holmes to Watson
in *The Reigate Squires*

11.1 the population regression model

The location of any area on our planet can be pinpointed at the intersection of three scores: latitude (degrees north or south of the equator), longitude (degrees east or west of the prime meridian in Greenwich, England), and elevation (distance above sea level) of the area. Suppose that we wish to study the relationship between location and temperature in the United States. From public records, we can obtain the average annual temperatures for most U.S. cities having populations in excess of 5,000. But collecting data on the many thousands of these jurisdictions would be a formidable task. Accordingly, we decide to base our relationship investigation on a random sample of only 30 cities.

The variables in our investigation, with their units of measurement, are

temp: (the dependent variable) The average annual temperature of a U.S. city in degrees Fahrenheit.
lat: The city's latitude in degrees north of the equator.
long: The city's longitude in degrees west of the prime meridian.

elev: The city's elevation in hundreds of feet above sea level.

Illustrative data for four cities in this sample appear in Chart 11-1.

the sample
regression
equation

In studying the relationship between location and temperature, we intend to answer such questions as

How much colder might we expect it to get as we move 1° north or climb an additional 100 feet in altitude?

In the United States, does east-to-west movement across a path of constant latitude and elevation systematically alter temperature levels?

These are questions about rates of response. In Chapter 7, we saw that one approach to answering this type of question is regression — more specifically, *multiple* regression, because we have defined more than a single predictor of *temp*. Accordingly, to describe the relationship between temperature and location in our sample of 30 cities, we can write the multiple regression equation

$$\widehat{temp} = b_0 + b_1\ lat + b_2\ long + b_3\ elev \qquad (11\text{-}1)$$

Using the least squares criterion to select the values for the *b*'s and a computer program to perform the necessary calculations, we obtain the results

$$\widehat{temp} = 98.420 - 1.402\ lat + 0.057\ long - 0.134\ elev$$

$$R^2 = 94.8\% \qquad\qquad (11\text{-}2)$$

chart 11-1

CITY	TEMP (Degrees Fahrenheit)	LAT (Degrees North)	LONG (Degrees West)	ELEV (Hundreds of Feet)
Portland, Maine	45.5°	43.7°	70.3°	0.63
Denver, Colorado	50.2	39.8	105.0	53.32
Nashville, Tennessee	59.6	36.2	86.8	6.05
Phoenix, Arizona	70.3	33.5	112.1	11.07

From Equation (11-2), we can predict the rates of response of *temp* to changes in location, as follows:

For each 1° movement north with no change in *long* and *elev*, we predict a *reduction* of 1.402° in *temp* (since $b_1 = -1.402$).

For each 1° movement west with no change in *lat* and *elev*, we predict an increase of 0.057° in *temp* (since $b_2 = 0.057$).

For every 100 feet climbed (1 unit of *elev* holding *lat* and *long* constant, we predict a 0.134° decline in *temp* (since $b_3 = -0.134$).

We also note that since $b_0 = 98.42$, the predicted temperature at a point on the equator (*lat* = 0), on the prime meridian (*long* = 0), and at sea level (*elev* = 0) is 98.42°. Of course, this point lies outside the United States, so that the value of b_0 (the intercept) is implicitly an extrapolation beyond the range of our data.

We also see that $R^2 = 94.8$ percent. R^2 represents the percentage of the total variation in *temp* that is explained by the regression equation. So the equation is a good predictor of *temp* for *this sample of 30 cities*. Only 5.2 percent of the variation in *temp* is unexplained by the locational variables.

from sample description to population inference

Having completed our description of the sample regression results, we now want to make inferential statements about the relationships between *temp* and the locational variables. Although our sample is limited to 30 cities in the United States, our area of interest is broader. Conceptually, our population of inference includes the temperature–location scores of *all* cities in the country.

We must remember that our sample evidence is subject to the chance variation inherent in random sampling. If the luck of the draw had produced a different group of 30 cities, our least squares regression procedure would have generated a different set of values for the *b*'s in Equation (11-1). To develop laws of chance variation for the *b*'s, we must first define the concept of a *population regression model*.

A population regression model is a theoretical statement
in the form of an equation about the manner in which a dependent variable responds to changes in its external environment.

Inevitably, the external environment includes numerous
variables. We identify some variables as *predictors*, because
we believe they have a measurable influence on the behavior
of the dependent variable. Other variables are not specifically
identified as predictors because we have no theoretical reason
to believe that they affect the dependent variable. We refer to
the latter group of variables as *noise*. Each predictor is believed to emit a "signal" that produces some response in the
dependent variable. Although some other components of the
external environment may also emit signals, they are assumed
to be weak and unrelated to each other. The composite of these
signals is just "noise."

Now let's consider the population regression model

$$temp = \beta_0 + \beta_1 \, lat + \beta_2 \, long + \beta_3 \, elev + noise \quad (11\text{-}3)$$

This model expresses several *beliefs* about the manner in
which *temp* responds to changes in its external environment.

1. *Temp* will respond in a systematic manner to changes
    in location that are measured by the three predictors
    lat, long, and *elev.* In other words, to understand why
    different areas have different average annual tempera-
    ture levels, we should consider differences in their lati-
    tude, longitude, and elevation.

2. The response in *temp* to a change in each locational
    variable is basically *linear,* so that the response rates
    can be represented by constants—the β's (betas) in
    Equation (11-3). For example, the population regres-
    sion model assumes that the extent of the change in
    temp resulting from movement north or south depends
    only on the extent of the change in *lat* and not on the
    location from which the change occurs (the initial level
    of *lat, long,* or *elev*). So the *rates* of response (the β's) do
    not change as the levels of the predictors change: The
    betas are constants.

3. By introducing the term *noise,* the model acknowledges
    that a unit change in one locational variable (holding

all others constant) will not *always* produce a response of the same size in *temp*. For example, if *lat* and *long* are fixed, a 100-foot increase in *elev* will not always change *temp* by exactly β_3 degrees. The exact rate of response in *temp* will vary from β_3 due to hidden changes in variables that have been excluded from the model (*noise*).

However, if *noise* is just a composite of weak and unrelated forces, the effects of hidden changes in *noise* on *temp* should average to 0. So if *lat* and *long* are fixed, a 100-foot increase in *elev* will change *temp* by β_3 degrees *on the average*. Accordingly, the β's can be interpreted as true *average* response rates.

β_1: The true rate at which *temp* responds on the average to changes in *lat*, holding *long* and *elev* constant.

β_2: The true rate at which *temp* responds on the average to changes in *long*, keeping *lat* and *elev* fixed.

β_3: The true rate at which *temp* responds on the average to changes in *elev*, if *lat* and *long* do not change.

Viewing the β's in the population regression model as true average rates of response, we refer to the values of the b's in Equation (11-1) as sample estimates of the true average response rates. So the change in Y *we would predict* from the sample regression equation per unit change in each X — our descriptive interpretation of each b — is really an estimate of the true average change in Y per unit change in X.

But, as we know, the b's will vary by chance from sample to sample. To draw inferences from the b's about the β's, therefore, we need to have laws of chance variation for sample regression coefficients.

EXERCISES

11-1   Does the population regression model $Y = \beta_0 + \beta_1 X_1 + \beta_2 X_2 + noise$ express the belief that:

(a) The scores on variable Y can be predicted without error from knowledge of the scores on X_1 and X_2?

(b) A one-unit increase in X_1, holding X_2 constant, will always change Y by exactly β_1 units?

(c) A one-unit increase in X_2, holding X_1 constant, will change Y by an average of β_2 units?

11-2   In a discussion concerning the types of variables that could explain why some graduates of a particular college achieved higher grade point averages, *gpa*, than their peers, one counselor suggested that measures of intellectual ability *IQ*, nonscholastic interests *diversity*, and income aspiration *ambition* should be considered.

(a) Express the counselor's suggestion in the form of a population regression model.

(b) Naturally, the mix of instructors differed from student to student. What does the model you constructed in (a) assume about the effect on *gpa* of differences in instructors and their grading standards?

11-3   A student enrolled in a statistics course plans to investigate the relationship between an individual's height and age by estimating the population regression model

$$height = \beta_0 + \beta_1 \, age + noise$$

Cite at least two reasons why this model seems to be an oversimplification of the relationship between *age* and *height*.

*11-4   Youths sentenced to federal prison are reviewed by a parole board within six months of confinement. The board then establishes a period of continuance—the time the offender must remain in prison—ranging from 0 months (the offender is paroled immediately) to 36 months (at which time another hearing is held).

A researcher hypothesizes that variation in period of continuance, *PC*, from case to case is a linear function of:

(1) *severity*, a measure of the severity of the offender's crime (higher score = more severe).

(2) *progress*, a measure of the offender's participation in prison programs (higher score = greater progress).

(3) *behavior*, a measure of the offender's behavior in prison (higher score = better behavior).

(4) *risk*, a measure of the risk that an offender will violate parole (higher score = greater risk).

(a) Express the researcher's hypothesis as a population regression model.

(b) Why did you include the term *noise* in your statement of the model?

(c) Interpret the meaning of each β coefficient and indicate the sign (+ or –) that you would expect each β coefficient to have.

*11-5   Consider the population regression model

$$cost = \beta_0 + \beta_1\ distance + noise$$

where

$$cost = \text{expense of an automobile trip}$$
$$\text{(for example, gas, food, and tolls)}$$
$$distance = \text{number of miles traveled}$$

(a) Name several variables that you would assume would be part of *noise.*
(b) What is the meaning of β_1?

11.2   the laws of chance variation for sample regression coefficients

When we studied chance variation in the possible values of a sample mean $\overline{X}_n$ about μ, we defined the concept of a sampling distribution of $\overline{X}_n$ and presented the set of laws that govern the sampling distribution of $\overline{X}_n$.

the sampling distribution of a regression coefficient

A sampling distribution for a sample regression coefficient b is defined in the same way that a sampling distribution for a sample mean is defined.

SAMPLING DISTRIBUTION OF b The relative frequency distribution of all the possible scores for b in random samples of n observations.

In our illustrative example, three sampling distributions are of interest:

1. *The sampling distribution of b_1 for $n = 30$.* The relative frequency distribution of all possible scores for the least squares coefficient on *lat* in random samples of 30 U.S. cities.
2. *The sampling distribution of b_2 for $n = 30$.* The relative frequency distribution of all possible scores for the least squares coefficient on *long* in random samples of 30 U.S. cities.

3. *The sampling distribution of b_3 for $n = 30$.* The relative frequency distribution of all possible scores for the least squares coefficient on *elev* in random samples of 30 U.S. cities.

As we did in the case of means, we will use several laws of chance variation to describe the essential characteristics of these sampling distributions. Again, we will present an Expected Value Law, a Standard Error Law, and a *t* Distribution Law.

the Expected
Value Law for
a sample
regression
coefficient

THE EXPECTED VALUE LAW FOR *b*   If the variables that are omitted as predictors from the population regression model (the *noise* variables) are uncorrelated with an included predictor, then the expected value of the sample regression coefficient *b* on that predictor equals the true average response rate β. This law holds for any sample size *n*.

Symbolically, the law permits us to write

$$\mu(b) = \beta \tag{11-4}$$

The Expected Value Law tells us that the average of all possible scores for the least squares regression coefficient b is β. So under the proviso stated in this law, a sample regression coefficient provides an unbiased estimate of the true average response rate of Y to changes in X.

However, the proviso is important. If the forces that are considered to be *noise* are correlated with the predictors in the model, then the least squares sample regression coefficients will tend to either underestimate or overestimate the true average response rates.

In Chapter 7, we noted this consequence of excluding predictors. We said that the hidden effects of missing predictors distort the measured response rates (the b's). Now we can clarify this point by stating that the distortion is in the form of a bias.

Suppose that in reality

$$Y = \beta_0 + \beta_1 X_1 + \beta_2 X_2 + \beta_3 X_3 + noise \tag{11-5}$$

whereas our sample regression equation

$$\hat{Y} = b_0 + b_1 X_1 + b_2 X_2 \qquad (11\text{-}6)$$

excludes the variable X_3 as a predictor, making X_3 a component of *noise*. If X_1 is *correlated* with X_3, then the omission of X_3 from Equation (11-6) will cause b_1 to provide a biased estimate of β_1. The value of b_1 will reflect the hidden effects that changes in X_3 produce in Y. On the other hand, if X_2 is *uncorrelated* with X_3, b_2 will still provide an unbiased estimate of β_2.

> The sample regression coefficient b provides an unbiased estimate of β if and only if X is uncorrelated with the forces that are omitted from the model (*noise*).

the Standard Error Law for a sample regression coefficient

We have defined the standard error of $\overline{X}_n$, $\sigma(\overline{X}_n)$, as the standard deviation of the sampling distribution of $\overline{X}_n$. The Standard Error Law for $\overline{X}_n$ states that the value of $\sigma(\overline{X}_n)$ is a function of the population standard deviation and sample size. Specifically, $\sigma(\overline{X}_n) = \sigma/\sqrt{n}$. From this, it follows that the standard error of $\overline{X}_n$ decreases as the sample size increases. We define the standard error of b, denoted by $\sigma(b)$, as the standard deviation of the sampling distribution of b. Again, we find that the standard error decreases as the sample size increases.

> **THE STANDARD ERROR LAW FOR b**   The size of the standard error of a sample regression coefficient *b* is influenced by a number of factors, but it always decreases as sample size increases.

Because the true average response rate β is the mean (expected value) of the sampling distribution of b, $\sigma(b)$—the standard deviation of the sampling distribution of b—measures the "average" spread of the scores for b about β. As such

> $\sigma(b)$ measures the "average" error we will make when we use the sample regression coefficient b to estimate the true average response rate β.

The Standard Error Law tells us that the larger the sample size is, the smaller $\sigma(b)$ will be. So the precision of our estimate of β will improve as sample size increases.

Like the value of $\sigma(X_n)$, the value of $\sigma(b)$ is seldom known in advance of sampling and therefore must be estimated from the sample data.[1] The estimate of $\sigma(b)$ will be denoted by $\hat{\sigma}(b)$ and will be called the *estimated standard error of b*.

the *t* distribution law for a sample regression coefficient

We have previously defined the *t* score of a sample mean as

$$t_{\overline{X}_n} = \frac{\overline{X}_n - \mu}{\hat{\sigma}(\overline{X}_n)}$$

where t_{X_n} tells us the number of estimated standard errors between the sample mean and the population mean.

The *t* score of a sample regression coefficient b is defined analogously as

$$t_b = \frac{b - \beta}{\hat{\sigma}(b)}$$

and represents the number of estimated standard errors between the sample regression coefficient and the population regression coefficient.

The *t Distribution Law* for b describes the distribution of possible *t* scores t_b in random sampling.

> *t* **DISTRIBUTION LAW for** *b*   **If** *noise* **is uncorrelated with the predictors, then as sample size** *n* **increases, the shape of the sampling distribution of** t_b **approaches the shape of a** *t* **distribution with degrees of freedom** *df* **equal to** *n* **minus the number of population regression coefficients (**β**'s) in the model.**

When inferences are made concerning a population mean μ, the sample size required for valid use of the *t* distribution

[1] To know the *true* standard errors, our data set would have to be the population of inference—not just a sample from that population.

depends on the shape of the population of inference. When inferences are made in regression analysis, the required sample size depends on the shape of the distribution of *noise* scores. We will discuss these points in greater detail later in the chapter. At this point, we should remember that in practice

> If all variables having measurable impact on the dependent variable are represented as predictors in the model (so that *noise* is merely a composite of weak, unrelated forces) and if the degrees of freedom (*df*) equal 15 or more, it is generally safe to use the *t* distribution to draw inferences about the β's from the *b*'s.

Returning to our illustrative example, the *t* Distribution Law tells us that

1. The *t* score of the sample regression coefficient on *lat*

$$t_{b_1} = \frac{b_1 - \beta_1}{\hat{\sigma}(b_1)}$$

is a score from a *t* distribution with $df = 26$. There are 30 cities ($n = 30$) and 4 parameters — β_0, β_1, β_2, and β_3 — in the model, so $df = 30 - 4 = 26$.

2. The *t* score of the sample regression coefficient on *long*

$$t_{b_2} = \frac{b_2 - \beta_2}{\hat{\sigma}(b_2)}$$

is a score from a *t* distribution with $df = 26$.

3. The *t* score of the sample regression coefficient on *elev*

$$t_{b_3} = \frac{b_3 - \beta_3}{\hat{\sigma}(b_3)}$$

is a score from a *t* distribution with $df = 26$.

Once we have obtained numeric values for the *b*'s and $\hat{\sigma}(b)$'s, we can utilize the *t* distribution with $df = 26$ both to form confidence intervals for the β's and to test hypotheses about the values of the β's.

EXERCISES

11-6   For each of the following cases, determine the number of degrees of freedom associated with the t scores of the sample regression coefficients:

	SAMPLE SIZE	SAMPLE REGRESSION EQUATION	df
(a)	$n = 20$	$\hat{Y} = b_0 + b_1X_1$	_____
(b)	$n = 25$	$\hat{Y} = b_0 + b_1X_1 + b_2X_2$	_____
(c)	$n = 60$	$\hat{Y} = b_0 + b_1X_1 + b_2X_2 + b_3X_3 + b_4X_4$	_____

*11-7   (*Continuation of Exercise 11-4, page 326*) The researcher randomly selects 44 cases from the parole board's records and uses these data to compute least squares estimates (the b's) of the population regression coefficients (the β's).

(a) Based on the Expected Value Law for b, what can we say about the b's as estimates of the β's?

(b) Based on the t Distribution Law for b, what can we say about the t score t_b of each regression coefficient?

*11-8   The correlation matrix given in Exercise 6-13 on page 156 is reproduced here, where

> $damage$ = Dollars of damage in a dorm per resident.
> gpa = The mean grade-point average of dorm residents.
> % $male$ = The percentage of dorm residents who are male.
> % $in\text{-}state$ = The percentage of dorm residents who were in-state students.

	DAMAGE	GPA	% MALE	% IN-STATE
DAMAGE	1.000			
GPA	−0.741	1.000		
% MALE	0.683	−0.656	1.000	
% IN-STATE	0.002	0.014	−0.027	1.000

(a) Suppose that you calculated the coefficients of the regression equation

$$\widehat{damage} = b_0 + b_1\, gpa$$

With % *male* excluded as a predictor, would you expect b_1 to provide an unbiased estimate of the true average rate of response of *damage* to changes in *gpa*? Support your answer.

(b) Now suppose that you performed the regression

$$\widehat{damage} = b_0 + b_1\, gpa + b_2\, \%\ male$$

As a consequence of the exclusion of % *in-state* as a predictor, would you expect b_1 and b_2 to provide seriously biased estimates of the true average response rates in *damage* to changes in *gpa* and % *male,* respectively?

11.3   confidence intervals for the true average response rates

Chart 11-2 supplies the computer output on the coefficients, b's, and on the estimated standard errors, $\hat{\sigma}(b)$'s, for our sample regression equation. Notice that instead of reading "Estimated Standard Error," the third column head reads "Standard Error." Because the true standard error is essentially never known, it is understood that when someone refers to a standard error, he or she is referring to an estimate. Throughout the remainder of the text, we will follow this convention.

With this information, we can construct confidence intervals for the true average response rates by using the same logic we employed to form a confidence interval for a population mean μ.

In the case of μ, a confidence interval is of the form

$$\overline{X}_n \pm t \cdot \hat{\sigma}(\overline{X}_n)$$

This interval is symmetric about the sample mean; the width on either side of the mean (that is, the margin for error) is the distance $t \cdot \hat{\sigma}(\overline{X}_n)$. The appropriate value of t is obtained from the t table (Appendix Table D) in the row given by $df = n - 1$ and the column that reflects the desired level of confidence.

chart 11-2

DEPENDENT VARIABLE: *temp*		
INDEPENDENT VARIABLE	REGRESSION COEFFICIENT b	STANDARD ERROR $\hat{\sigma}(b)$
Lat	−1.402	0.066
Long	0.057	0.027
Elev	−0.134	0.027
Intercept	98.420	

In the case of β, we have

$$\text{Confidence interval for } \beta = b \pm t \cdot \hat{\sigma}(b) \qquad (11\text{-}7)$$

which is a symmetric interval about the sample regression coefficient whose margin for error is the distance $t \cdot \hat{\sigma}(b)$. Again, the appropriate value of t is obtained from the t table in the row given by the number of degrees of freedom and the column that reflects the desired level of confidence.

PROBLEM 1 Compute and interpret a 95-percent confidence interval for β_1, the true average response rate of *temp* to changes in *lat*.

1. $b_1 = -1.402$ (from Chart 11-2)
2. $\hat{\sigma}(b) = 0.066$ (from Chart 11-2)
3. t at $df = 26$ and 95 percent confidence (two-tailed P value $= 0.05) = 2.056$

So

$$b_1 \pm t \cdot \hat{\sigma}(b_1) = -1.402 \pm 2.056 \cdot 0.066$$

$$= \{-1.266 \text{ to } -1.538\} \text{ degrees}$$

Interpretation: We are 95 percent confident that the average response to a 1° movement north, holding *long* and *elev* constant, is a reduction in *temp* of between 1.266° and 1.538°.

PROBLEM 2 Compute and interpret a 90-percent confidence interval for β_2, the true average response rate of *temp* to changes in *long*.

1. $b_2 = 0.057$
2. $\hat{\sigma}(b_2) = 0.027$
3. t at $df = 26$ and 90 percent confidence (two-tailed P value $= 0.10) = 1.706$

So

$$b_2 \pm t \cdot \hat{\sigma}(b_2) = 0.057 \pm 1.706 \cdot 0.027$$

$$= \{0.011 \text{ to } 0.103\} \text{ degrees}$$

Interpretation: We are 90 percent confident that the average response to a 1° movement west in *long*, holding *lat* and *elev* constant, is an increase in *temp* of between 0.011° and 0.103°.

PROBLEM 3 Compute and interpret a 99-percent confidence interval for β_3, the true average response rate of *temp* to changes in *elev*.

1. $b_3 = -0.134$
2. $\hat{\sigma}(b_3) = 0.027$
3. t at $df = 26$ and 99 percent confidence (two-tailed P value $= 0.01$) $= 2.779$

So

$$
\begin{aligned}
b_3 \pm t \cdot \hat{\sigma}(b_3) &= -0.134 \pm 2.779 \cdot 0.027 \\
&= \{-0.059 \text{ to } -0.209\} \text{ degrees}
\end{aligned}
$$

Interpretation: We are 99 percent confident that the average response to a 100-ft climb (1 unit of *elev*), holding *lat* and *long* constant, is a reduction in *temp* of between 0.059° and 0.209°.

EXERCISES

11-9

(*Continuation of Exercise 11-8, page 332*) The following model ultimately was proposed:

damage $= \beta_0 + \beta_1$ *gpa* $+ \beta_2$ % *male* $+ \beta_3$ % *in-state* $+$ *noise*

The least squares estimates of the β's and the estimated standard errors for a sample of 23 dormitories follow:

DEPENDENT VARIABLE: *damage*

INDEPENDENT VARIABLE	REGRESSION COEFFICIENT	STANDARD ERROR
Gpa	$b_1 = -10.889$	$\hat{\sigma}(b_1) = 3.832$
% *Male*	$b_2 = 0.050$	$\hat{\sigma}(b_2) = 0.022$
% *In-state*	$b_3 = 0.115$	$\hat{\sigma}(b_3) = 0.059$
Intercept	$b_0 = 29.091$	

(a) Interpret the value of b_1 as an estimate of β_1. *Note:* A one-unit change in *gpa* = one-tenth of a point; for example 3.0 to 3.1.

(b) Compute a 95-percent confidence interval for β_1. Interpret your result.
(c) Compute a 99-percent confidence interval for β_2. Interpret your result.
(d) Compute a 90-percent confidence interval for β_3. Interpret your result.

*11-10   Consider the population regression model

$$food = \beta_0 + \beta_1\ income + noise$$

where

food = Annual expenditures by U.S. families on food products.

income = Annual disposable personal income of U.S. families.

Both variables are measured in hundreds of millions of dollars. When the model is estimated based on the six-year time period 1971–1976, the least squares regression results are

$$\widehat{food} = -0.086 + 0.200\ income \qquad R^2 = 0.997$$
$$(0.008)$$

The number in parentheses represents the estimated standard error of the regression coefficient on *income*.

(a) What can you report with 99 percent confidence about the portion of additional *income* that tends to be spent on *food*?
(b) Comment on the validity of the method used to form the confidence interval in (a).

11-11   In Exercise 7-20 on page 200, we reported some results of a regression investigation of the TV viewing habits of American families. That investigation was based on the model

$$hours = \beta_0 + \beta_1\ income + \beta_2\ size + noise$$

where

hours = Number of TV hours watched during a test month.

income = Family income in thousands of dollars.

size = Number of individuals in the household.

The model was estimated from data collected in a random sampling of 64 families, with the following results (standard errors are in parentheses):

$$\widehat{hours} = 120 - 4.0\ income + 28.2\ size$$
$$\qquad\qquad (1.6) \qquad\qquad (5.8)$$

(a) Report a 90-percent confidence interval for β_1. What does your result indicate about the effect of an increase of \$1,000 in family income?
(b) Report and interpret a 90-percent confidence interval for β_2.

11-12   (*Continuation of Exercise 11-5, page 327*) Based on a sample of 12 trips and assuming the t distribution is applicable, confidence intervals for β_1 are:

90% confidence interval for $\beta_1 = \{19.564$ to $30.436\}$ cents

99% confidence interval for $\beta_1 = \{15.493$ to $34.507\}$ cents

where *cost* is measured in cents; *distance*, in miles.
(a) Interpret the 90-percent and 99-percent confidence intervals for β_1.
(b) Obtain the value of b_1, the point estimate of β_1, from either confidence interval.
(c) Determine the margin for error in each of the two confidence intervals and break down each margin for error into its two components: (1) the t value required for the specified level of confidence, and (2) the estimated standard error of b_1.
(d) Report a 95-percent confidence interval for β_1.

11.4   hypothesis tests on regression coefficients

Frequently, researchers want to test hypotheses rather than estimate true average response rates.

A researcher may want to know if a particular predictor does, in fact, have a measurable influence on the behavior of the dependent variable. In this case, the null hypothesis is $\beta = 0$ and the alternative hypothesis is $\beta \neq 0$. To claim that $\beta = 0$ is to claim that the true average response rate of Y to changes in X is 0, or more intuitively, that changes in X do not produce measurable changes in Y on the average.

The alternative hypothesis implies a two-tailed test because the direction of the variable's influence has not been specified. Alternatively, the researcher may have a theoretical reason to believe that a predictor influences the dependent variable in a specific direction, either positively or negatively. Again, the null hypothesis will be that $\beta = 0$, but the alternative hypothesis will be that $\beta > 0$ *or* that $\beta < 0$ (a one-tailed test).

We can imagine both situations arising in a study of the relationships between *temp* and locational variables.

one-tailed tests

We would expect *temp* to respond *negatively* to increases in *elev*, holding *lat* and *long* constant. (The higher the elevation, the colder the climate.) The sample results show $b_3 = -0.134$, which is consistent with this belief. But it can be argued that this result is due to chance variation in random sampling and that, in fact, $\beta_3 = 0$. To verify that β_3 is less than 0, we have to discredit the chance explanation.

To do so, we begin by stating the null hypothesis $\beta_3 = 0$. Then we follow the same logical procedure we applied to hypothesis testing regarding a population mean μ.

To test the null hypothesis, we attempt to determine whether or not chance is a reasonable explanation of the discrepancy between (1) the observed results and (2) the results that are expected to occur if the null hypothesis is true. We observed a sample regression coefficient b_3 equal to -0.134. If the null hypothesis that $\beta_3 = 0$ is true, according to the Expected Value Law for b we would expect to find that b_3 is close to 0. Therefore, we wish to measure the probability (the one-tailed P value) that b_3 can assume a value as far below 0 as $b_3 = -0.134$ when, in fact, $\beta_3 = 0$.

To determine the requisite P value, we begin, as in the t test for μ, by calculating a t score. This time we calculate the t score of the value found for b_3:

$$t_{b_3} = \frac{b_3 - \beta_3}{\hat{\sigma}(b_3)} = \frac{-0.134 - 0}{0.027} = \frac{-0.134}{0.027} = -4.963$$

The result tells us that our sample estimate $b_3 = -0.134$ lies nearly 5 standard errors—that is, nearly 5 $\hat{\sigma}(b)$—below 0. The probability that a value will lie this much below 0 when $\beta = 0$ is represented by the blue area in Chart 11-3.

Referring to the t table we see that the largest numeric entry in the table at $df = 26$ is 2.779. So a value of t that is more

Chart 11-3
t Distribution
with *df* = 26

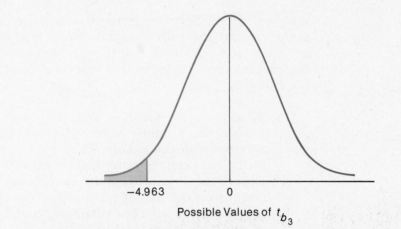

Possible Values of t_{b_3}

negative than -2.779 cannot occur by chance more than 5 in 1,000 times (one-tailed P value = 0.005). Because t_{b_3} is well below (more negative than) this value, we can conclude:

> If β_3 is equal to 0, we face substantially less than a 0.5-percent chance of obtaining a value for b_3 as far below 0 as $b_3 = -0.134$. The results are significant at the 0.5-percent level of significance. Accordingly, we can reject the claim $\beta_3 = 0$ with less than a 0.5 percent risk of rejecting a true claim. Holding *lat* and *long* constant, increases in *elev* tend to reduce *temp* on the average.

The t_b values for evaluating claims of the form $\beta = 0$ typically are supplied on computer printouts, as shown in Chart 11-4 for the data in our example. Each t_b listed is the number of

chart 11-4

DEPENDENT VARIABLE: *temp*			
INDEPENDENT VARIABLE	REGRESSION COEFFICIENT b	STANDARD ERROR $\hat{\sigma}(b)$	t SCORE t_b
Lat	−1.402	0.065	−21.372
Long	0.059	0.027	2.139
Elev	−0.134	0.027	−4.963
Intercept	98.420		
$R^2 = 0.948$			

standard errors between b and 0, since the computer always assumes that the null hypothesis is $\beta = 0$ unless otherwise specified. If $\beta = 0$, each $t_b = (b - \beta)/\hat{\sigma}(b) = b/\hat{\sigma}(b)$ (the ratio of the coefficient to its estimated standard error). So to evaluate the claim $\beta_3 = 0$ we:

1. Read t_{b_3}. $t_{b_3} = -4.963$.
2. Locate the numeric value of t_{b_3} (4.963) in the appropriate row of the t table. At $df = 26$, t_{b_3} is larger in absolute value than the entry for one-tailed P value = 0.005.
3. Report the P values of the entries bounding t_{b_3} as the bounds on the risk of rejecting a true claim. The claim $\beta_3 = 0$ can be rejected with less than a 0.5-percent risk of rejecting a true claim.

Turning to the relationship between *temp* and *lat*, we would expect an increase in *lat* to cause a reduction in *temp*, all other factors being equal. Since the sample regression coefficient on *lat*, $b_1 = -1.402$, lies more than 21 standard errors below 0 ($t_{b_1} = -21.372$), we can reject the claim $\beta_1 = 0$ and report that holding *long* and *elev* constant, increases in *lat* tend to reduce *temp* on the average.

a two-tailed test

The direction of response of temperature to increases in longitude (westerly movements) across the United States is not intuitively obvious. The evidence from our sample regression equation is that $b_2 = +0.057$; that is, among our sample of 30 cities, each 1° increase in *long* will increase *temp* by a predicted 0.057°, all other factors being equal. To see if this result could be due to chance, we will evaluate the claim $\beta_2 = 0$. This time our alternative hypothesis will be $\beta_2 \neq 0$, implying a two-tailed test. We will reject the null hypothesis if we find that b_2 is significantly different from 0 in either direction.

To evaluate the claim $\beta_2 = 0$, we:

1. Read t_{b_2}. $t_{b_2} = +2.139$. Our sample regression coefficient lies an estimated 2.139 standard errors above 0.
2. Locate the numeric value of t_{b_2} in the appropriate row of the t table. At $df = 26$, the value of t_{b_2} (2.139) lies between the entries for a two-tailed P value of 0.05 and a two-tailed P value of 0.02.

3. Report the P values of the entries bounding t_b as the bounds on the risk of rejecting a true claim. We can reject the claim $\beta_2 = 0$ with less than a 5-percent but more than a 2-percent chance of rejecting a true claim. Using a 5-percent level of significance, we are able to report that increases in *long* tend to produce systematic changes in *temp* on the average.

EXERCISES
11-13

The population regression model underlying the multiple regression investigation of *robbery rate* in Chapter 7 is

$$robbery\ rate = \beta_0 + \beta_1 \%\ low\ income + \beta_2\ density + noise$$

Least squares estimates of the model for data from 69 police precincts in New York City follow:

DEPENDENT VARIABLE: *robbery rate*

INDEPENDENT VARIABLE	REGRESSION COEFFICIENT	STANDARD ERROR	t SCORE
% *Low income*	3.9	1.3	3.000
Density	0.009	0.004	2.250
Intercept	−12.9		

(a) You wish to test the null hypothesis $\beta_1 = 0$ against the alternative hypothesis $\beta_1 \neq 0$. Explain what the null and alternative hypotheses assert.

(b) What is the meaning of the t score of 3.000 for the % *low income* coefficient? How would you have calculated this t score if its value had not been provided?

(c) Can you reject the null hypothesis without facing more than a 5-percent risk of rejecting a true claim? Support your answer.

(d) Is the coefficient on % *low income* significantly different from 0 at the 1-percent level of significance?

(e) Now you wish to test the null hypothesis that increases in *density* will not affect robbery rates on the average against the alternative hypothesis that increases in *density* will increase robbery rates on the average. State the null and alternative hypotheses symbolically.

(f) If you reject the null hypothesis in (e), what is the risk that you have rejected a true hypothesis? (Determine the one-tailed P value.)

11-14 Referring to the estimate of the food–income model in Exercise 11-10 on page 336:

(a) Determine the t score of the estimate of β_1 (that is, determine the t score of $b_1 = 0.200$) implied under the null hypothesis $\beta_1 = 0$.

(b) Assuming that the t distribution is applicable, what is the probability that we would find a sample regression coefficient as high as $b_1 = 0.200$ if *food* and *income* were uncorrelated?

(c) As a rule of thumb, families tend to spend another $0.22 on food products for each additional dollar of disposable personal income. State this rule of thumb as a null hypothesis.

(d) Calculate the t score of $b_1 = 0.200$ implied under the null hypothesis in (c).

(e) The alternative hypothesis is that families spend other than $0.22 on food products for each additional dollar of disposable personal income. State this hypothesis symbolically. Does the alternative hypothesis imply a one-tailed or a two-tailed test?

(f) Is the discrepancy between the sample evidence and the null hypothesis you stated in (c) statistically significant at the 10-percent level?

11-15 The Sudden Infant Death Syndrome *sids* is a term used to describe infant deaths of undetermined cause. Using 1958 data available from 35 states in the United States, researchers reported the following regression results:

For whites:

$$\widehat{sids\ rate} = -0.433 + 0.024\ low\ income + 0.219\ low\ birth\ weight$$
$$(P < 0.05) \qquad\qquad (P < 0.01)$$

For blacks:

$$\widehat{sids\ rate} = \ \ 1.711 + 0.098\ low\ income - 0.062\ low\ birth\ weight$$
$$(P < 0.01) \qquad\qquad (P > 0.20)$$

where

$sids\ rate$ = Number of sudden infant deaths per 1,000 live births.

$low\ income$ = Percent of families with incomes below $3,500.

> low birth weight = Percent of infants born who weighed less than 2,500 g (~5.5 lb).

The two-tailed P values are given in parentheses for each sample coefficient.
(a) What conclusions can you draw regarding the statistical significance of the effects of *low income* and *low birth weight* on *sids rate?*
(b) Is the *sids rate* of whites or blacks more sensitive to variation in *low income* among the states? Explain the relevant evidence.
(c) What additional information would you need to form a confidence interval for any one of the population regression coefficients?

11-16   (*Continuation of Exercises 11-4 and 11-7*) The least squares estimates of the model are presented below.

DEPENDENT VARIABLE: *PC* (measured in months)

INDEPENDENT VARIABLE	REGRESSION COEFFICIENT	STANDARD ERROR	t SCORE
Severity	3.140	1.224	2.565
Progress	−1.431	0.620	−2.308
Discipline	−0.465	0.383	−1.214
Risk	0.393	0.250	1.572
Intercept	5.457		

$R^2 = 55.0\%$

For each of the four predictors, let the null hypothesis be that $\beta = 0$.
(a) Based on your answer to Exercise 11-4(c) on page 326, state the alternative hypothesis symbolically.
(b) In each case, how plausible is the chance explanation for the discrepancy between the estimated value of β and the null value of 0? (Determine the appropriate P value.)
(c) *Severity* is measured on a scale of 1 to 6, where 6 represents the most severe offense. *PC* is measured in number of months. What can you say with 90 percent confidence about the change in *PC* on the average for each one-unit increase in *severity*, holding *progress, discipline,* and *risk* constant?
(d) Estimate the difference in *PC* for two offenders who have identical scores on *progress, discipline,* and *risk* if Offender 1 has committed a crime with a severity rating of 1 and Offender 2 has committed a crime with a severity rating of 6.

Caveat: The Limitations of the *t* Distribution in Regression Analysis

Our ability to use the *t* distribution to make inferences concerning a population mean depends on the shape of the distribution of the scores of the population of inference. Our ability to use the *t* distribution to make inferences in regression depends on the shape of the distribution of *noise* scores.

The *t* distribution yields exact probabilities (*P* values) only if the relevant population distribution is *normal*. If *noise* consists of many independent forces, each exerting a small influence on the dependent variable, then we can expect the distribution of *noise* scores to be approximately *normal*. In fact, these are the conditions we described in Chapter 4 when we were discussing why many distributions appear to be *normal* in shape.

A variable that is not explicitly mentioned in the regression model is, by definition, part of *noise*. Therefore if a variable that has a major impact on the dependent variable is ignored, *noise* will not be a composite of *small* forces and the distribution of *noise* scores may be very *non-normal*.

If a major variable has been excluded from the regression model, a very large sample size may be required before the *t* distribution can be employed.

More importantly, if the missing variable is correlated with one or more independent variables, the least squares regression coefficients (the *b*'s) will provide biased estimates of the true average response rates (the β's). In this case, use of the *t* distribution is unwarranted no matter how large sample size is.

Clearly, the usefulness of the *t* distribution for inference in regression is in doubt until we can make some statement about the composition of *noise* and especially about the possibility that it may include an important missing variable. Methods for analyzing the composition of *noise*, called *residual analysis,* are presented in textbooks that specialize in regression analysis.[2] However, our best insurance against invalid inferences lies in our theoretical understanding of the behavior of the dependent variable.

> All forces that are expected to influence the dependent variable should be included in the regression equation as predictors, even if their effects on the dependent variable are not specifically of interest to us.

COMING
TO
TERMS

Expected Value of b, $\mu(b)$ The average of all the possible scores for a sample regression coefficient in random samples of n observations.

Noise A term for the group of variables that are not included as predictors in a regression model because there is no theoretical reason to believe they exert a measurable influence on the dependent variable.

Number of Degrees of Freedom df The difference between sample size n and the number of β's in the population regression model.

Population Regression Coefficient β The true average rate at which the dependent variable responds to changes in the independent variable, holding all other independent variables constant.

Population Regression Model A theoretical statement in the form of an equation indicating the manner in which a dependent variable responds to changes in its external environment.

Predictor A variable that is considered to exert a measurable influence on the dependent variable.

Sampling Distribution of b The relative frequency distribution of all the possible scores for b that can result from random samples of n observations.

Standard Error of b, $\sigma(b)$ The standard deviation of the sampling distribution of b. The "average" error we will make when we use b to estimate β.

t Score of b, t_b The number of estimated standard errors between b and β.

---

[2] See, for example, John Neter and William Wasserman, *Applied Linear Statistical Models* (Homewood, Ill.: Richard D. Irwin, 1974).

1. The *temp* model (Equation 11-3 on page 324) used as the illustrative example in this chapter has been criticized for failing to include *windspeed* (units of miles per hour) as a predictor. "After all, don't people feel colder when the wind is blowing faster?"

   (a) Based on the belief that variation in *windspeed* will have a measurable effect on *temp,* revise the regression model given by Equation (11-3) to include *windspeed* as a predictor.

   (b) Assuming that *windspeed* does belong in the model as a predictor, discuss two distinct problems that would arise if the results of Equation (11-2) on page 322 were used to draw inferences about the true average response rates in *temp* to changes in the locational variables.

   (c) Applying the revised model to our random sample of 30 U.S. cities produces the following regression results:

   DEPENDENT VARIABLE: *temp*

INDEPENDENT VARIABLE	REGRESSION COEFFICIENT	STANDARD ERROR	*t* SCORE
Lat	−1.388	0.070	−19.683
Long	0.056	0.029	1.931
Elev	−0.131	0.028	−4.630
Windspeed	−0.186	0.313	−0.593
Intercept	100.230		

   $R^2 = 0.948$

   Based on these results, comment on the validity of the criticism made of the original *temp* model.

2. The County Transportation Authority (CTA) hired a consulting firm to develop a model suitable for forecasting the demand for bus transportation in the county. In one phase of the study, the consultant gathered annual data for the 20-year time period 1951–1970 on each of the following variables:

   tokens: The average daily number of bus tokens sold.
   pop: The population of the county.
   gas: The average price in cents of a gallon of gasoline.
   charge: The average charge in cents for a bus token.

   Over the period 1951–1970, both county population and bus-token charges increased dramatically, but gasoline prices remained remarkably stable.

(a) An important issue of the study was the sensitivity of bus ridership to increases in the charge for a bus token. All other factors being equal, an increase in *charge* was expected to result in a decline in *tokens*. But a simple regression of *tokens* on *charge* revealed that

$$\widehat{tokens} = 2{,}000 + 10.5\ charge \qquad (1)$$
$$(7.0)$$

where the standard error is in parentheses. Comment on the credibility of the estimated effect of *charge* on *tokens*.

(b) When *gas* and *pop* were included as predictors in addition to *charge*, the result was

$$\widehat{tokens} = 812 + 0.15\ pop + 0.01\ gas - 7.30\ charge \qquad (2)$$
$$(0.06) \qquad (0.02) \qquad (3.65)$$

Interpret the estimates of the respective effects of *pop*, *gas*, and *charge* on *tokens*.

(c) Using the results of Equation (2), report a 90-percent confidence interval for the true β coefficient on *pop*.

(d) Using the results of Equation (2), can you conclude that increases in *charge* will result in a statistically significant reduction in *tokens* on the average?

(e) Equation (2) produced reasonably accurate forecasts of the average daily number of tokens sold during 1971 and 1972, but in the face of a very sharp increase in gasoline prices in the following year, seriously underestimated the demand for bus transportation in 1973. Provide a possible explanation for this result.

PETITIONS FILED

Petitions containing 156,000 signatures were turned over to the Board of Canvassers yesterday by a group seeking to remove the gasoline surtax imposed during last year's session of the state legislature. The Board has 30 days to certify that the petition contains more than 124,800 valid signatures. To be valid, a signature can be no more than 90 days old at the time the petitions are filed and must be of a person registered to vote. If the petition is certified, the Secretary of State must call a statewide referendum within 60 days.

inferences based on a yes–no variable 12

"When you follow two separate chains of thought, Watson, you will find some point of intersection which should approximate to the truth."

The Disappearance of
Lady Frances Carfax

12.1 estimating a proportion

The headline on the opposite page describes the legal course of a petition drive in a midwestern state. A State Board of Canvassers is responsible for determining whether the petitions filed contain the minimum number of valid signatures required by state law to force a statewide referendum. In this case, 156,000 petitions were filed and the petition is to be certified if more than $124,800/156,000 = 0.80$, or 80 percent, of these signatures are valid.

The Board had to accomplish a massive task—validating 156,000 signatures—within the short time period of 30 days. Realizing that it would be impossible to check every signature on the petition against state voter registration lists, the Board adopted a sampling plan.

Utilizing a random digits table, the Board selected a simple random sample of 2,720 signatures from the population of the 156,000 signatures on the petition ($n = 2,720$; $N = 156,000$). It then completed the signature validation sheet shown in Chart 12-1, assigning each valid signature a score of 1 and each invalid signature a score of 0.

chart 12-1
Signature
Validation
Sheet

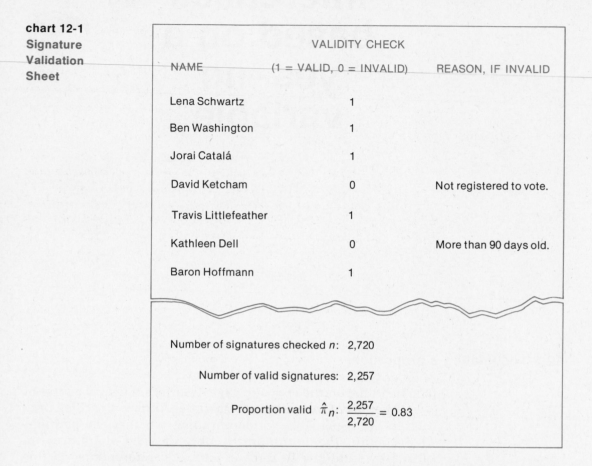

NAME	VALIDITY CHECK (1 = VALID, 0 = INVALID)	REASON, IF INVALID
Lena Schwartz	1	
Ben Washington	1	
Jorai Catalá	1	
David Ketcham	0	Not registered to vote.
Travis Littlefeather	1	
Kathleen Dell	0	More than 90 days old.
Baron Hoffmann	1	

Number of signatures checked n: 2,720

Number of valid signatures: 2,257

Proportion valid $\hat{\pi}_n$: $\dfrac{2,257}{2,720} = 0.83$

We will use the symbol π (lower-case Greek pi) to denote the true proportion of valid signatures and, more generally, to denote the true proportion of 1's in a population of inference where the possible values are 0 and 1. In our example, the population of inference is comprised of 156,000 signature scores and the petition is to be certified if the Board concludes that $\pi > 0.80$.

We will use the symbol $\hat{\pi}_n$ ("pi-hat sub n") to denote the proportion of 1's in a simple random sample of n scores where the possible values are 0 and 1. As shown in Chart 12-1, 2,257 of the sample of 2,720 signature scores that were checked for validity were found to be valid (1's). So

$$\hat{\pi}_n = \hat{\pi}_{2,720} = \frac{2,257}{2,720} = 0.83 \text{ (or 83.0\%)}$$

A sample proportion of 1's provides an estimate of the population proportion of 1's. Having found that $\hat{\pi}_{2,720} = 0.83$ and making due allowance for chance variation in random sampling, the Board inferred that the true proportion of valid signatures π was greater than 0.80. Consequently, the petition was certified.

We said that the Board made due allowance for chance variation. Like any other sample statistic, $\hat{\pi}_n$ is subject to chance variation from sample to sample. Even if π is less than 0.80, chance may produce a sample for which $\hat{\pi}_n > 0.80$. The Board, however, heeded the laws of chance variation in $\hat{\pi}_n$ in making its decision.

EXERCISES

12-1   In Chittenden County in 1974, 29 cases of rape were reported to the police. Of these, 9 offenders were prosecuted and 8 convictions were obtained. Determine the proportion of:
(a) Reported rape cases that resulted in prosecution.
(b) Reported rape cases that resulted in conviction.
(c) Reported rape cases that did *not* result in conviction.
(d) Prosecutions for rape that resulted in conviction.

12-2   (*Continuation of Exercise 12-1*) National law enforcement authorities estimate that only 1/5 to 1/20 of all occurrences of rape are reported to the police. Assuming that these fractions apply to Chittenden County, estimate the total number of rapes that occurred in this county during 1974.

*12-3   A simple random sample of the personnel records of 100 employees of a company employing a total of 22,000 people was audited to estimate the proportion of employees that exceeded the company's sick-day allotment last year. Let the value 1 denote a case of excessive sick leave and the value 0 denote a case of not excessive sick leave.
(a) What is the value of the population size N?
(b) What is the value of the sample size n?
(c) What would π denote in this problem?
(d) What would $\hat{\pi}_n$ denote in this problem?

12-4   A sample of the audience at a Jackson Browne concert during the summer of 1978 was asked: "Which of Jackson's five albums currently in circulation do you consider his best?" Here is a tabulation of the responses:

ALBUM	FAVORITE OF
"Saturate Before Using" (1972)	23
"For Everyman" (1973)	79
"Late for the Sky" (1974)	215
"The Pretender" (1976)	183
"Running on Empty" (1977)	200
No preference	100

(a) Let π denote the proportion of the entire audience who preferred "Running on Empty" to Jackson Browne's other albums. Estimate π from the tabulated responses.

(b) Let π denote the proportion of the entire audience who actually had a preference. Determine $\hat{\pi}_n$ from the tabulated responses.

(c) Estimate the proportion of those having a preference who preferred "Running on Empty."

12.2   laws of chance variation for sample proportions

In Chapter 2, we used (the term *yes–no variable* to define a variable whose scores fall into two categories.) This is true of the signature scores. Each signature is either valid (*yes*) or invalid (*no*). The Board of Canvassers has chosen the value 1 to denote a valid signature and the value 0 to denote an invalid signature, thereby compiling a distribution of 0, 1 scores.

(The key point we learned about a *yes–no* variable in Chapter 2 (page 36) is that

The proportion of 1's is the mean of a *yes–no* variable.

So $\hat{\pi}_n$, the proportion of 1's in a random sample of n scores, is just a special case of a sample mean $\overline{X}_n$, and all the laws of chance variation for $\overline{X}_n$ must apply to $\hat{\pi}_n$ as well.)

EXPECTED VALUE LAW FOR $\hat{\pi}_n$ For any sample size n, the expected value of the sample proportion of 1's, denoted by $\mu(\hat{\pi}_n)$, is equal to the population of 1's. Symbolically

$$\mu(\hat{\pi}_n) = \pi \quad) \qquad (12\text{-}1)$$

$\Big($ So $\hat{\pi}_n$ provides an unbiased estimate of π. The average of all the possible scores for the proportion of 1's in random samples of n observations will be equal to the true population proportion of 1's. $\Big)$

$\Big($ **THE STANDARD ERROR LAW FOR** $\hat{\pi}_n$ **The standard error of** $\hat{\pi}_n$, **denoted by** $\sigma(\hat{\pi}_n)$, **is equal to the standard deviation of the population of inference divided by the square root of the sample size, or**

$$\sigma(\hat{\pi}_n) = \frac{\sigma}{\sqrt{n}} \quad) \qquad (12\text{-}2)$$

$\Big($ So as sample size increases, $\hat{\pi}_n$ provides a more and more precise estimate of π. As in the case of $\overline{X}_n$, quadrupling the sample size will reduce the standard error to half its former value. $\Big)$

$\Big($ **THE LAW OF NORMALITY FOR** $\hat{\pi}_n$ **As sample size increases, the shape of the sampling distribution of** $\hat{\pi}_n$ **approaches the shape of a** *normal* **curve. If** π **lies between 0.2 and 0.8, the sampling distribution of** $\hat{\pi}_n$ **will be virtually** *normal* **in shape for** $n \geq 25.$ $\Big)$

The classic *yes–no* variable. The expected value of the proportion of *heads* (or tails) resulting from *n* flips of the coin is 0.5.

In Chapters 10 and 11, we were forced to use the t distribution instead of the *normal* distribution, because we did not know the value of the population standard deviation σ and, therefore, the value of the standard error. In the case of a sample proportion, however, the standard error has a special relationship to π — a relationship that will enable us to avoid using the t distribution.

the standard error of $\hat{\pi}$

To see how we might expect the standard error to be related to π, let's consider a few special cases.

$\Big($ Suppose that $\pi = 1$ (every signature in the population is valid). Then no matter what sample is chosen, all the signatures in the sample will be valid and $\hat{\pi}$ will be equal to 1. This implies that there is no variability in the sampling distribution of $\hat{\pi}_n$. The standard error of $\hat{\pi}_n$ will be 0. The same is true if $\pi = 0$ (all signatures are invalid).)

(If π is close to 1 (or 0), then $\hat{\pi}_n$ will vary, but the *yes* (or *no*) responses will dominate most samples and the standard error will not be large. On the other hand, if $\pi = 0.5$ (50 percent valid; 50 percent invalid), a wide variety of possibilities are likely and there is a good chance that $\hat{\pi}_n$ will under- or overestimate π by a sizable amount. In this case, the standard error is relatively large.)

(So intuitively we expect the size of the standard error to increase as the true proportion of 1's becomes closer to 0.5. The exact relationship is

$$\text{Standard error of } \hat{\pi}_n = \sigma(\hat{\pi}_n) = \sqrt{\frac{\pi(1-\pi)}{n}} \quad\big) \quad (12\text{-}3)$$

(If π is equal to either 0 or 1, then $\sigma(\hat{\pi}_n) = 0$ as expected. Also, by using Equation (12-3) to find $\sigma(\hat{\pi}_n)$ for a series of values for π, we can verify that $\sigma(\hat{\pi}_n)$ increases as π gets closer to 0.5.)

In the next section, we will learn how the Board of Canvassers applied these laws of chance variation in making its decision to certify the gasoline surtax petition.

EXERCISES

12-5   (*Continuation of Exercise 12-3, page 351*) In the context of the sick-leave study:

(a) Explain the meaning of the concepts denoted by $\mu(\hat{\pi}_{100})$ and $\sigma(\hat{\pi}_{100})$.

(b) Complete the following chart:

If $\pi =$	Then $\mu(\hat{\pi}_{100}) =$	and $\sigma(\hat{\pi}_{100}) =$
0.5	_____	_____
0.4	_____	_____
0.3	_____	_____
0.2	_____	_____
0.1	_____	_____

12-6   (a) Calculate the "average" error you would make if you used $\hat{\pi}_n$ to estimate π when, in fact, $\pi = 0.5$ and

(1) $n = 25$

(2) $n = 100$

(3) $n = 10,000$

(b) If $\pi = 0.5$, how large a sample size would you have to obtain to make $\sigma(\hat{\pi}_n) = 0.01$ (that is, one percentage point)?

12-7

(a) Calculate the value of $\sigma(\hat{\pi}_n)$ if:
  (1) $\pi = 0.25$ and $n = 64$
  (2) $\pi = 0.75$ and $n = 64$
(b) Why are the two results you obtained in (a) identical?
(c) To what fraction of its value in (a) is $\sigma(\hat{\pi}_n)$ reduced if sample size is increased ninefold from 64 to 576?

12-8

The National Auto Theft Bureau wishes to estimate π, the proportion of car thefts recorded last year that involved unlocked cars.
(a) If $\pi = 0.2$, calculate $\sigma(\hat{\pi}_n)$, the "average" error that would be made if π were estimated on the basis of a simple random sample of:
  (1) $n = 400$ car thefts
  (2) $n = 1,600$ car thefts
(b) Within one set of axes, sketch the frequency curves of:
  (1) the sampling distribution of $\hat{\pi}_{400}$.
  (2) the sampling distribution of $\hat{\pi}_{1,600}$.
  Indicate the location of π ($\pi = 0.2$) and the distances represented by $\sigma(\hat{\pi}_{400})$ and $\sigma(\hat{\pi}_{1,600})$.

12-9

Of a random sample of senior citizens 65 years and older, 70 percent were opposed to a mandatory retirement law.
(a) Can you conclude from the Expected Value Law for $\hat{\pi}_n$ that *exactly* 70 percent of the population of senior citizens are opposed to mandatory retirement? Explain your answer.
(b) Can you conclude from the Standard Error Law for $\hat{\pi}_n$ that the "average" error made in using the sample proportion $\hat{\pi}_n = 0.70$ to estimate π would be reduced by half if the sample size were doubled? Explain your answer.
(c) Can you conclude from the Law of Normality for $\hat{\pi}_n$ that the distribution of scores for $\hat{\pi}_n$ is *normal* in shape? Explain your answer.

12.3   a hypothesis test for π

Recall that for certification of the petition, the Board must establish that π, the proportion of valid signatures, is more than 0.80. Its sample evidence is $\pi_{2,720} = 0.83$. But can a value for $\hat{\pi}_{2,720}$ as high as 0.83 arise merely by chance when π is actually no more than 0.80? To determine if this possibility could be ruled out beyond a reasonable doubt, the Board performed a hypothesis test.

Null hypothesis: $\pi \leq 0.80$
Alternative hypothesis: $\pi > 0.80$

Method: Determine the probability (the one-tailed P value) that $\hat{\pi}_{2,720} \geq 0.83$ if the null hypothesis is true. To do so, assume that $\pi = 0.80$, the cutoff point between the null and alternative hypotheses.

If $\pi = 0.80$, then the possible values of $\hat{\pi}_n$ form a *normal* distribution (Law of Normality) centered on $\pi = 0.80$ (Expected Value Law), with a standard error of

$$\sqrt{\frac{0.80\,(1.00 - 0.80)}{n}} \qquad \text{(Standard Error Law)}$$

These implications of the laws of chance variation are illustrated in Chart 12-2, where the P value of interest – the probability that $\hat{\pi}_{2,720} \geq 0.83$ if $\pi = 0.80$ – is represented by the blue area under the *normal* curve.

To implement the laws of chance variation, the Board must

1. Compute $\sigma(\hat{\pi}_{2,720})$.
2. Determine the Z score of $\hat{\pi}_{2,720}$.

chart 12-2

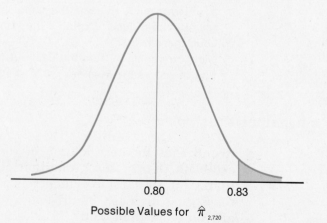

Possible Values for $\hat{\pi}_{2,720}$

1. Computation of $\sigma(\hat{\pi}_{2,720})$:

$$\sigma(\hat{\pi}_{2,720}) = \sqrt{\frac{0.80\,(1.00 - 0.80)}{2,720}}$$

$$= 0.008$$

2. With this knowledge, the Z score for $\hat{\pi} = 0.83$ can be calculated as

$$Z_{\hat{\pi}_n} = \frac{\hat{\pi}_n - \pi}{\sigma(\hat{\pi}_n)}$$

$$= \frac{0.83 - 0.80}{0.008}$$

$$= 3.75$$

The *normal* table tells us that the probability of obtaining a Z score as high as 3.50 is 0.0002 (0.5000 − 0.4998). So a Z score as high as 3.75 can arise by chance less than 0.02 percent of the time. Even at the 0.02 percent level of significance, the null hypothesis could be rejected.

If, in fact, $\pi = 0.80$, then there is less than a 2 in 10,000 chance that the $\hat{\pi}$ value of a random sample of 2,720 signatures will be as large as $\hat{\pi} = 0.83$. The null hypothesis that $\pi \leq 0.80$ can be rejected with only a negligible risk of rejecting a true claim.

$\big($ Let's review the steps we just took to test a hypothesis about π: $\big)$

$\big($ 1. State the null and alternative hypotheses. $\big)$

Null: $\pi \leq 0.80$
Alternative: $\pi > 0.80$

$\big($ 2. Calculate $\sigma(\hat{\pi}_n)$ at the cutoff point of the null hypothesis. $\big)$
The cutoff point is 0.80 and, with $n = 2,720$

$$\sigma(\hat{\pi}_{2,720}) = \sqrt{\frac{0.80\,(1.00 - 0.80)}{2,720}} = 0.008$$

3. Calculate the sample proportion of 1's and the Z score of the sample proportion of 1's under the null hypothesis:

$$\hat{\pi}_{2,720} = 0.83 \qquad Z_{0.83} = \frac{0.83 - 0.80}{0.008} = 3.75$$

4. Determine the appropriate P value as an area under the *normal* curve:

$$P \text{ value} < 0.0002$$

5. Report the test results. The sample results $\hat{\pi} = 0.83$ are statistically significant at the 0.02-percent level. The null hypothesis is rejected.

determination of sample size

The Board of Canvassers selected a simple random sample of $n = 2{,}720$ signatures. Why was this number chosen instead of a larger or a smaller sample size? The answer lies in the risks the Board was willing to take that it would make an incorrect decision based on the sample evidence.

The Board was concerned about the risks of making two types of errors:

1. The risk of certifying a petition that actually contained fewer than the minimum number of valid signatures required by state law. This is equivalent to the risk of concluding that $\pi > 0.80$ when, in fact, $\pi \leq 0.80$. In terms of the hypothesis test presented in the previous section, this risk is the risk of Type I error: rejecting the null hypothesis $\pi \leq 0.80$ when it is true.
2. The risk of failing to certify a petition that actually contained a legally sufficient number of valid signatures. This risk is the risk of Type II error: failing to reject the null hypothesis $\pi \leq 0.80$ when it is false.

As we learned in Chapter 10, one way to reduce the risks of both Type I and Type II errors is to increase sample size. Although the procedure used to calculate the sample size required to achieve acceptable levels of Type I and Type II errors is beyond the scope of this text, the Board computed this sample size to be $n = 2{,}720$ signatures. With a smaller sample size, the risks of reaching incorrect decisions would be unacceptably high; with a larger sample size, time and effort would be spent unnecessarily.

12-10   Television networks typically project a winner in a political contest long before the votes have been completely tabulated. In the current election, the Democrat is favored to win over the Republican opponent. You are watching the broadcast of the evening election returns, but you would like to go to sleep as soon as you are confident of the outcome of the race.

(a) State the assertion that the Democrat will *not* win as a null hypothesis about π, the final proportion of votes for the Democrat. Then state the alternative hypothesis. (*Note:* It takes more than 50 percent of the vote to win.)

(b) At 9 P.M., a newscaster announces that 100 ballots have been counted and that the Democrat has received 55 percent of the votes. If you conclude that the Democrat has won—that is, if you reject the null hypothesis at this point and go to sleep—what risk do you face of waking up the next morning to find that the Republican has been elected? (Assume that the 100 ballots counted are a simple random sample of the ballots cast.)

(c) At 10 P.M., the newscaster announces that 2,500 ballots have been counted and that 51.6 percent of the votes favor the Democrat. How probable is it that these results would arise if the null hypothesis were true?

(d) At 11 P.M., you hear the newscaster say that 10,000 votes have been tabulated, but you do not hear what proportion of these votes has been cast for the Democrat. What proportion $\hat{\pi}_{10,000}$ would lead you to reject the null hypothesis and go to sleep at this point, if you are willing to accept a 1-percent risk of waking up to find that the Republican has been elected?

*12-11   A recent advertisement proclaimed that the majority of the Vermonters who tried new Groovy Griddles Syrup—a mixture of sugars, corn syrup, artificial flavoring, and artificial coloring—said they preferred its taste to that of "real maple syrup." More specifically, 54 out of a sample of 100 Vermont residents allegedly expressed a preference for Groovy Griddles Syrup. Assume that these 100 Vermonters represent a random sample of "real maple syrup" users (the ad's implication).

(a) If π denotes the true proportion of "real maple syrup" users who actually prefer the taste of Groovy Griddles Syrup, does the sample evidence $\hat{\pi}_{100} = 0.54$ permit you to reject the null hypothesis that $\pi \leq 0.50$ in favor of the alternative hypothesis that $\pi > 0.50$?

(b) How large must $\hat{\pi}_{100}$ be for you to reject the null hypothesis that $\pi \leq$ 0.50 without facing more than a 1-percent risk of rejecting a true claim?

12-12   Consider the null hypothesis that $\pi = 0.40$ and the alternative hypothesis that $\pi \neq 0.40$ (implying a two-tailed test). In each of the following cases, indicate whether the discrepancy between the sample evidence and the null hypothesis is:
(1) Statistically significant at the 1-percent level.
(2) Statistically significant at the 5-percent level but not at the 1-percent level.
(3) Statistically significant at the 10-percent level but not at the 5-percent level.
(4) Not statistically significant at the 10-percent level.

(a) $\hat{\pi}_{25} = 0.60$
(b) $\hat{\pi}_{25} = 0.25$
(c) $\hat{\pi}_{64} = 0.47$
(d) $\hat{\pi}_{225} = 0.31$

12-13   A candy store advertises that its nut mixture is 80 percent peanuts and 20 percent cashews (the more expensive type of nut). In preparing the mixture, the proprietor faces two concerns. If the proportion of cashews π is lower than advertised, he risks being accused of cheating the public. If the proportion of cashews π is higher than advertised, he will earn a lower than expected profit on the sale of the nut mixture. Each morning, the proprietor takes a scoop of nuts (selects a simple random sample) from the bin and counts the number of each type of nut in the sample.
(a) Suppose that one morning the scoop contains 27 cashews and 54 peanuts. Should this finding lead the proprietor to conclude that $\pi \neq 0.20$? (Conduct a two-tailed test.)
(b) Describe the potential harm that could result if the proprietor commits a Type I error.

12-14   Company management decides to increase promotional expenditures if its product Clean Up appears to be used in less than 30 percent of all households. A sample survey of 400 households is commissioned to obtain evidence for testing the hypothesis that the true proportion of households using Clean Up is at least 30 percent ($\pi \geq 0.30$).
(a) Can management reject the null hypothesis on finding that 150 of the 400 households sampled are using Clean Up?

(b) Can management reject the null hypothesis on finding that 100 of the 400 households sampled are using Clean Up?

(c) Suppose that as a result of the test performed, the company commits a Type II error. What harm would be done?

(d) Suppose that as a result of the test performed, the company commits a Type I error. What harm would be done?

12.4 confidence intervals for π

$\Big($We have seen how the laws of chance variation in $\hat{\pi}_n$ are applied in a hypothesis test concerning the value of π. These laws also give us the necessary information to calculate a confidence interval for π when appropriate, as it is in the following problem.$\Big)$

Union organizers were trying to decide whether it would be worthwhile to lobby for a faculty union at a certain state university. They anticipated that an informational campaign would attract support from faculty members who currently were not in favor of unionization. But if too large a proportion of the faculty were opposed to unionization, then the union's effort would be better spent at another institution.

To estimate the proportion of faculty members currently favoring the union, a simple random sample of 100 faculty members was interviewed. Of the 100, 42 responded *yes* and 58 responded *no*, resulting in $\hat{\pi}_{100} = 0.42$. Recognizing that the actual proportion π was not likely to be 0.42, union officials decided to calculate a 99-percent confidence interval for π.

$\Big($If n is sufficiently large, the laws of chance variation for $\hat{\pi}_n$ permit us to express a confidence interval for π as

$$\hat{\pi}_n \pm Z \cdot \sigma(\hat{\pi}_n) \qquad (12\text{-}4)$$

where $Z \cdot \sigma(\hat{\pi}_n)$ is the margin for error. The value of Z can be determined from the *normal* table.$\Big)$ For 99 percent confidence, we find $Z = 2.58$. $\Big($From Equation (12-3), we know that $\sigma(\hat{\pi}_n) = \sqrt{\pi(1-\pi)/n}$. In hypothesis testing, we used this equation and the value for π assumed to be true under the null hypothesis to determine $\sigma(\hat{\pi}_n)$. In confidence interval estimation, no value for π is specified in advance, so we need a new approach. First we'll consider the *conservative* confidence interval for π.$\Big)$

(When we introduced the relationship between $\sigma(\hat{\pi}_n)$ and π, we noted that the standard error becomes larger as π approaches 0.5.

For any sample size, the standard error of $\hat{\pi}_n$, $\sigma(\hat{\pi}_n)$, reaches a maximum value when π is equal to 0.5. This maximum value is

$$\sqrt{\frac{0.5(1-0.5)}{n}} = \frac{0.5}{\sqrt{n}.}$$

In this problem, $n = 100$, so the maximum value of $\sigma(\hat{\pi}_n)$ is $0.5/\sqrt{n} = 0.5/\sqrt{100} = 0.05.$ (Respecifying Equation (12-4), substituting $0.5/\sqrt{n}$ for $\sigma(\hat{\pi}_n)$, we obtain

$$\hat{\pi}_n \pm Z \cdot \frac{0.5}{\sqrt{n}} \qquad (12\text{-}5)$$

Equation (12-5) is called a *conservative confidence interval for* π, because it assumes that the standard error $\sigma(\hat{\pi}_n)$ is at its maximum value.

To form a conservative confidence interval for π, let $\sigma(\hat{\pi}_n) = 0.5/\sqrt{n}$, and determine the Z score from the *normal* table that supplies the desired level of confidence. The *actual* level of confidence will be *at least* the confidence level desired.)

In the faculty union example, a conservative 99-percent confidence interval for the true proportion of the faculty that currently plans to vote *yes* is

$$\hat{\pi}_{100} \pm 2.58 \cdot \frac{0.5}{\sqrt{100}} = 0.42 \pm 0.129$$
$$= \{0.291 \text{ to } 0.549\}$$

"We can be *at least* 99 percent confident," the pollster reported, "that between 29.1 percent and 54.9 percent of the faculty would have voted *yes* on the issue if the vote had been taken at the time of the survey."

CHIEF JUSTICE BURGER ESTIMATES THAT 20-50% OF TRIAL LAWYERS ARE INCOMPETENT. CHECK THE PROFESSIONS WHERE THAT IS NOT TRUE.

□ MECHANIC   □ JUDGE   □ POLITICIAN   □ PROFESSOR   □ JOURNALIST   □ BUSINESSMAN

□ ATHLETE   □ DOCTOR   □ CHEF   □ CRIMINAL   □ CARTOONIST   □ OTHER

an approximate confidence interval for π

(The standard error component of the conservative confidence interval for π is found by letting $\pi = 0.5$ and computing $\sigma(\hat{\pi}_n) = 0.5/\sqrt{n}$. Alternatively, once we select our sample and compute $\hat{\pi}_n$, we can *estimate* the standard error, substituting $\hat{\pi}_n$ for π. The *estimated standard error* $\hat{\sigma}(\hat{\pi}_n)$ is then

$$\hat{\sigma}(\hat{\pi}_n) = \sqrt{\frac{\hat{\pi}_n(1 - \hat{\pi}_n)}{n}} \ ) \tag{12-6}$$

Following this procedure, the union officials would find

$$\hat{\sigma}(\hat{\pi}_{100}) = \sqrt{\frac{0.42\,(0.58)}{100}} = 0.049$$

The 99-percent confidence interval is then

$$0.42 \pm 2.58 \cdot 0.049 = 0.42 \pm 0.126$$
$$= \{0.294 \text{ to } 0.546\}$$

This result is slightly more precise than the conservative 99-percent confidence interval {0.291 to 0.549}, but this time the confidence level is only *approximately* 99 percent, rather than *at least* 99 percent. For this reason, we call a confidence interval based on an estimated standard error an *approximate confidence interval.*

precision and
sample size

One advantage of the conservative confidence interval is that it permits us to determine the required sample size for desired confidence–precision goals *in advance of sampling.*

For example, suppose that the union officials had wanted to obtain a confidence interval for π that:

1. carries the 99-percent confidence level and
2. allows a margin for error of, at most, 5 percentage points.

When dealing with means in general, we can find the required sample size by solving the following equation for n:

$$\text{Margin for error} = Z \cdot \frac{\sigma}{\sqrt{n}} \qquad (12\text{-}7)$$

We introduced this equation in Section 9.4 in the context of determining a confidence interval for μ.

When dealing with the mean of a *yes–no* variable — that is, with a proportion of 1's — we use the analagous equation

$$\text{Margin for error} = Z \cdot \sqrt{\frac{\pi(1 - \pi)}{n}} \qquad (12\text{-}8)$$

or, conservatively

$$\text{Margin for error} = Z \cdot \frac{0.5}{\sqrt{n}}$$

Solving this equation for n gives us

$$n = \left(\frac{0.5 \cdot Z}{\text{Margin for error}} \right)^2 \qquad (12\text{-}9)$$

For 99 percent confidence, $Z = 2.58$ and the margin for error is set at 0.05 (5 percentage points). Substituting these values into Equation (12-9), we obtain

$$n = \left(\frac{0.5 \cdot 2.58}{0.05}\right)^2$$

$$n = 666$$

To be 99 percent sure of attaining a $\hat{\pi}_n$ within 5 percentage points of π, the union organizers would have had to interview 666 faculty members.

<table>
<tr><td>a correction
factor for
small populations</td><td>(Sometimes the sample size that is required to achieve the desired mix of confidence and precision is so large that the sample may include a substantial part of the scores in the population of inference. Previously, we have not concerned ourselves with this possibility. In situations involving inferences about means and regression coefficients, our samples usually are composed of less than 1 percent of the population.)</td></tr>
</table>

Even the Board of Canvassers' enormous sample of 2,720 signatures represented only 1.7 percent of the 156,000 signatures on the petition.

However, if the surveyors of faculty opinion are to achieve their confidence and precision goals, the sample size they require will be quite large in relation to the population size. To attain a margin for error of, at most, 5 percentage points, a sample of 666 faculty members would be necessary. Only the largest universities employ more than 2,000 faculty members. Even at this size, a sample of 666 would represent 33 percent of the population of scores.

When sample size n is more than a small fraction of population size N, a *correction factor* should be used in calculating $\sigma(\hat{\pi}_n)$. As a rule of thumb, the correction factor should be used when the sample size is more than 5 percent of the population size—that is, when $n/N > 0.05$.

Equation (12-10) supplies the value of the standard error using this correction factor:

$$\sigma(\hat{\pi}_n) = \sqrt{\frac{\pi(1-\pi)}{n}} \cdot \sqrt{\left(1 - \frac{n}{N}\right)} \tag{12-10}$$

(The expression $\sqrt{1 - (n/N)}$ is called the *small population correction factor.* Let's see how it works.

1. Suppose that the sample size is one-third of the population size. Then $1 - (n/\text{N}) = 1 - 1/3 = 2/3$. The square root of 2/3 is approximately 0.816. So the standard error will actually be only 81.6 percent of the value of $\sqrt{\pi(1-\pi)/n}$.
2. Suppose that the sample size is one-tenth of the population size. Then $1 - (n/\text{N}) = 1 - 1/10 = 9/10$. The square root of 9/10 is approximately 0.949. So $\sigma(\hat{\pi}_n)$ will be only 94.9 percent of the value of $\sqrt{\pi(1-\pi)/n}$.
3. If n is very small in relation to N, then $1 - (n/N)$ will be very close to 1 and the correction factor can be ignored.)

(The important point to remember is that using the correction factor, when it is applicable, *reduces* the standard error. In turn, the reduced standard error leads to a narrower (more precise) confidence interval for π. Alternatively, a lower standard error reduces the sample size required to achieve specific confidence and precision goals.)

The calculation of the required sample size, employing the correction factor, is a complex procedure and tables are available for this purpose. In our illustrative example, the faculty opinion pollster — to be 99 percent confident of obtaining a value for $\hat{\pi}_n$ within 5 percentage points of π from a population of $N = 2,000$ — would have to interview not quite 500 faculty members, rather than the 666 indicated earlier.

For if $n = 500$, we obtain from Equation (12-10),

$$\sigma(\hat{\pi}_n) = \sqrt{\frac{0.5^2}{500}} \cdot \sqrt{1 - \frac{500}{2,000}} = 0.019$$

For 99 percent confidence, $Z = 2.58$, leaving

$$\text{Margin for error} = 2.58 \cdot 0.019$$
$$= 0.049, \text{ or } 4.9 \text{ percentage points}$$

which is just within the allowable margin for error of 5.0 percentage points.

EXERCISES

12-15   (*Continuation of Exercise 12-11, page 359*) Only 64 out of the sample of 100 Vermonters had used "real (100 percent) maple syrup." The other 36 actually had used a blend of maple and other syrups. Of the former 64, 24 expressed a preference for Groovy Griddles Syrup. Of the latter 36, 30 expressed a preference for Groovy Griddles Syrup.

(a) Calculate $\hat{\pi}_{64}$ and $\hat{\pi}_{36}$. Explain what the values of these statistics estimate.

(b) Calculate a conservative 95-percent confidence interval based on the value you obtained for $\hat{\pi}_{64}$. Interpret your result.

(c) Calculate a conservative 95-percent confidence interval based on the value you obtained for $\hat{\pi}_{36}$. Interpret your result.

(d) Use your results in (b) and (c) to criticize the advertisement described in Exercise 12-11.

12-16   Last month the Bureau of Labor Statistics estimated that the nation's unemployment rate was 6.5 percent, with a margin for error of 0.2 percent at the 99-percent confidence level. Use this information to form a confidence interval for π, the nation's true unemployment rate last month.

12-17   During a survey of 144 randomly selected households receiving food stamps, 36 of the households were discovered to be ineligible for the federal food-subsidy program. What can you report with 90-percent confidence about the true proportion of households receiving food stamps that are ineligible for the program?

12-18   Of 64 people who claimed to be chronic insomniacs and who reported taking over an hour to fall asleep, actual tests showed that

28 fell asleep within 30 min.
16 fell asleep within 30 to 60 min.

(a) If π denotes the true proportion of self-declared chronic insomniacs who actually take more than an hour to fall asleep, determine $\hat{\pi}_{64}$.

(b) Report an approximate 90-percent confidence interval for π.

(c) If π is the true proportion of self-declared chronic insomniacs who take at least 30 min to fall asleep, what is the approximate 90-percent confidence interval for π?

12-19   From interviews with a random sample of 625 pregnant women, researchers learned that 250 had been cigarette smokers prior to pregnancy but that 100 had given up cigarette smoking during pregnancy.

(a) Compute a conservative 90-percent confidence interval for π, the proportion of women smokers who give up smoking during pregnancy. Interpret your result.

(b) Compute an approximate 90-percent confidence interval for π as defined in (a). Interpret your result.

(c) Compute an approximate 90-percent confidence interval for the proportion of women who smoke during pregnancy. Interpret your result.

(d) Rank the confidence intervals you obtained in (a), (b), and (c) from narrowest (most precise) to widest (least precise). Explain the reasons for this result.

12-20   The Drug Enforcement Administration (DEA) wishes to estimate π, the proportion of heroin buys in which the purity level of the compound could be rated "high." How many buys must the DEA make (that is, determine the required sample size) to insure with 95-percent confidence that:

(a) Its estimate of π will be within 3 percentage points of π.

(b) Its estimate of π will be within 1 percentage point of π.

12-21   A simple random sample of 100 *yes–no* scores is selected from a population of N *yes–no* scores, resulting in $\hat{\pi}_{100} = 0.5$. For each of the following values of N, calculate:

(1) $\dfrac{n}{N}$

(2) The value of the small population correction factor.

(3) An estimate of $\sigma(\hat{\pi}_{100})$.

(a) $N = 200$

(b) $N = 500$

(c) $N = 2,000$

(d) $N = 10,000$

12-22   When a scout master asked 25 of his cub scouts whether their fathers also had been cub scouts, 17 answered "yes."

(a) Assuming that the 25 scouts represent a random sample from a very large scout troop, form a conservative 90-percent confidence interval for π, the true proportion of scouts in the troop who have followed in their fathers' footsteps.

(b) In fact, the scout troop has a membership of $N = 100$. Recompute the conservative 90-percent confidence interval for π, using the small population correction for $\sigma(\hat{\pi}_n)$.

(c) What effect does the small population correction factor have on the precision of the 90-percent confidence interval for π?

12.5 estimating a total number

(Frequently, our principal interest is not in estimating a *proportion* of 1's but in estimating the *total number of 1's.*)

When the Board of Canvassers validates a petition, it actually follows a two-stage process. First, even before sampling, signatures are checked for "face validity." For example, a signature without an address is considered invalid on its face. Second, a sample of the signatures that pass the test of face validity are checked against the state's voter registration lists. We discussed only this second stage in our examination of hypothesis testing in Section 12.3.

A statistical consultant to the Board suggested that time could be saved by checking only a *sample* of signatures for face validity instead of checking all signatures. This sampling procedure was adopted for a subsequent petition containing a total of 200,000 signatures.

Selecting a simple random sample of 1,600 signatures ($n = 1,600$), the Board found that 1,200 were face valid—a proportion of $\frac{1,200}{1,600} = 0.75$, or 75 percent. From this result, the Board wished to calculate a conservative 90-percent confidence interval for the *total number of face-valid signatures.* Here is how the Board proceeded.

1. It computed a conservative 90-percent confidence interval for the true *proportion* of the 200,000 signatures that were face valid:

$$0.75 \pm 1.64 \, \frac{0.5}{\sqrt{1,600}} = \{0.73 \text{ to } 0.77\}$$

So the Board could be at least 90 percent confident that between 73 percent and 77 percent of the signatures on the petition were face valid. The margin for error of 2 percentage points appears to be quite small.

2. The Board then extended this result to a conservative 90-percent confidence interval for the *total number* of face-valid signatures by multiplying the lower and upper bounds of the confidence interval for the *proportion* of face-valid signatures by the total number of signatures on the petition:

$$0.73 \cdot 200{,}000 = 146{,}000$$

$$0.77 \cdot 200{,}000 = 154{,}000$$

The Board could be at least 90 percent confident that between 146,000 and 154,000 signatures were face valid. Or—which is the same—the Board could be at least 90 percent confident that the total number of face-valid signatures fell within the interval of 150,000 ± 4,000.

Note that a margin for error of 2.0 percentage points in the confidence interval for the *proportion* of face-valid signatures becomes a margin for error of 4,000 signatures $(0.02 \cdot 200{,}000 = 4{,}000)$.

What seems like a negligible margin for error in a confidence interval for a proportion of 1's may in fact imply an unacceptably imprecise estimate of the total number of 1's if the population size is large.

EXERCISES

12-23   From a sample survey of senior citizens in a state, the following inference was drawn: "We can be 95 percent confident that between 41.2 and 46.4 percent of the eligible population (aged 65 and over) will file for a tax rebate under the provisions of the Property Tax Relief Act." Census data indicate that there are 850,000 senior citizens in the state. Form a conservative 95-percent confidence interval for the total number of senior citizens who plan to file for a tax rebate.

12-24   The Municipal Power Authority (MPA) wishes to estimate the number of households within its jurisdiction that would request installation of a "peak-load" meter, which automatically cuts off electrical power in the home for several hours during each day. In return for this inconvenience, the electricity rates of peak-load meter users would be reduced during nonpeak hours. From a random sample of 144 households, the MPA determined that 54 households would request meter installation.
(a) Compute an approximate 99-percent confidence interval for π, the true proportion of households that would request meter installation.
(b) Given 50,000 households in the community, compute an approximate 99-percent confidence interval for the total number of requests for the peak-load meter.

Caveats on the Use of the *Normal* Distribution

We have shown how the *normal* distribution can be applied both to test hypotheses and to form confidence intervals for π, a population proportion of 1's. However, to use the *normal* distribution we must obtain a certain minimum sample size.

Earlier, we said that if π is between 0.20 and 0.80, then a sample size of $n \geq 25$ permits us to apply the *normal* distribution safely. This statement is based on the more general theorem presented here.

The *normal* distribution provides a close approximation to the sampling distribution of $\hat{\pi}_n$ based on a simple random sample, provided that both:
(1) $n \cdot \pi \geq 5$
(2) $n \cdot (1 - \pi) \geq 5$

If π is within the range of 0.20 to 0.80, both conditions of this theorem will be satisfied when $n \geq 25$. For example, when $\pi = 0.20$ and $n = 25$

$$n \cdot \pi = 25 \cdot 0.20 = 5$$

and

$$n(1 - \pi) = 25 \cdot 0.80 = 20$$

To use the *normal* distribution when π is outside the range of 0.20 to 0.80, we must obtain a sample size larger than 25. For example, if $\pi = 0.10$, condition (1) in the theorem becomes

$$n \cdot 0.10 \geq 5$$

implying that

$$n \geq \frac{5}{0.10}$$

$$n \geq 50$$

When the sample size is less than the value of n required to use the *normal* distribution, inferences about π can still be made, but in place of the *normal* table, we should use tables of the *binomial* distribution. The *binomial* tables provide exact P values (and confidence levels) for inferences about π, regardless of sample size.[1]

In certain cases, the size of the simple random sample we must obtain to use the *normal* distribution is prohibitively large. Such situations occur commonly when the incidence of a rare event is to be estimated.

Suppose that we wish to estimate the current proportion of American males who are afflicted with cancer of the bladder. Past records indicate that cancer of the bladder affects only 8 out of every 100,000 males, or a proportion of 0.00008. It is a *rare event*.

If π is approximately equal to 0.00008, then the size of the simple random sample required to use the *normal* distribution is

$$n \cdot 0.00008 \geq 5$$

or

$$n \geq 62,500$$

It is usually not possible to obtain such a large sample. For this reason, when we wish to estimate the incidence of a rare event, generally we do not use a simple random sampling procedure to select our data. Such sampling techniques as *stratified random sampling* are used instead, because they permit valid inferences to be made about π from much smaller samples.

EXERCISE

The federal government periodically audits state welfare files. If an audit reveals that more than 3 percent of current welfare recipients are ineligible for state aid (usually because they exceed income limitations), the state must take corrective action or risk a cutback in federal funding. In one state, a random sample of 225 files ($n = 225$) was audited from more than 20,000 current records in the state. Out of the 225 files, 9 ineligible cases were discovered, resulting in $\hat{\pi}_{225} = 9/225 = 0.04$. State

---

[1] For more information on the binomial distribution, see George W. Snedecor and William G. Cochran, *Statistical Methods*, Sixth Edition (Ames, Iowa: Iowa State University Press, 1967).

officials wished to conduct a hypothesis test to determine the necessity of imposing corrective action. If the null hypothesis is that $\pi \leq 0.03$ and the alternative hypothesis is that $\pi > 0.03$:

(a) Can state officials use the *normal* distribution to perform this test? Explain your answer.

(b) If your response to (a) is "yes," determine the appropriate P value.

(c) Should the state take corrective action?

COMING
TO
TERMS

Approximate Confidence Interval for π A confidence interval for π that uses the estimated standard error for $\hat{\pi}_n$

$$\hat{\sigma}(\hat{\pi}_n) = \sqrt{\frac{\hat{\pi}_n(1 - \hat{\pi}_n)}{n}}$$

The resulting confidence interval is narrower than the confidence interval that is obtained if the conservative standard error is used, but the confidence level is only approximate.

Binomial Distribution   The distribution that can be used to obtain exact probabilities for the sampling distribution of $\hat{\pi}_n$, regardless of sample size.

Conservative Confidence Interval for π A confidence interval for π that uses the maximum possible standard error for $\hat{\pi}_n$, which is $0.5/\sqrt{n}$. Using a conservative confidence interval assures us that the confidence level is *at least as great* as the stated level.

Cutoff Point   In conducting a one-tailed hypothesis test about a population proportion, the cutoff point is the value that divides the null and the alternative hypotheses. The values of π on one side of the cutoff point are of no consequence—the null hypothesis. The values of π on the other side of the cutoff point are of consequence—the alternative hypothesis.

Population Proportion π In a population in which the observations can be represented by only two values, labeled 0 and 1, π is the number of 1's in the population divided by the population size N.

Sample Proportion $\hat{\pi}_n$ When a sample of n observations is taken on a variable whose possible values are 0 and 1, $\hat{\pi}_n$ is the number of 1 scores divided by the sample size n.

Small Population Correction Factor for the Standard Error   A multiplication factor to be used in calculating the standard error of $\hat{\pi}_n$ when the sample is more than 5 percent of the population size. If the correction factor is not used, the standard error of $\hat{\pi}_n$ will be overstated.

Standard Error of $\hat{\pi}_n$ The standard deviation of the sampling distribution of $\hat{\pi}_n$, or the "average" error that is made when $\hat{\pi}_n$ is used to estimate π. For a given n, $\sigma(\hat{\pi}_n)$ reaches its maximum when $\pi = 0.5$.

ISSUES FOR ANALYSIS

1. From a simple random sample of 1,600 scores representing lawyer fees charged last year for uncontested divorces, the following summary statistics were reported:

    Median: $800

    Interquartile range: ($1,200 − $500) = $700

    (a) Estimate the proportion of lawyers' fees for uncontested divorces that are:
       (1) $\geq$ $800
       (2) $\leq$ $500
       (3) $\leq$ $1,200
    (b) Calculate an approximate 95-percent confidence interval based on each proportion you reported in (a). Interpret each result.
    (c) Is the sample evidence consistent with the null hypothesis that less than 50 percent of the fees for uncontested divorces are higher than $500? Conduct the appropriate hypothesis test.

2. The manufacturer of shelf board for do-it-yourself bookcases (see Chapter 3, page 63) has adopted the following *quality control* procedure.

    From a production run of $N = 10,000$ boards, a simple random sample of $n = 100$ boards is selected and measured.
    (1) If less than 10 of the 100 boards fail to meet specifications, the full production run of 10,000 boards is stamped "Acceptable for Shipment." (The manufacturer considers a failure rate of less than 10 percent acceptable.)
    (2) If 10 or more of the 100 boards fail to meet specifications, the full production run of 10,000 boards is measured (an expensive process).
    (a) Suppose that a production run of $N = 10,000$ boards actually contains 500 boards that fail to meet specifications. What is the probability that the manufacturer will measure the full production run based on a sample of 100 boards?
    (b) Suppose that a production run of $N = 10,000$ boards actually contains 1,500 boards that fail to meet specifications. What is the probability that the manufacturer will stamp the run "Acceptable for Shipment" based on a sample of 100 boards?

The quality control procedure can be formulated as a hypothesis testing problem. Let the null hypothesis be that the actual failure rate is < 10 percent ($\pi < 0.10$, where π denotes the proportion of $N = 10,000$ boards in a production run that fail to meet specifications).

(c) Describe a decision that represents a Type I error. Describe a decision that represents a Type II error.

(d) What can you conclude about the manufacturer's risks of committing Type I and Type II errors under the conditions given in (a) and (b)?

4 comparing groups: the analysis of differences

Do Calculators Belong in Schools? Yes, Says Study; No, Say Parents

WASHINGTON — Should pocket calculators be used in the nation's elementary school classrooms?

Parents and mathematicians have been arguing the matter ever since minicalculator prices dropped down to the affordable $5–$10 range three years ago. Now, researchers are providing a tentative answer.

Dr. Jesse A. Rudnick, director of Temple University's Mathematics Education Action Research Center, has just finished compiling the results of a year-long study of the impact of adding calculators to the existing teaching curriculum of 350 seventh graders in West Chester, Pa. public schools. This group was compared with 350 seventh graders who did not have calculators.

"The students who used the calculators in the classroom were denied their use during tests," said Rudnick. "At the end of the year, we found no significant difference in the children's ability to calculate. . . . In fact, there was a little gain in favor of the experimental group, but it was not statistically significant.

"The conclusion we reached was, since there was no damage, the time is now ripe for adjusting the curriculum to take advantage of the machine."

Source: Carol R. Richards, "Do Calculators Belong in Schools? Yes, Says Study; No, Say Parents." Reproduced by permission of Gannett News Service, Washington, D.C.

two-group comparisons 13

13.1 drawing inferences from comparisons

Several years ago, a study was conducted to assess the performance of policewomen on the beat by comparing the felony arrest records of male and female officers during a 6-month time period. The results revealed that the policewomen had made fewer felony arrests on the average than the policemen, but that the policewomen had handled instances of violence as competently as the policemen.

A school district hired a statistician to evaluate the effectiveness of its experimental open classroom program. Among other factors, the consultant was asked to compare the standard test scores of children participating in this program with the test scores of children participating in the traditional, structured classroom program.

A razor blade manufacturer wished to test the durability of a newly developed blade—the Diamondedge. One group of men was asked to try Diamondedge, and a second group of men was asked to shave with a popular stainless steel blade. The test results revealed that on the average the men using the Diamondedge shaved twice as many times with each blade as the men using the stainless steel blade.

Each of these studies compared the *responses* (scores) of two groups of test subjects to evaluate the effect of some distinguishing feature. Classically, such a feature is called a *treatment.* The open classroom and the structured classroom are distinctive educational treatments, and the use of Diamondedge and the use of stainless steel razor blades are different shaving treatments. (The use of the term *treatment* is also extended to inherent subject characteristics.) So the sexes (male and female) of the police personnel whose arrest records were compared are termed *treatments* as well.

(Using this terminology, we can say that each of these studies was planned in the hope of drawing meaningful inferences about the differences between the effects of two treatments.)

the need
for unbiased
comparisons

In each case, however, the purpose of the comparison was defeated because the treatment difference was not the only feature that distinguished the two groups of test subjects.

A spokeswoman for the policewomen asserted that their poorer felony arrest record should not necessarily be attributed to sex—the treatment under comparison—but to experience. She indicated that the policewomen in the study averaged less than 7 months prior experience on the job, whereas the policemen in the study averaged more than 4 years prior experience. The groups differed inherently not only in sex but also in prior experience.

The school district consultant pointed out that a comparison of open and structured classroom programs was invalid because both the children and the teachers had been allowed to choose their own programs. The brighter students and the better teachers had selected the open classroom program. So the difference between the programs—the treatments—was not the only inherent difference between the two groups of test subjects.

The razor blade manufacturer wished to advertise the test results that made Diamondedge appear twice as durable as the stainless steel blade. The Federal Trade Commission (FTC) prohibited publication of the test results on discovering that the men who shaved with the stainless steel blade typically had heavier beards than the men who shaved with the Diamondedge blade. The manufacturer, the FTC alleged, had biased the test results in favor of the Diamondedge blade.

Two treatments.

(To provide a meaningful comparison of the effects of two treatments, two groups of test subjects must be selected who do not differ inherently in any manner that can alter their relative response to treatment.)

random sampling
and random
assignment

The school district consultant suggested that for the next year each child and teacher be *randomly assigned* to open or structured classroom programs instead of being allowed to choose between the two. Similarly the FTC told the razor blade manufacturer that subjects must be assigned randomly to treatments (Diamondedge versus stainless steel blades) before the test results could be included in advertising.

(The random assignment of test subjects to treatments is an extension of random sampling. Each subject is placed in one group or the other according to a chance mechanism—the luck of the draw—rather than according to the judgment of the researcher or the preferences of the subjects themselves.)

Subjects can be randomly assigned by many mechanisms. For example, each person can draw a card that specifies either Treatment A or Treatment B. Whatever mechanism is used, (the purpose of random assignment is to let chance spread similar subjects evenly between groups.)With random assignment, it is unlikely that tough beards will be concentrated in the stainless steel treatment group or that all the brighter students will be enrolled in the open classroom program.

(If test subjects are randomly assigned to treatment groups, the differences observed in group responses can be more readily attributed to treatment effects, since the groups should be similar in all other respects.)

Random assignment is not always possible. In the case of the felony–arrest comparison, the treatments under consideration are *being men versus being women*, and the researchers can merely observe the responses (arrests made) to treatment.

However, individuals still can be randomly selected from each group. If the policemen and policewomen have equivalent distributions of years of experience on the job, random samples of policemen and policewomen should be comparable in terms of prior experience. Unfortunately, at the time of the study, most policewomen were new members of the force, so that random sampling in itself would not have eliminated the potential bias due to varying job experience. In this case, in addition to random sampling, subjects would have to be matched by duration of experience. We will examine the technique of matched pairs in Section 13.6.

EXERCISES

13-1   A controversial issue has been whether or not capital punishment (the death penalty) is an effective deterrent to homicide. In several studies, homicide rates (the number of homicides per 100,000 people) in states that impose the death penalty were compared with homicide rates in states that do not impose the death penalty.

(a) What are the treatments being compared?

(b) What variable is being used to measure response to treatment?

13-2   To compare the nutritional value of two types of cereals for human consumption, 100 rats are randomly assigned to two groups. One group is fed Cereal A; the other, Cereal B.

(a) What are the treatments being compared?

(b) What are the test subjects?

(c) Why are the test subjects randomly assigned to treatment? How could random assignment be accomplished in this case?

(d) Given the purpose of the study, what assumption must be made to justify the choice of test subjects?

13-3   Is hospitalization in a private hospital more expensive on the average than hospitalization in a community hospital? To compare the costs per patient day between private and community hospitals, random samples of both types of institutions were selected.

(a) What are the treatments being compared?

(b) What are the test subjects?

(c) What variable is being used to measure response to treatment?

(d) Why is random assignment of subjects to treatment not possible in this study?

(e) Suppose that costs per patient day are highly correlated with hospital size. Could this correlation bias the proposed comparison of the two treatments?

13-4   In the midst of a severe regional drought, people were asked to adopt voluntary water conservation measures in their homes. To evaluate the effectiveness of the voluntary conservation program, a random sample of households was selected. Daily water-usage data for these households were acquired and compared with data on household daily water usage that were available from a random sample taken prior to the drought. What findings would you expect if the voluntary conservation program had been successful?

13-5   My neighbor and coworker claimed to have discovered a new route between our homes and office that would shorten our commute. Taking our usual route, the trip typically lasts 20 to 30 minutes. Taking the "shortcut" one morning, we timed the trip at 34 minutes, but in fairness, an automobile accident slowed us up. We wish to make a scientific comparison between the two routes. Each workday next week we will drive the usual route to work in the morning and the new route home in the evening. We will record the time of each trip and compare the average times.

Comment on this so-called scientific comparison.

13.2   the structure of hypothesis testing for comparisons

Throughout the 1960s, a common treatment for individuals suffering from hypertension (high blood pressure) was a drug compound called guanethidine sulfate. While successful in inducing antihypertensive effects (reducing high blood pressure), the drug caused adverse physical side effects, including nausea and diarrhea. Also in the 1960s, clinical trials were conducted on a new antihypertensive agent called guanadrel sulfate, which failed to produce these adverse side effects in animals. Medical researchers then sought to compare the effectiveness of guanadrel and guanethidine in the relief of hypertension in humans.

In one study, 40 patients being treated in a hypertension clinic at a major hospital were selected to produce a cross section of different ages, sexes, and severity of hypertension.

These patients were assigned to two equal-sized groups *at random.* One group was treated with guanethidine; the other, with guanadrel. The percentage change in each patient's blood pressure was recorded after about 4 months of therapy. To maintain objectivity, a *double blind* experiment was used: No patient knew which drug he or she was receiving (the first blind), and no researcher recording the responses knew which drug a patient was being given (the second blind).

Chart 13-1 lists the blood pressure responses recorded during the trial. A plus sign indicates an *increase* in blood pressure level.

The blood pressure of the patients treated with guanethi-

**chart 13-1
Responses to
Treatment**

GROUP A: GUANETHIDINE-TREATED		GROUP B: GUANADREL-TREATED	
PATIENT NO.	PERCENT CHANGE IN BLOOD PRESSURE X_A	PATIENT NO.	PERCENT CHANGE IN BLOOD PRESSURE X_B
1	0	21	−18
2	−12	22	− 9
3	− 4	23	+15
4	−29	24	+ 4
5	− 3	25	+ 9
6	−16	26	+ 2
7	−12	27	−15
8	−15	28	−24
9	−20	29	0
10	− 7	30	− 8
11	−12	31	−23
12	−11	32	−23
13	+ 9	33	+ 1
14	+ 6	34	− 8
15	−11	35	−18
16	−25	36	−16
17	− 3	37	−16
18	+ 1	38	−16
19	−22	39	+17
20	− 5	40	0

Summary Statistics

$\overline{X}_A = -9.55\%$ $\overline{X}_B = -7.30\%$

$\hat{\sigma}_A = 10.02\%$ $\hat{\sigma}_B = 12.64\%$

dine decreased by an average of 9.55 percent, whereas the blood pressure of the patients treated with guanadrel decreased an average of 7.3 percent. In this study, guanadrel was somewhat less effective on the average than guanethidine in treating high blood pressure. But can we infer from these findings that guanadrel will be less effective in general?

Not necessarily. The test subjects were randomly assigned to treatment. But (although random assignment reduces the threat of *bias*, sample responses can still differ due to *chance variation*.) The guanadrel-treated patients may have responded less favorably to treatment merely by chance.

(To infer that the two treatments differ in effectiveness, first we must rule out chance variation as an explanation for the observed differences. We must be able to reject the null hypothesis of no inherent difference between treatments.)

the null
hypothesis

When we introduced the concept of the null hypothesis in Chapter 10, we said that (*null* implies *of no consequence*. When we compare two treatments, therefore, we adopt the null hypothesis that any difference between the treatments is of no consequence.)

(Logically, the null hypothesis implies that if we administer treatments to two randomly acquired groups of patients, we should not expect to observe systematic differences between the responses of Group A and the responses of Group B.)

We should think of the blood pressure responses of the 20 guanethidine-treated patients as a random sample of scores from a population of potential responses to guanethidine. Similarly, we should think of the blood pressure responses of the 20 guanadrel-treated patients as a random sample of scores from a population of potential responses to guanadrel. In these terms, we can state the null hypothesis in the formal manner:

(**NULL HYPOTHESIS** The two samples of responses are random samples from populations of inference with identical distributions. If the frequency distributions of potential responses to guanethidine

chart 13-2
Distributions
of Potential
Responses to
Guanethidine
and Guanadrel:
The Null
Hypothesis

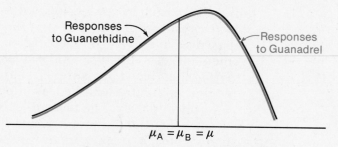

Responses to Guanethidine

Responses to Guanadrel

$$\mu_A = \mu_B = \mu$$

Percent Decrease in Blood Pressure

and guanadrel are superimposed on one another, the two distributions will appear as one (see Chart 13-2).)

the alternative hypothesis

The alternative to the null hypothesis, in its most general form, is that the difference between treatments is of consequence: The responses of the test subjects will differ systematically, depending on which treatment they receive.)

(ALTERNATIVE HYPOTHESIS   The two samples of responses are random samples from populations of inference whose distributions differ in some respect. If the frequency distributions of potential responses to guanethidine and guanadrel are superimposed on one another, two distinct frequency distributions will appear.)

(To test the null hypothesis statistically, the alternative hypothesis must specify the particular respect in which the treatments allegedly differ. In the guanadrel–guanethidine comparison, the alternative hypothesis might be that the effects of the treatments differ *on the average* — that is, differ in the average (mean) percentage by which they lower blood pressure — but that the variability and shape of the distributions are identical) This is illustrated in Chart 13-3.

chart 13-3
Distributions of Potential Responses to Guanethidine and Guanadrel: The Alternative Hypothesis

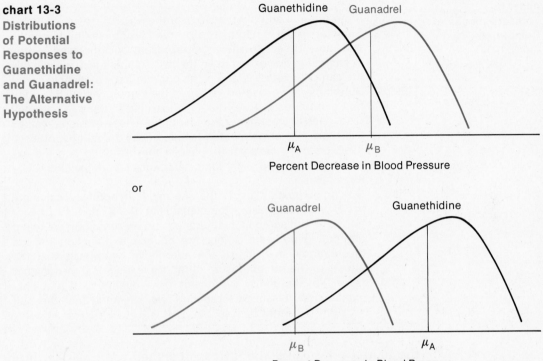

or

the testing
procedure

    In the one-sample hypothesis tests in Chapters 10–12, we followed a common logical process. We compared the observed value of a sample statistic to the value we would expect if the null hypothesis were true. If the discrepancy between the observed and the expected results favored the alternative hypothesis, we decided whether chance was a plausible explanation for the discrepancy by determining a P value. If chance was not a plausible explanation—if the P value was small—then we rejected the null hypothesis. If chance was a plausible explanation or if the discrepancy did not favor the alternative hypothesis, then we did not reject the null hypothesis.

    Tests to compare two treatments are performed in a very similar manner. The four comparison tests described in this chapter—the two-sample t test, the Wilcoxon rank sum test,

**chart 13-4
Logical Process of
Comparison Tests**

Step 1. State the null and alternative hypotheses.

Step 2. Summarize our samples of observed responses by summary statistics. These statistics will represent our observed results in sampling. For example, in the two-sample t test, we will use the sample means $\overline{X}_A$ and $\overline{X}_B$ as our summary statistics.

Step 3. Determine the expected values of the summary statistics if the null hypothesis is true (that is, if the distributions of potential responses to treatments are identical). These expected values will represent our expected results under the null hypothesis.

Step 4. Measure the discrepancy between the observed and the expected results and see if the discrepancy favors the alternative hypothesis.

Step 5. If the discrepancy favors the alternative hypothesis, calculate the P value to measure the plausibility of the chance explanation for the discrepancy. If chance can be ruled out, we reject the null hypothesis (reject the chance explanation for the discrepancy) in favor of the alternative hypothesis that the effects of the treatments do differ. If the discrepancy does not favor the alternative hypothesis or if chance *is* a plausible explanation for the discrepancy, we conclude that the null hypothesis cannot be rejected.

the paired t test, and the Wilcoxon signed rank test—will all follow the logical process outlined in Chart 13-4. Become familiar with these steps before reading further.

EXERCISES

13-6   (*Continuation of Exercise 13-1, page 382*) For the issue presented in Exercise 13-1:

(a) State the null hypothesis.

(b) State the alternative hypothesis.

(c) Does the alternative hypothesis warrant a one-tailed or a two-tailed test?

13-7   (*Continuation of Exercise 13-2, page 382*) Suppose that weight gain is the measure of nutritional value: The more weight that is gained, the more nutritious the cereal is assumed to be.

(a) State the null hypothesis in terms of the issue under analysis.

(b) State the alternative hypothesis in terms of the issue under analysis.

(c) Restate the alternative hypothesis symbolically.

(d) Does the alternative hypothesis warrant a one-tailed or a two-tailed test?

13-8   (*Continuation of Exercise 13-3, page 382*) Designating the private hospitals Group P and the community hospitals Group C:
(a) State the null hypothesis.
(b) State the alternative hypothesis symbolically and explain what it asserts.
(c) Does the alternative hypothesis dictate a one-tailed or a two-tailed test?

*13-9   (*Continuation of Exercise 13-4, page 383*) Designating the population of household water usages *before* the conservation request Group B and the population of household water usages *after* the conservation request Group A:
(a) State the null hypothesis.
(b) State the alternative hypothesis in words and symbols.
(c) If the goal is a reduction in average household water usage of 25 gallons per day and the null hypothesis is that the goal is not met, state the alternative hypothesis symbolically.

13.3   the two-sample *t* test

We will now perform a *two-sample t test* to compare the effectiveness of guanethidine and guanadrel by following the steps outlined in Chart 13-4.

(1. State the null and alternative hypotheses.)
Null Hypothesis (illustrated in Chart 13-2): The two drug treatments produce identical distributions of blood pressure reductions.
Alternative Hypothesis (illustrated in Chart 13-3): The responses to the two treatments differ on the average; $\mu_A \neq \mu_B$, or $\mu_A - \mu_B \neq 0$.

(2. Summarize our samples of observed responses by summary statistics.)
Since we are conducting a test to determine if potential differences exist between population means μ_A and μ_B we represent the observed responses to treatment by the sample means $\overline{X}_A$ and $\overline{X}_B$.
From Chart 13-1:

$$\overline{X}_A \text{ (guanethidine)} = -9.55\%$$
$$\overline{X}_B \text{ (guanadrel)} \quad = -7.30\%$$

3. Determine the expected values of the summary statistics if the null hypothesis is true.

The Expected Value Law for Sample Means (see page 253) tells us that the expected value of the sample mean is equal to the mean of the population of inference. We are considering two sample means $\overline{X}_A$ and $\overline{X}_B$ and two population means μ_A and μ_B. The Expected Value Law then implies that $\mu(\overline{X}_A) = \mu_A$ and that $\mu(\overline{X}_B) = \mu_B$.

If the null hypothesis is true, $\mu_A = \mu_B = $ a common value μ. Accordingly, we can write that $\mu(\overline{X}_A) = \mu$ and that $\mu(\overline{X}_B) = \mu$. Therefore, $\mu(\overline{X}_A) = \mu(\overline{X}_B)$.

> If the null hypothesis is true, we can expect $\overline{X}_A$ to be close to $\overline{X}_B$ or, equivalently, we can expect $\overline{X}_A - \overline{X}_B$ to be close to 0.

4. Measure the discrepancy between the observed and the expected results and see if the discrepancy favors the alternative hypothesis.

Under the null hypothesis we would expect $\overline{X}_A$ to be close to $\overline{X}_B$. So any difference between these two values $(\overline{X}_A - \overline{X}_B \neq 0)$ represents a discrepancy between the observed and the expected results. In our example

$$\overline{X}_A - \overline{X}_B = (-9.55) - (-7.30) = -2.25 \text{ percentage points}$$

After administering each drug treatment to 20 patients, a discrepancy of -2.25 percentage points was observed. The mean of the blood pressure reductions of patients treated with guanethidine (Group A) is 2.25 percentage points larger (the minus sign indicates a larger *reduction*) than the mean of the blood pressure reductions of patients treated with guanadrel.

Since the alternative hypothesis $\mu_A - \mu_B \neq 0$ implies a two-tailed test, any discrepancy—any value for $\overline{X}_A - \overline{X}_B$ that differs from 0—favors the alternative hypothesis. But before we can reject the null hypothesis, we must see if *chance* can be ruled out as an explanation for this discrepancy.

5. If the discrepancy favors the alternative hypothesis, calculate the P value to measure the plausibility of the chance explanation for the discrepancy.

This final step requires us to determine the probability (a two-tailed P value in this example) that a discrepancy between $\overline{X}_A$ and $\overline{X}_B$ of at least 2.25 percentage points can occur by chance(Our method of finding the appropriate P value is based on three laws of chance variation: the *Expected Value Law for* $\overline{X}_A - \overline{X}_B$, the *Standard Error Law for* $\overline{X}_A - \overline{X}_B$ and the *t Distribution Law for* $\overline{X}_A - \overline{X}_B$.)

THE EXPECTED VALUE LAW FOR $\overline{X}_A - \overline{X}_B$ If $\overline{X}_A$ and $\overline{X}_B$ are the means of two samples selected at random from populations of inference that have equal means $\mu_A = \mu_B$, then the expected value of $\overline{X}_A - \overline{X}_B$, denoted by $\mu(\overline{X}_A - \overline{X}_B)$, is equal to 0. This law prevails for any pair of sample sizes n_A and n_B, where n_A and n_B are not necessarily equal. Symbolically

$$\mu(\overline{X}_A - \overline{X}_B) = 0 \qquad (13\text{-}1)$$

(The Expected Value Law for $\overline{X}_A - \overline{X}_B$ is the formal justification for our choice of $\overline{X}_A - \overline{X}_B$ as the measure of discrepancy between the observed and the expected results.)

THE STANDARD ERROR LAW FOR $\overline{X}_A - \overline{X}_B$ If $\overline{X}_A$ and $\overline{X}_B$ are the means of two samples selected at random from populations of inference that have equal standard deviations ($\sigma_A = \sigma_B =$ a common value σ), then the standard error of $\overline{X}_A - \overline{X}_B$, denoted by $\sigma(\overline{X}_A - \overline{X}_B)$, is given by

$$\sigma(\overline{X}_A - \overline{X}_B) = \sigma\sqrt{\frac{1}{n_A} + \frac{1}{n_B}} \qquad (13\text{-}2)$$

(We may interpret the standard error $\sigma(\overline{X}_A - \overline{X}_B)$ as the "average" amount by which $\overline{X}_A - \overline{X}_B$ will deviate from its expected value by chance. Notice that the standard error depends on the population standard deviation σ and on sample sizes n_A and n_B.)

(In general, the value of σ—the common standard deviation of the populations of inference—is unknown and must be estimated from the sample observations. Under the null hypothesis, both samples are taken from populations with identical distributions, which implies that $\sigma_A = \sigma_B = $ a common value σ. So both sample standard deviations $\hat{\sigma}_A$ and $\hat{\sigma}_B$ provide estimates of σ.)

(We take advantage of both these estimates by combining the two into what we call a *pooled estimate of* σ, denoted by $\hat{\sigma}_{pool}$ ("sigma-hat pool"). Symbolically

$$\hat{\sigma}_{pool} = \sqrt{\frac{(n_A - 1)\hat{\sigma}_A^2 + (n_B - 1)\hat{\sigma}_B^2}{n_A + n_B - 2}} \qquad (13\text{-}3)$$

The pooling procedure given by Equation (13-3) generates a kind of weighted average, weighting the squares of the values of the two estimates of σ by the relative size (the number of degrees of freedom) of the two samples. Therefore, if $n_A > n_B$, the estimate $\hat{\sigma}_A$ is weighted more heavily because it is based on more information.

Substituting $\hat{\sigma}_{pool}$ for σ in Equation (13-2) gives us the *estimated standard error of* $\overline{X}_A - \overline{X}_B$, denoted by $\hat{\sigma}(\overline{X}_A - \overline{X}_B)$:

$$\hat{\sigma}(\overline{X}_A - \overline{X}_B) = \hat{\sigma}_{pool}\sqrt{\frac{1}{n_A} + \frac{1}{n_B}} \qquad (13\text{-}4)$$

If the sample sizes are equal ($n_A = n_B = n$), we can simplify Equation (13-4) to

$$\text{for } n_A = n_B = n, \; \hat{\sigma}(\overline{X}_A - \overline{X}_B) = \sqrt{\frac{\hat{\sigma}_A^2 + \hat{\sigma}_B^2}{n}} \qquad (13\text{-}5)$$

Since the two drugs guanethidine and guanadrel were administered to an equal number of patients, $n_A = n_B = 20$, we should use Equation (13-5) to calculate an estimate of the standard error of $\overline{X}_A - \overline{X}_B$:

$$\hat{\sigma}(\overline{X}_A - \overline{X}_B) = \sqrt{\frac{\hat{\sigma}_A^2 + \hat{\sigma}_B^2}{n}} = \sqrt{\frac{\hat{\sigma}_A^2 + \hat{\sigma}_B^2}{20}}$$

From Chart 13-1, $\hat{\sigma}_A = 10.02$ percent and $\hat{\sigma}_B = 12.64$ percent. Completing the calculation

$$\hat{\sigma}(\overline{X}_A - \overline{X}_B) = \sqrt{\frac{(10.02)^2 + (12.64)^2}{20}} = 3.61 \text{ percentage points}$$

This result tells us that the possible scores for $\overline{X}_A - \overline{X}_B$ in random sampling vary from their expected value by an estimated "average" of 3.61 percentage points. So finding a discrepancy as large as ±3.61 percentage points between the two sample means would certainly not be unusual.

Because the discrepancy of 2.25 percentage points between $\overline{X}_A$ and $\overline{X}_B$ is even smaller than the "average" discrepancy, we should surmise that the size of our discrepancy can be readily explained by chance. (To calculate the precise *P* value, however, we need to know the shape of the sampling distribution of $\overline{X}_A - \overline{X}_B$, which is derived from the *t* Distribution Law for $\overline{X}_A - \overline{X}_B$.)

(**THE *t* DISTRIBUTION LAW FOR $\overline{X}_A - \overline{X}_B$ If $\overline{X}_A$ and $\overline{X}_B$ are the means of two random samples from identical populations of inference and if sample sizes are large, then the standard score**

$$t_{\overline{X}_A - \overline{X}_B} = \frac{\overline{X}_A - \overline{X}_B}{\hat{\sigma}(\overline{X}_A - \overline{X}_B)}$$ (13-6)

will be a point from a *t* distribution with *df* = $n_A + n_B - 2$.)

(As in the one-sample test, a *t* score represents the distance between the observed value of a statistic and its expected value in units of estimated standard errors. $\mu(\overline{X}_A - \overline{X}_B)$, the expected value of $\overline{X}_A - \overline{X}_B$, is equal to 0 under the null hypothesis. So $t_{\overline{X}_A - \overline{X}_B}$ is simply the ratio of the discrepancy $\overline{X}_A - \overline{X}_B$ to its estimated standard error $\hat{\sigma}(\overline{X}_A - \overline{X}_B)$.)

(The number of degrees of freedom is the sum of the degrees of freedom associated with each of the two estimates of σ that are combined to form $\hat{\sigma}_{pool}$, or $(n_A - 1) + (n_B - 1) = n_A + n_B - 2$. The number of degrees of freedom is the denominator of $\hat{\sigma}_{pool}$.)

(Once again, the applicability of the *t* distribution depends on population shape and sample size. If the common shape of both populations of inference is *normal*, then the *t* distribution can be applied to any set of sample sizes. More generally, unless the common population distribution is badly skewed, we can assume that the t distribution can be applied when each sample supplies at least 15 degrees of freedom so that total $df \geq 30.$

The t Distribution Law tells us that the distribution of the t scores for $\overline{X}_A - \overline{X}_B$ will follow a t distribution with $df = n_A + n_B - 2.$ So the law directs us to:

1. Compute the t score of the value obtained for $\overline{X}_A - \overline{X}_B$.
2. Obtain the appropriate P value from the t table at $df = 38$.

$$n_A = 20 \text{ and } n_B = 20, \text{ so } df = n_A + n_B - 2 = 38$$

We know that the t score of $\overline{X}_A - \overline{X}_B$ is defined as the ratio of $\overline{X}_A - \overline{X}_B$ to its standard error $\hat{\sigma}(\overline{X}_A - \overline{X}_B)$, or

$$t_{\overline{x}_A - \overline{x}_B} = \frac{\overline{X}_A - \overline{X}_B}{\hat{\sigma}(\overline{X}_A - \overline{X}_B)}$$

We've already determined that $\overline{X}_A - \overline{X}_B = -2.25$ percentage points and that $\hat{\sigma}(\overline{X}_A - \overline{X}_B) = 3.61$ percentage points. Completing the calculation for $t_{\overline{x}_A - \overline{x}_B}$

$$t_{\overline{x}_A - \overline{x}_B} = \frac{-2.25}{3.61} = -0.623$$

The t score is -0.623, which tells us that the distance of 2.25 percentage points between $\overline{X}_A$ and $\overline{X}_B$ is a distance of 0.623 estimated standard errors. The sum of the blue areas in Chart 13-5 depicts the two-tailed P value we must determine.

Checking the t table, we find no row at $df = 38$. Consequently, we check the row closest to $df = 38$, which is $df = 40$. The lowest entry on this row is 0.681, the entry associated with a two-tailed P value of 0.50. Our numeric value for $t_{\overline{x}_A - \overline{x}_B}$ (0.623) is even smaller than this entry. So the two-tailed P value for $t_{\overline{x}_A - \overline{x}_B} = -6.23$ exceeds 0.50.

There is more than a 50-percent chance that $\overline{X}_A - \overline{X}_B$ will differ from 0 (its expected value under the null hypothesis) by 0.623 standard errors. Consequently, if we reject the null hypothesis, we face more than a 50-per-

chart 13-5

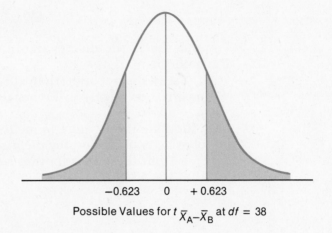

−0.623   0   + 0.623

Possible Values for $t_{\overline{X}_A - \overline{X}_B}$ at $df = 38$

cent risk of rejecting a true hypothesis—an excessive
risk by any standards. So we cannot conclude from the
evidence gathered in this study that guanadrel's effectiveness differs from guanethidine's effectiveness. We
cannot reject the null hypothesis.

to review the
two-sample *t* test

1. State the null and alternative hypotheses)
   Null hypothesis: The two treatments produce identical distributions of blood pressure reductions.
   Alternative hypothesis: The responses to the two
treatments differ on the average. Specifically, $\mu_A - \mu_B$
$\neq 0$.
2. Calculate the value of the summary statistics $\overline{X}_A$ and
$\overline{X}_B$.)
   From Chart 13-1
$$\overline{X}_A = -9.55$$
$$\overline{X}_B = -7.30$$
3. Determine the expected values of $\overline{X}_A$ and $\overline{X}_B$ if the null
hypothesis is true.
   Under the null hypothesis, $\mu(\overline{X}_A) = \mu(\overline{X}_B) =$ a common value μ. So we would expect $\overline{X}_A$ to be close to $\overline{X}_B$,
or equivalently, we would expect $\overline{X}_A - \overline{X}_B$ to be close to 0.)
4. Measure the discrepancy between the observed and the
expected results (that is, measure $\overline{X}_A - \overline{X}_B$):)

$$\overline{X}_A - \overline{X}_B = (-9.55) - (-7.30) = -2.25 \text{ percentage points}$$

and see if the discrepancy favors the alternative hypothesis.

(Any discrepancy that differs from 0, whether it is positive or negative, favors the alternative hypothesis $\mu_A - \mu_B \neq 0$.)

(5. Calculate the P value to measure the plausibility of the chance explanation for the discrepancy.

From the t Distribution Law for $\overline{X}_A - \overline{X}_B$, we know that the t score

$$t_{\overline{x}_A - \overline{x}_B} = \frac{\overline{X}_A - \overline{X}_B}{\hat{\sigma}(\overline{X}_A - \overline{X}_B)}$$

will be a point from a t distribution with $df = n_A + n_B - 2$.)

((a) Compute $\hat{\sigma}(\overline{X}_A - \overline{X}_B)$:

Since $n_A = n_B$, we use Equation (13-5) to compute $\hat{\sigma}(\overline{X}_A - \overline{X}_B)$.

$$\hat{\sigma}(\overline{X}_A - \overline{X}_B) = \sqrt{\frac{\hat{\sigma}_A^2 + \hat{\sigma}_B^2}{n}} = \sqrt{\frac{(10.02)^2 + (12.64)^2}{20}}$$
$$= 3.61 \text{ percentage points}$$

((b) Compute $t_{\overline{x}_A - \overline{x}_B}$:

$$t_{\overline{x}_A - \overline{x}_B} = \frac{\overline{X}_A - \overline{X}_B}{\hat{\sigma}(\overline{X}_A - \overline{X}_B)} = \frac{-2.25}{3.61} = -0.623$$

((c) Obtain the appropriate P value:)

At $df = n_A + n_B - 2 = 20 + 20 - 2 = 38$, the two-tailed P value > 0.50.

EXERCISES

13-10   A population of inference is known to have a virtually normal distribution with a mean μ and a standard deviation σ. A random sample of n scores is selected, and the sample mean is denoted by $\overline{X}_A$. Another random sample, also of n scores, is selected, and the mean is denoted by $\overline{X}_B$. So we have two samples of *equal size* from the *same population*. Complete the following chart:

LAWS OF CHANCE VARIATION	FOR $\overline{X}_A$	FOR $\overline{X}_B$	FOR $\overline{X}_A - \overline{X}_B$ (Equal Sample Sizes)
Expected Value Law	$\mu(\overline{X}_A) = \mu$	$\mu(\overline{X}_B) = $ _____	$\mu(\overline{X}_A - \overline{X}_B) = $ _____
Standard Error Law	$\sigma(\overline{X}_A) = \dfrac{\sigma}{\sqrt{n}}$	$\sigma(\overline{X}_B) = $ _____	$\sigma(\overline{X}_A - \overline{X}_B) = $ _____
t Distribution Law	$t_{\overline{X}_A} = \dfrac{\overline{X}_A - \mu}{\hat{\sigma}_A/\sqrt{n}}$ follows a *t* distribution with $df = n - 1$	$t_{\overline{X}_B} = $ _____ _____ _____ _____	$t_{\overline{X}_A - \overline{X}_B} = $ _____ _____ _____ _____

13-11 A random sample of 26 scores is selected from each of two populations that are assumed to have equal standard deviations ($\sigma_A = \sigma_B = $ a common value σ). The sample standard deviations are, respectively, $\hat{\sigma}_A = 8$ units and $\hat{\sigma}_B = 12$ units. Use Equation (13-5) on page 392 to calculate $\hat{\sigma}(\overline{X}_A - \overline{X}_B)$, the standard error of $\overline{X}_A - \overline{X}_B$.

13-12 A random sample is selected from each of two populations that are assumed to have equal standard deviations ($\sigma_A = \sigma_B = $ a common value σ). The standard deviations are $\hat{\sigma}_A = 4$ units and $\hat{\sigma}_B = 6$ units, and the sample sizes are $n_A = 28$ and $n_B = 38$.
   (a) Use Equation (13-3) on page 392 to calculate $\hat{\sigma}_{pool}$, the pooled estimate of σ.
   (b) State one reason why the value you calculated for $\hat{\sigma}_{pool}$ in (a) is closer to $\hat{\sigma}_B$ than it is to $\hat{\sigma}_A$.
   (c) Use Equation (13-4) on page 392 to calculate $\hat{\sigma}(\overline{X}_A - \overline{X}_B)$.
   (d) Whereas $\hat{\sigma}_{pool}$ provides an estimate of σ, the standard deviation of the distribution of scores in either population of inference, $\hat{\sigma}(\overline{X}_A - \overline{X}_B)$ provides an estimate of the standard deviation of *what* distribution?

*13-13 Is time perception affected by the color of light in an environment? A group of 50 volunteers participated in the following experiment. After drawing a chip from a bowl, each subject was assigned to one of two groups. Subjects in Group A were placed in a room bathed in warm (reddish) tones; subjects in Group B were placed in a room bathed in cool (bluish) tones. Each subject was asked to signal when a period of exactly 10 minutes had elapsed. (Obviously, the test subjects were asked to leave their watches outside.)

The actual time that elapsed before each subject signaled was recorded. The summary statistics follow.

	GROUP A (Warm Tones)	GROUP B (Cool Tones)
Mean Time (in minutes)	$\overline{X}_A = 8.96$	$\overline{X}_B = 10.46$
Standard Deviation	$\hat{\sigma}_A = 1.46$	$\hat{\sigma}_B = 1.54$
Sample Size	$n_A = 25$	$n_B = 25$

(a) State the null hypothesis.
(b) State the alternative hypothesis to imply a two-tailed test.
(c) Measure the discrepancy $\overline{X}_A - \overline{X}_B$ and interpret its value.
(d) What is the expected value of $\overline{X}_A - \overline{X}_B$ if the null hypothesis is true?
(e) Calculate $\hat{\sigma}(\overline{X}_A - \overline{X}_B)$ and interpret your result.
(f) Calculate $t_{\overline{X}_A-\overline{X}_B}$ and interpret your result.
(g) Is it reasonable to assume that the score obtained for $t_{\overline{X}_A-\overline{X}_B}$ is a point from a t distribution? Briefly justify your answer.
(h) If you answered yes to (g), sketch a frequency curve of the appropriate t distribution and shade the area(s) that depict the P value of interest.
(i) Determine the appropriate P value from the t table and state your conclusion about the issue under analysis.

*13-14 Random samples of 16 male and 16 female students at a community college revealed the following information about the quantity of beer consumed over a one-week period:

	GROUP A (Male Students)	GROUP B (Female Students)
Mean Consumption (in liters)	$\overline{X}_A = 7.5$	$\overline{X}_B = 6.0$
Standard Deviation	$\hat{\sigma}_A = 2.2$	$\hat{\sigma}_B = 1.9$
Sample Size	$n_A = 16$	$n_B = 16$

Perform a two-sample t test of the null hypothesis that the amount of beer consumed does not differ inherently between male and female students. Let the alternative hypothesis be that $\mu_A \neq \mu_B$. Your answer should include the result calculated for $t_{\overline{X}_A-\overline{X}_B}$, the appropriate P value, and a statement of your conclusion about the issue under analysis.

*13-15   The manufacturer of the Diamondedge blade randomly assigned 50 test subjects to Treatment A (Diamondedge) or Treatment B (stainless steel blade). However, 6 of the test subjects assigned to Treatment A and 2 of the test subjects assigned to Treatment B did not complete the experiment, leaving $n_A = 19$ and $n_B = 23$. Summary statistics follow.

NUMBER OF SHAVES PER BLADE

	DIAMONDEDGE	STAINLESS STEEL
Mean	$\overline{X}_A = 23.7$	$\overline{X}_B = 22.8$
Standard Deviation	$\hat{\sigma}_A = 2.8$	$\hat{\sigma}_B = 2.4$
Sample Size	$n_A = 19$	$n_B = 23$

(a) State the null hypothesis verbally.

(b) State the alternative hypothesis symbolically to imply a one-tailed test.

(c) Calculate the discrepancy $\overline{X}_A - \overline{X}_B$ and interpret your result.

(d) Use Equations (13-3), (13-4), and (13-6) to calculate $\hat{\sigma}_{pool}$, $\hat{\sigma}(\overline{X}_A - \overline{X}_B)$, and $t_{\overline{X}_A - \overline{X}_B}$.

(e) Is the discrepancy you calculated in (c) statistically significant at the 1-percent level?

(f) State your conclusion about the relative durability of the Diamondedge blade.

13-16   (*Continuation of Exercise 13-5, page 383*) Ultimately the two-route comparison was based on 10 commutes: 5 in the morning and 5 in the evening. Before each commute, a coin was tossed: Heads indicated the new route would be followed; tails, the usual route. When 5 commutes were completed on one route, the remaining commutes were completed on the other route. The resulting data follow:

COMMUTING TIMES
(in minutes)

New Route	23.7,	17.6,	19.1,	18.5,	20.3
Usual Route	22.4,	25.3,	20.6,	18.8,	21.5

(a) Would the results of a two-sample *t* test permit you to conclude that there is a difference on the average in commuting times between the usual route and the new route? (Perform the test and report the *P* value.)

(b) Is the two-sample *t* test a justifiable procedure in this case?

13.4   a confidence interval for $\mu_A - \mu_B$

The results of our two-sample t test show that the mean blood pressure reductions $\overline{X}_A$ and $\overline{X}_B$ do not differ by an amount that is statistically significant. Of course, this result does not *prove* that the two population means do not differ. We may have made a Type II error.

Let's suppose we have made a Type II error—that, in fact, $\mu_A \neq \mu_B$. Could the difference between μ_A and μ_B be large enough to be medically important? To estimate how large the mean difference $\mu_A - \mu_B$ conceivably might be, we could form a confidence interval for $\mu_A - \mu_B$.

In Chapter 10, we learned that a confidence interval for the mean of a single population μ has the form

$$\overline{X}_n \quad \pm \quad t \cdot \hat{\sigma}(\overline{X}_n)$$
$$\begin{array}{ccc} \text{Sample} & \pm & \text{Margin} \\ \text{mean} & & \text{for error} \end{array}$$

The confidence interval for $\mu_A - \mu_B$ has the same structure:

$$(\overline{X}_A - \overline{X}_B) \pm t \cdot \hat{\sigma}(\overline{X}_A - \overline{X}_B)$$

So the steps to form a 99-percent confidence interval for $\mu_A - \mu_B$ are:

1. Calculate $\overline{X}_A - \overline{X}_B$:
   As before, $\overline{X}_A - \overline{X}_B = -2.25$.

2. Calculate $\hat{\sigma}(\overline{X}_A - \overline{X}_B)$:
   As before, $\hat{\sigma}(\overline{X}_A - \overline{X}_B) = 3.61$.

3. Find the appropriate t value:
   For a 99-percent confidence interval, we check the column for a two-tailed P value $= 0.01$. At $df = 40$ (since $df = 38$ is not included in the t table), $t = 2.704$.

So the 99-percent confidence interval for $\mu_A - \mu_B$ becomes

$$(\overline{X}_A - \overline{X}_B) \pm t \cdot \hat{\sigma}(\overline{X}_A - \overline{X}_B)$$
$$= -2.25 \pm 2.704 \cdot 3.61$$
$$= \{-12.01 \text{ to } +7.51\} \text{ percentage points}$$

We can be 99 percent confident that the difference in the mean blood pressure effects of the two drugs is between -12.01 and $+7.51$ percentage points. A mean difference $\mu_A - \mu_B = -12.01$ implies that, in general, patients who take guanadrel will be less responsive to treatment than patients who take guanethidine: Guanadrel patients may experience 12.01 percentage points less reduction in blood pressure on the average. A mean difference $\mu_A - \mu_B = +7.51$ implies that in general guanadrel patients will respond better to treatment than guanethidine patients, experiencing 7.51 percentage points more reduction in blood pressure on the average. (The confidence interval includes the value $\mu_A - \mu_B = 0$. So the data are consistent with the belief that $\mu_A = \mu_B$, which is the same conclusion we drew from the two-sample t test.) However, the possibility that a difference of as much as 12.01 percentage points exists between μ_A and μ_B may be important enough on medical grounds to warrant further experiments with the two drugs.

EXERCISES

13-17   Referring to the sample evidence from the time-perception study given in Exercise 13-13 (page 397), compute a 90-percent confidence interval for $\mu_A - \mu_B$ and interpret your result in terms of the issue under analysis.

13-18   Referring to the sample evidence from the beer-consumption study given in Exercise 13-14 (page 398), compute a 95-percent confidence interval for $\mu_A - \mu_B$ and interpret your result in terms of the issue under analysis.

13-19   (*Continuation of Exercise 13-9, page 389*) The data from the two random samples follow.

	GROUP B	GROUP A
$\bar{X}$	365 gallons	330 gallons
$\hat{\sigma}$	20.6 gallons	18.7 gallons
n	100	100

Estimate the average drop in water usage $\mu_B - \mu_A$ at the 99-percent confidence level. Does the city appear to have met its goal of an average reduction of at least 25 gallons?

13.5   the Wilcoxon rank sum test

(We have learned how to use the t distribution to compare the effects of two treatments. But the two-sample t test will provide valid inferences only if certain sample size requirements are met. To review these requirements:

1. If the populations of inference (the populations of potential responses to guanadrel and guanethidine) are both *normal* in shape, then the t table will provide the correct probabilities for any pair of sample sizes.
2. If the populations of inference are not badly skewed, the t table will provide approximately correct P values as long as n_A and n_B are each greater than 15. For badly skewed populations, even larger samples are necessary.)

Too often, however, the t distribution is applied to very small samples, even when the shape of the population is not known. This can result in potentially misleading probabilities and thus faulty inferences.

(When comparisons are based on only a few responses, one effective alternative to the two-sample t test is the *Wilcoxon rank sum test*.[1] Unlike the two-sample t test, the Wilcoxon rank sum test does not require us to make an assumption about the shapes of the populations of inference. For this reason, it is called a *distribution-free test*.[2])

an illustrative example

A major medical concern today is the prevention of asthmatic attacks. Frequently, asthma sufferers experience these attacks when exercising. One method of testing the effectiveness of a new drug is to administer the drug to individual patients, ask them to exercise, and then measure the severity of their asthmatic attacks. This procedure is called an "exercise challenge test." Since each test subject is almost certain to experience an asthmatic attack, few patients are willing to volunteer, and so sample sizes are typically quite small.

[1] Another test frequently suggested is the Mann–Whitney test. Because the two tests are equivalent, we will only discuss the Wilcoxon procedure here.
[2] Such a test is also called a *nonparametric test*.

In one such study, 10 volunteers were randomly assigned to two treatment groups of equal size. Each of the 5 patients in Group A received an experimental drug identified by the proposed trade name Breathe-Ease. Each of the 5 patients in Group B received a placebo that appeared to be identical to Breathe-Ease. A placebo—pronounced "pleh-see-boh"—contains no medicine and permits the researcher to conduct a double-blind study. Thus neither the test patients nor the researcher can be influenced subconsciously by the knowledge that the drug being administered is supposed to reduce the severity of the asthmatic attack. All patients were asked to exercise on a treadmill (illustrated below) until their pulse rates increased to a designated level.

chart 13-6

BREATHE-EASE, TREATMENT A		PLACEBO, TREATMENT B	
PATIENT NO.	ASTHMATIC RESPONSE INDEX*	PATIENT NO.	ASTHMATIC RESPONSE INDEX*
1	45	6	55
2	13	7	40
3	62	8	17
4	24	9	31
5	8	10	64

*Higher index numbers indicate more severe asthmatic responses.

At this point, the severity of each asthmatic attack induced by the exercise was measured according to an asthmatic response index. The data are presented in Chart 13-6.

(With samples of only 5 scores apiece, the two-sample t test would be hard to justify, so the *Wilcoxon rank sum test* was used.)

the Wilcoxon procedure

Again we follow the steps outlined in Chart 13-4 on page 388.

(1. State the null and alternative hypotheses.)

The null hypothesis is that the treatments do not differ inherently. Since one of the treatments is a placebo, which, by definition, should produce no physiological effects, an equivalent statement of the null hypothesis is that Breathe-Ease is not effective in reducing exercise-induced asthmatic attacks. The alternative hypothesis is that Breathe-Ease decreases the average severity of asthmatic response more than the placebo does.

(2. Summarize our samples of observed responses by summary statistics.)

(In the two-sample t test, we use the sample means $\overline{X}_A$ and $\overline{X}_B$ as summary statistics. In the Wilcoxon rank sum test, we use a summary statistic known as the *rank sum* W. This statistic is based on the *ranks* of the responses to treatment, not on the *values* of the responses.)

We will use the asthmatic-response data given in Chart 13-6 to illustrate the computation of this summary statistic.

chart 13-7

RESPONSE	RANK R	TREATMENT
8	1	A
13	2	A
17	3	B
24	4	A
31	5	B
40	6	B
45	7	A
55	8	B
62	9	A
64	10	B
	$\Sigma R = 55$	

We begin by placing the responses of both groups into a single array, as in the first column of Chart 13-7. We then list the ranks of the responses (1 = smallest), as in the second column of Chart 13-7.

(To compute W, we sum the ranks of the group that contains *fewer* subjects. (If $n_A = n_B$, either group can be used.))Selecting Group A, we find the result $W = 1 + 2 + 4 + 7 + 9 = 23$.

(3. Determine the expected value of the summary statistic if the null hypothesis is true.)

(If the two groups contain the same number of patients and if the severity of response is not inherently affected by the treatment used, then W — the rank sum for Treatment A — will not be expected to differ from the rank sum of Treatment B; that is, W will be expected to be equal to one-half of the rank sum of the combined group of patients, or $(1/2)\Sigma R$.)

(In general, the expected value of W depends on the total sample size and on the proportion of the sample who received Treatment A. The specific relationship is given by the *Expected Value Law for W.*)

(**EXPECTED VALUE LAW FOR W**   If the null hypothesis is true, then the expected value of the rank sum for Treatment A, denoted by $\mu(W)$, is

$$\mu(W) = \left(\frac{n_A}{n_A + n_B} \right) \cdot \Sigma R$$
)

Since our combined sample totals ten scores, ΣR is simply the sum of the digits 1 through 10, or 55. So if asthmatic response is not relieved by Breathe-Ease, we would expect W to be close to $(1/2)\Sigma R = 55/2 = 27.5$.

4. Measure the discrepancy between the observed and the expected results and see if the discrepancy favors the alternative hypothesis.)

If Breathe-Ease effectively reduces asthmatic response — if the alternative hypothesis is true — most of the lower ranks should be associated with Group A. Hence, W should be small. The smallest possible value for W is $W = 15$, which results if the five smallest ranks are in the group receiving Treatment A ($W = 1 + 2 + 3 + 4 + 5 = 15$).

Our observed result is $W = 23$, and we know that the expected value of this statistic under the null hypothesis is $(1/2)\Sigma R = 27.5$. Since 23 is less than 27.5, the discrepancy favors the alternative hypothesis that Treatment A (Breathe-Ease) is more effective than the placebo. Now we must see if chance can be ruled out as an explanation for this discrepancy; that is, we need to determine the appropriate P value.

5. Calculate the P value to measure the plausibility of the chance explanation for the discrepancy.)

For selected sample sizes, Appendix Table E, reproduced here as Chart 13-8, provides W scores associated with a range of P values. Let's use this table in our drug comparison.

(First, we select the sample-size combination (n_A, n_B).) Each group contains five subjects, giving us the combination (5, 5) and the following associated information:

(5, 5)

TWO-TAILED P VALUE	ONE-TAILED P VALUE	L	U
0.20	0.10	20	35
0.10	0.05	19	36
0.05	0.025	17	38
0.02	0.01	16	39

The L entries provide cutoff points for the lower (left) tail of the W distribution. We refer to the L entries and to the one-tailed P values if our alternative hypothesis implies that W will be *lower* than its expected value under the null hypothesis. If our alternative hypothesis implies that W will be higher than the mean (in the *upper* tail of the distribution), we refer to the U entries and to the one-tailed P values.

Our alternative hypothesis is that Breathe-Ease is more effective in treating asthmatic attacks than the placebo.

chart 13-8

Table of the
Rank Sum W

Sample Sizes:		(3, 3)		(3, 4)		(3, 5)		(3, 6)	
P Value Two-Tailed	One-Tailed	L	U	L	U	L	U	L	U
0.20	0.10	7	14	7	17	8	19	9	21
0.10	0.05	6	15	6	18	7	20	8	22
0.05	0.025	*	*	*	*	6	21	7	23
0.02	0.01	*	*	*	*	*	*	*	*

Sample Sizes:		(4, 4)		(4, 5)		(4, 6)		(4, 7)	
P Value Two-Tailed	One-Tailed	L	U	L	U	L	U	L	U
0.20	0.10	13	23	14	26	15	29	16	32
0.10	0.05	11	25	12	28	13	31	14	34
0.05	0.025	10	26	11	29	12	32	13	35
0.02	0.01	*	*	10	30	11	33	11	37

Sample Sizes:		(5, 5)		(5, 6)		(5, 7)		(5, 8)	
P Value Two-Tailed	One-Tailed	L	U	L	U	L	U	L	U
0.20	0.10	20	35	22	38	23	42	25	45
0.10	0.05	19	36	20	40	21	44	23	47
0.05	0.025	17	38	18	42	20	45	21	49
0.02	0.01	16	39	17	43	18	47	19	51

Sample Sizes:		(6, 6)		(6, 7)		(6, 8)		(6, 9)	
P Value Two-Tailed	One-Tailed	L	U	L	U	L	U	L	U
0.20	0.10	30	48	32	52	34	56	36	60
0.10	0.05	28	50	29	55	31	59	33	63
0.05	0.025	26	52	27	57	29	61	31	65
0.02	0.01	24	54	25	59	27	63	28	68

* Scores with *P* values this small do not exist for this sample size.

Adapted from W.J. Dixon and F.J. Massey, Jr., *Introduction to Statistical Analysis*, Second Edition, page 445. Copyright © McGraw-Hill Book Company, 1957. Used with permission of McGraw-Hill Book Company.

Since we have chosen W to represent the rank sum of the scores for patients using Breathe-Ease, W should be lower than $\mu(W)$. So we need to compare the observed value for W ($W = 23$) with the L entries. (If we had chosen W to represent the rank sum of the placebo patients' scores, then we would compare W with the U entries.)

At sample size (5, 5), the entry $L = 16$ tells us that if the null hypothesis is true, no more than 1 percent (a one-tailed P value of 0.01) of the possible values for W will be ≤ 16. This entry also tells us that any score for $W > 16$ will have a P value > 0.01. We have found that $W = 23$, so now we know that the P value for $W = 23$ is more than 0.01.

Proceeding up the L column, we find that our observed value for W exceeds the entry $L = 20$ associated with a one-tailed P value of 0.10. So the P value associated with $W = 23$ must be greater than 0.10.[3]

> If we reject the null hypothesis, we face greater than a 10-percent risk of rejecting a true claim. (The discrepancy between the observed and the expected values for W is not statistically significant at the 10-percent level.) Breathe-Ease has not been found to be more effective than the placebo in reducing the severity of asthmatic response.

modifications of
the Wilcoxon test

(We have learned to perform a Wilcoxon rank sum test when (1) the sample size permits us to use a table of the distribution of possible values for W (Chart 13-8), and (2) the responses can be uniquely ranked (that is, no two responses are equal). But how do we conduct this test under other conditions?)

(1. The *Normal* Approximation for W)

Suppose $n_A = 10$ and $n_B = 20$. On one hand, (the small number of responses in sample A makes the use of the two-sample t test problematic: For this test, both df_A and df_B should be at least 15, unless the distributions of the populations of

---

[3] If our asthmatic response comparison had required a two-tailed test, we would have focused on the two-tailed P values. For example, based on Chart 13-8, we would reject the null hypothesis for a two-tailed test with no more than a 5-percent risk (at 5, 5) only if W were ≤ 17 (L) or ≥ 38 (U).

potential responses are virtually *normal.* On the other hand, tables of the possible values for W do not generally include such large sample size combinations. To resolve our problem, we can apply the *Law of Normality for W.*)

LAW OF NORMALITY FOR *W* If sample size is large (conservatively, both n_A and n_B should be at least 10), then under the null hypothesis, *W* (the sum of the ranks in sample A) will follow an approximately *normal* distribution with an expected value

$$\mu(W) = \frac{n_A(n_A + n_B + 1)}{2}$$

and a standard error

$$\sigma(W) = \sqrt{\frac{n_A n_B(n_A + n_B + 1)}{12}}$$

To test a null hypothesis, therefore, we calculate the Z score

$$Z_W = \frac{W - \mu(W)}{\sigma(W)}$$

and obtain the appropriate P value from the *normal* table.)

2. Dealing with Ties

(When scores are tied, we assign each of the tied observations an average rank.) To see what this means, let's consider the following data:

TREATMENT A	TREATMENT B
12	17
13	17
17	18
	19
	20

The smallest two scores are 12 and 13, so ranks 1 and 2 are assigned to Treatment A. But then there are three scores of 17. Since the next three available ranks are 3, 4, and 5,

$\Big($the rule is to assign the mean of these ranks to each score$\Big)$ that is, each score of 17 is given a rank of 4.

Calculating W, we have

$$W = 1 + 2 + 4 = 7$$

Referring to Chart 13-8 at sample sizes (3, 5), we see that 7 has a one-tailed P value $= 0.05$. The chance of the value of W being ≤ 7 is no more than 5 percent. $\Big($Because Chart 13-8 was developed assuming *no ties*, the P values are only approximate when scores are tied. The approximations will be good as long as the number of ties is small relative to the total number of observations.$\Big)$

a final comment

$\Big($We have said that the Wilcoxon rank sum test is more flexible than the two-sample t test because the Wilcoxon test provides valid inferences for much smaller samples. In the two-sample t test, each sample must have $df \geq 15$, unless the populations of potential responses are virtually *normal* in shape. The Wilcoxon rank sum test can be used even when sample size is as small as 3 responses per group$\Big)$ $\Big($Another advantage of the Wilcoxon test is that it is applied to the *ranks* of the responses rather than to the values of the responses, so that the Wilcoxon test can be used when the original data result from a ranking process—that is, when all the data are ranks, not measurements.$\Big)$ For example, applicants for a graduate program in psychology are ranked by a faculty committee from strongest to weakest based on many factors. If we wished to compare the qualifications of male and female applicants, we would have to use a procedure for comparing ranks, such as the Wilcoxon rank sum test.[4]

EXERCISES

13-20   If there is no inherent difference between two treatments A and B, what can you report about the probability that:
(a) $W \leq 25$ if $n_A = 6$ and $n_B = 7$.
(b) $W \leq 22$ if $n_A = 5$ and $n_B = 8$.

[4] For a more detailed discussion of the Wilcoxon rank sum test, see E.L. Lehmann, *Nonparametrics: Statistical Methods Based on Ranks* (San Francisco: Holden-Day, Inc., 1975).

(c) $W \geq 38$ if $n_A = 5$ and $n_B = 5$.
(d) $W \leq 12$ or $W \geq 32$ if $n_A = 4$ and $n_B = 6$.
where W always refers to the rank sum of the smaller sample.

*13-21   Two county justices alternately presided over the Family Court, where all divorce hearings were held. It seemed that counsel for the husband always tried to schedule the case before Judge Bloom, whereas counsel for the wife sought to bring the case before Judge Artis. Curious as to why, a new lawyer in the county determined the alimony settlements in a random sample of 6 divorce decisions made by each justice. The alimony settlements, expressed in dollars per month, were ranked as follows:

RANK	ALIMONY (per month)	JUDGE (A = Artis; B = Bloom)
1	$ 0	B
2	10	A
3	20	B
4	25	B
5	50	B
6	100	A
7	110	B
8	115	A
9	140	B
10	220	A
11	300	A
12	650	A

The Wilcoxon rank sum test was selected to test the null hypothesis of no inherent difference in the distributions of alimony settlements against the alternative hypothesis that Judge Artis's settlements are larger.
(a) Compute the rank sum statistic W. (Let W be the rank sum of Group A.)
(b) What value would you expect W to have if the null hypothesis is true?
(c) Does the discrepancy between W and $\mu(W)$ favor the alternative hypothesis?
(d) Can you reject the chance explanation for the discrepancy? (Determine the appropriate P value.)

13-22   Is there a difference in "ease of adjustment" experienced by wearers of soft versus hard contact lenses? Eleven prospective contact lens wearers were used as test subjects, and the following "ease of adjustment" rankings were recorded.
(a) State the null and alternative hypotheses.
(b) Compute W.

EASE OF ADJUSTMENT RANKING (1 = easiest)	TREATMENT (H = hard; S = soft)
1	S
2	S
3	H
4	S
5	H
6	S
7	S
8	H
9	H
10	H
11	H

(c) Is the value of W statistically significant at the 10-percent level? (two-tailed P value ≤ 0.10).

(d) Calculate $\mu(W)$ — the expected value of W if the null hypothesis is true. Does the discrepancy between W and $\mu(W)$ seem to indicate that it is easier to adjust to soft lenses?

13-23   The following annual salary data were collected from random samples of male and female faculty members at a liberal arts college:

MALE	FEMALE
$12,600	$ 9,800
$14,100	$11,900
$14,500	$12,600
$18,300	$16,200

Do these data support the argument that female faculty members tend to earn lower salaries than male faculty members, or can the observed differences be attributed to chance in random sampling?

13-24   The following times (in seconds) were recorded on a slalom run for students from the advanced ski clinics at two resorts:

RESORT A	RESORT B
90	67
85	66
82	65
71	64
68	64
64	63

Assuming that the ability of the students entering the two clinics is the same, can Resort B use these data to claim it has a superior training program?

13-25 *(Continuation of Exercise 13-14, page 398)* The rank sum of the female students' scores was $W = 211$ in the beer-consumption comparison. Retest the null hypothesis, applying the Law of Normality for W. Compare your finding with the results of the two-sample t test you performed in Exercise 13-14.

13.6 the matched-pairs design

⌠The two-sample t test and the Wilcoxon rank sum test are methods for comparing the effects of two treatments. Both tests are based on the same experimental design—a design that calls for random sampling or the random assignment of test subjects to the two treatments being compared. The randomized design precludes bias and allows the researcher to use the laws of chance variation to calculate the probability that differences between group responses have arisen by chance.⌡

Matched pairs.

The randomized design, however, may not provide the most *precise* inferences possible. Precision is a requirement of both confidence intervals and hypothesis tests. A precise confidence interval for a population parameter contains a small margin for error at the specified confidence level. A precise or, synonymously, a *powerful* hypothesis test detects treatment differences when they exist; that is, in a precise hypothesis test, there is little risk of making a Type II error. We know that increasing sample size improves the precision of inferences. Using homogeneous subject groups also helps.

Look again at Chart 13-1 (page 384), which displays the responses of hypertensive patients to the two drugs guanadrel and guanethidine. *Within* each group, patient scores vary considerably. This much variability among patients receiving the same medication clouds the difference *between* the two groups.

For a fixed sample size, the larger the standard deviation of the responses to each treatment, the larger the standard error of the difference between group means $\sigma(\overline{X}_A - \overline{X}_B)$. A larger standard error implies a less precise inference about the difference between treatments.

The patients in the guanadrel–guanethidine study differed in age, sex, general health, life style, and other characteristics —any of which could be expected to affect their response to treatment. If a more homogeneous group of subjects had been tested, the responses within each group would probably have been more consistent, implying a smaller standard deviation for each group.

In many cases, however, the researcher cannot find a homogeneous group of subjects or, as in the guanadrel-guanethidine comparison, does not wish to be limited to making a statement about a specialized group. In such cases, another modification of the experimental design—the *matched-pairs design*—can often improve the precision of comparisons between treatments.

MATCHED-PAIRS DESIGN An experimental design in which test subjects are paired so that, as nearly as possible, characteristics believed to affect the response to treatment are the same for both members of a pair. One member of each pair is randomly assigned to Treatment A; the other, to Treatment B.

In comparing the open classroom to the standard classroom, students could be matched by intelligence and then one student in each pair could be randomly assigned to each program. In the study comparing the arrest records of male and female police officers, matched male–female pairs could be formed on the basis of equivalent job experience.

Sometimes the same effect that results from matching test subjects can be achieved by making *repeated measurements* on the *same* test subject. Since the effects of antihypertensive drugs such as guanadrel and guanethidine are temporary, each patient could be treated with both drugs if an appropriate interval of time were allowed to elapse between treatments. The two responses of each patient then would form a pair.

The matched-pairs design improves the precision of inferences by eliminating differences in subject characteristics within each pair. Through matching, we hope to make the response of each member of a pair identical except for any differences due to the effects of the treatments.

To take full advantage of the matched-pairs design, however, we need to apply appropriate statistical procedures— procedures that are based on measurements of the response differences between members of pairs. In the remainder of this chapter, we will present two of these techniques: the *paired t test* and the *Wilcoxon signed rank test.*

EXERCISES

13-26   What type of matched-pairs design would you consider using if you wished to analyze the capital punishment issue presented in Exercise 13-1 (page 382)? Why?

13-27   (*Continuation of Exercise 13-3, page 382*) What variables would you consider in forming matched pairs of private and community hospitals for the purpose of comparing costs per patient day?

13-28   Name at least two variables that you should consider in forming matched pairs to make the alimony settlement comparison in Exercise 13-21 (page 411)?

13-29   *(Continuation of Exercise 13-23, page 412)* What variables would you consider in forming matched pairs of male–female faculty members?

13.7 the paired *t* test

Some years ago, when the authors of this book were teaching at the same university, we decided to compare our effectiveness as lecturers in introductory statistics. Two sections of Statistics 111 were scheduled to meet at the same time, and 25 students were enrolled in each section. At enrollment, the students did not know whether Tashman would teach Lecture A and Lamborn would teach Lecture B, or vice versa. So, in effect, the students were randomly assigned to the two lecturers.

Effectiveness, we agreed, would be measured in part by the students' scores on a common exam. The exam itself was prepared and graded by a third statistician who was thoroughly familiar with the course objectives and content. The statistician did not know whether the students were assigned to Lecture A or Lecture B, and we (the lecturers) did not know the contents of the exam questions.

To compare our performances, we formulated a matched-pairs design, pairing students on the basis of grade-point averages at the beginning of the semester, areas of specialization, and numeric scores on the Standard Aptitude Test. Each matched pair contained one student in Lecture A and one student in Lecture B. Some students were unique, so that no appropriate match could be made. In all, 20 pairs were formed. Their scores are listed in the first two columns of Chart 13-9.

(Since the subjects in the two groups were paired, we focused on the difference between the grades of each pair. These differences — referred to as *difference scores*, or *d scores*, $d = X_A - X_B$ — appear in the third column of Chart 13-9.)

test procedure
for the
paired *t* test

(1. State the null and alternative hypotheses.)
The null hypothesis is that the lecturers are equally effective. A reasonable alternative hypothesis is that *on the average* one lecturer is more effective.

(2. Summarize our samples of observed responses by summary statistics.)

chart 13-9

SCORE IN LECTURE A X_A	SCORE IN LECTURE B X_B	$d = X_A - X_B$
98	95	3
48	45	3
92	90	2
47	47	0
86	85	1
51	52	−1
80	80	0
59	55	4
71	70	1
60	57	3
79	75	4
62	60	2
65	65	0
91	92	−1
63	62	1
87	87	0
65	67	−2
84	82	2
77	77	0
79	72	7
$\overline{X}_A = 72.20$	$\overline{X}_B = 70.75$	$\overline{d} = 1.45$
$\hat{\sigma}_A = 15.26$	$\hat{\sigma}_B = 15.22$	$\hat{\sigma}_d = 2.14$

To compare the effectiveness of the two lecturers, we determine the difference score for each pair of students and calculate the mean of these differences $\overline{d}$. This is our *summary statistic.* From Chart 13-9, we see that $\overline{d} = 1.45$ points. Lecture A students outscored their partners in Lecture B by an average of 1.45 points.

3. Determine the expected value of the summary statistic if the null hypothesis is true.

If the two lecturers were equally effective and we look at the differences in the grades of the student pairs, the distribution of difference scores should be centered at 0 ($\mu_d = 0$) and symmetric in shape, with large positive differences (a higher score for the member of the pair who attended Lecture A) occurring about as frequently as negative differences of the same size. What does this tell us about the expected value of $\overline{d}$?

$\bigl(\;\bar{d}$ is a sample mean, so the laws of chance variation for sample means govern the sampling distribution of $\bar{d}$. From the Expected Value Law for Sample Means we know that the expected value of $\bar{d}$, $\mu(\bar{d})$, is equal to the population mean. Under the null hypothesis, the population mean is 0, so if the null hypothesis is true

$$\mu(\bar{d}) = 0 \;\bigl)$$

We now have formulated the same type of problem that we faced in Chapter 10 when we conducted the one-sample t test. The d entries in the third column of Chart 13-9 give us one sample of difference scores from a single population of possible difference scores, and we want to determine whether or not our observed result, $\bar{d} = 1.45$, is consistent with a value of 0 for the population mean, $\mu_d = 0$.

$\bigl($ Once the test subjects have been matched and the difference scores have been obtained, the paired t test is conducted as if it were a one-sample t test on a distribution of difference scores. $\bigr)$

$\bigl($ 4. Measure the discrepancy between the observed and the expected results and see if the discrepancy favors the alternative hypothesis. $\bigr)$

Our discrepancy is the difference between our observed value $\bar{d} = 1.45$ and the expected value of 0. Since the alternative hypothesis did not indicate which lecturer was considered to be more effective, but only that $\mu_d \neq 0$, any discrepancy – any $\bar{d}$ value other than 0 – favors the alternative hypothesis.

$\bigl($ 5. Calculate the P value to measure the plausibility of the chance explanation for the discrepancy. $\bigr)$

We have calculated $\bar{d} = 1.45$. We need to know the probability that this large a deviation from the expected value of 0 will occur by chance. Since $\bar{d}$ is a sample mean, the t Distribution Law for Sample Means tells us that $t_{\bar{d}}$ is a point from a t distribution with $df = n - 1 = 20 - 1 = 19$.

By definition

$$t_{\bar{d}} = \frac{\bar{d} - \mu_d}{\hat{\sigma}(\bar{d})}$$

Under the null hypothesis, $\mu_d = 0$. From Chart 13-9, we know that $\hat{\sigma}_d = 2.14$. So our estimated standard error for $\overline{d}$ is

$$\hat{\sigma}(\overline{d}) = \frac{\hat{\sigma}_d}{\sqrt{n}} = \frac{2.14}{\sqrt{20}} = 0.479$$

The t statistic is then

$$t_{\overline{d}} = \frac{\overline{d}}{\hat{\sigma}(\overline{d})} = \frac{1.45}{0.479} = 3.027$$

Our value for $\overline{d}$ lies an estimated 3.027 standard errors from 0. The appropriate P value for this result is represented by the blue areas in Chart 13-10.

At $df = 19$, the largest numeric entry in that table is 2.861, which is associated with a two-tailed P value of 0.01. Since our t score is farther from 0 than ± 2.861, the blue areas in Chart 13-10 will sum to less than 1 percent. Accordingly, we can reject the null hypothesis with less than a 1-percent risk of rejecting a true claim. The effectiveness of the two instructors did differ significantly.[5]

chart 13-10

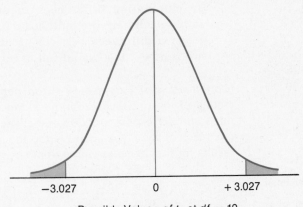

-3.027 0 $+3.027$

Possible Values of $t_{\overline{d}}$ at $df = 19$

[5] We are not going to tell you whether Tashman or Lamborn appeared to be the *more effective* lecturer. Nor can you infer this from noting that Tashman is still a professor at the University of Vermont but that Lamborn is not. (Why can't you?)

the paired t test versus the two-sample t test

(When the original responses of the two groups, X_A and X_B, are available—as they are in the example illustrated in Chart 13-9—we may be tempted to use the two-sample t test to compare the treatments. To do this would be not only inappropriate (the t table would not provide the correct P values) but foolish, because the two-sample t test would fail to utilize the matched-pairs design that has been created. The two-sample t test ignores the consideration that the variability of matched-pair differences, σ_d, in response to treatment will be smaller than the variability among the responses of individual subjects to treatment. So a discrepancy, $\bar{d} = \overline{X}_A - \overline{X}_B$, that is statistically significant according to the paired t test may be found *not statistically significant* if the two-sample t test is used.

For example, if we had applied the two-sample t test to the data in Chart 13-9, we would have calculated

$$t_{\overline{X}_A - \overline{X}_B} = \frac{\overline{X}_A - \overline{X}_B}{\hat{\sigma}(\overline{X}_A - \overline{X}_B)} = 0.301$$

a result that would *not* have permitted us to say that the discrepancy is statistically significant even at the 50-percent level of significance.

By eliminating differences in subject characteristics within each pair, the matched-pairs design enables us to detect a difference between treatment effects that the two-sample t test would fail to reveal.)

EXERCISES
13-30

(*Continuation of Exercise 13-15 on page 399*) Reluctant to accept the results of the previous experiment, the razor blade manufacturer instituted another experiment. This time, each test subject was asked to use one blade of each type and to report the number of shaves per blade. The results for the 16 males who participated in this experiment follow.

Subject Number	1	2	3	4	5	6	7	8	9	10	11	12	13	14	15	16
Number of Shaves with Diamondedge	18	19	20	20	21	22	23	23	23	24	25	25	26	26	27	29
Number of Shaves with Stainless Steel	18	18	19	21	22	20	21	24	22	21	25	25	24	27	24	28
Difference Score (Diamondedge − Stainless Steel)	—	—	—	—	—	—	—	—	—	—	—	—	—	—	—	—

A paired *t* test is to be performed. The null hypothesis that there is no in-herent difference in the durability of the two types of blades is to be tested against the alternative hypothesis that Diamondedge gives more shaves per blade on the average.

(a) Complete the table in this exercise by calculating the *d* score for each subject.

(b) Calculate $\bar{d}$ and interpret the result.

(c) What value would you expect $\bar{d}$ to have if the null hypothesis is true?

(d) Does the discrepancy between the observed and the expected values of $\bar{d}$ favor the alternative hypothesis?

(e) To determine the *P* value, which measures the plausibility of the chance explanation for the discrepancy, begin by calculating:

(1) $\hat{\sigma}_d$

(2) $\hat{\sigma}(\bar{d})$

(3) $t_{\bar{d}}$

and then determine the appropriate *P* value. Note:

$$\hat{\sigma}_d = \sqrt{\frac{\Sigma(d^2) - n\bar{d}^2}{n - 1}}$$

(f) To permit the manufacturer to advertise that Diamondedge is superior in durability, evidence favoring Diamondedge must be statistically significant at the 5-percent level. Can the manufacturer now adver-tise the superiority of Diamondedge?

(g) How could you explain the difference in test conclusions between the paired *t* test performed in this exercise and the two-sample *t* test you performed in Exercise 13-15? (Note that $\bar{d}$, the mean of the dif-ference scores here, is almost identical to the difference $\bar{X}_A - \bar{X}_B$ in Exercise 13-15.)

*13-31   To compare the effectiveness of the open classroom program and the traditional, structured program, children about to enter Grade 2 were paired on the basis of their Grade 1 performances. One member of each of 25 matched pairs was randomly assigned to the open classroom sec-tion (group O); the other member of each pair was placed in a struc-tured section (group S). At the end of the year, the children's achieve-ment scores were compared. The summary statistics follow.

	OPEN CLASSROOM SECTION	STRUCTURED SECTION	*d* SCORE (Open classroom section − Structured section)
Sample Mean	$\bar{X}_O = 86.2$	$\bar{X}_S = 82.2$	$\bar{d} = 4.0$
Sample Standard Deviation	$\hat{\sigma}_O = 7.6$	$\hat{\sigma}_S = 7.1$	$\hat{\sigma}_d = 3.0$

(a) State the null hypothesis and then state the alternative hypothesis to imply a two-tailed test.
(b) What is the expected value of $\bar{d}$ if the null hypothesis is true?
(c) Is the discrepancy between the observed and the expected values for $\bar{d}$ statistically significant?

13-32   *(Continuation of Exercise 13-31)* When the summary statistics presented in Exercise 13-31 were initially reported, no mention was made of the fact that the children had been matched on the basis of their performances in Grade 1. The omission of this point may have led us to conclude that the data should be analyzed by conducting a two-sample *t* test.

(a) Show that if a two-sample *t* test had been performed, the null hypothesis could not have been rejected with less than a 5-percent risk of rejecting a true claim.
(b) Which test results do you believe: the results of the paired *t* test in Exercise 13-31, or the results of the two-sample *t* test in (a) in this exercise? Explain your answer.

13.8 the Wilcoxon signed rank test

When data are unmatched, the Wilcoxon rank sum test presents an alternative approach to the two-sample *t* test for small samples. Now we introduce an alternative to the paired *t* test for small samples, which is called the *Wilcoxon signed rank test.*

To illustrate the Wilcoxon signed rank test, we will reconsider the asthmatic response data given in Chart 13-6. Recall that in using the Wilcoxon rank sum test, we failed to reject the null hypothesis that Breathe-Ease and the placebo produce equivalent effects. In a follow-up experiment, the patients switched treatments: Patients who previously received the placebo were given Breathe-Ease, and vice versa. The results appear in Chart 13-11, along with the difference score for each patient. In the chart, *d* represents the difference in severity of the asthmatic response of the patient after receiving the placebo minus the asthmatic response of the same patient after receiving Breathe-Ease.

1. State the null and alternative hypotheses.

As in our earlier analysis, the null hypothesis is that the two treatments do not differ inherently and the alternative

chart 13-11

PATIENT NO.	BREATHE-EASE TREATMENT A ASTHMATIC RESPONSE INDEX	PLACEBO TREATMENT B ASTHMATIC RESPONSE INDEX	d = DIFFERENCE IN ASTHMATIC RESPONSE INDEX (DRUG B − DRUG A)
1	45	56	11
2	13	23	10
3	62	67	5
4	24	21	−3
5	8	2	−6
6	55	55	0
7	36	40	4
8	10	17	7
9	22	31	9
10	69	64	−5

hypothesis is that Breathe-Ease is more effective than the placebo in reducing the severity of the asthmatic response.

(2. Summarize our samples of observed responses by summary statistics.)

(In both Wilcoxon tests, ranks are used in place of the actual numeric values. In this case, we rank the differences d in order of increasing size, *ignoring the sign of the difference;* that is, we rank the *absolute* differences. The differences, along with the ranks of the absolute differences (1 = lowest rank), are listed in Chart 13-12.)

chart 13-12

PATIENT NO.	d = DIFFERENCE IN SEVERITY OF ASTHMATIC RESPONSE	RANK
1	11	9
2	10	8
3	5	3.5
4	−3	1
5	−6	5
6	0	−
7	4	2
8	7	6
9	9	7
10	−5	3.5

(We can think of differences of 0, or $d = 0$, as having the rank of 0 and essentially can ignore them in the following computations. In cases of ties between difference scores, we assign the average rank to each score, as we did in the rank sum test.)

(In the Wilcoxon rank sum test, we summarize the ranks with a rank sum W. In the Wilcoxon signed rank test, our summary statistics consist of two rank sums: the sum of the ranks associated with the positive differences, $\Sigma R+$, and the sum of the ranks associated with the negative differences, $\Sigma R-$.)

For the data in Chart 13-12, we obtain

$$\Sigma R+ = 2 + 3.5 + 6 + 7 + 8 + 9 = 35.5$$
$$\Sigma R- = 1 + 3.5 + 5 = 9.5$$

(3. Determine the expected value of the summary statistics if the null hypothesis is true.)

(In this case, we want to determine the expected value of the rank sum of both the positive and the negative differences. If the two treatments do not affect the severity of asthmatic response differently, then we would expect the positive differences to be balanced by the negative differences so that the sum of the positive difference ranks, $\Sigma R+$, would be equal to the sum of the negative difference ranks, $\Sigma R-$. In turn, this implies that either $\Sigma R+$ or $\Sigma R-$ would be equal to one-half of the sum of all the ranks, ΣR.

In this example, $(1/2)\Sigma R = 45/2 = 22.5$. (Since the sixth subject has a difference score and, therefore, a rank of 0, ΣR is simply the sum of the first nine digits, or 45.) So the expected value of both $\Sigma R+$ and $\Sigma R-$ is 22.5 under the null hypothesis.)

(4. Measure the discrepancy between the observed and the expected results and see if the discrepancy favors the alternative hypothesis.)

If Breathe-Ease reduces the severity of asthmatic response, we would expect most of the difference scores to be positive, indicating more severe responses when the patients are treated with the placebo. Likewise, we would expect only a few difference scores to be negative and those that are negative to be small and therefore of low rank. So under the alternative hypothesis of reduced severity in the Breathe-Ease group, $\Sigma R-$ will be the smaller of the two rank sums. In this example, $\Sigma R-$ is actually the smaller rank sum, having a value of 9.5 rather than the expected value of 22.5. The discrepancy does favor the alternative hypothesis.

5. Calculate the P value to measure the plausibility of the chance explanation for the discrepancy.

To determine the P value, we need to define the *signed rank statistic V.*

> Denote by V the rank sum ($\Sigma R+$ or $\Sigma R-$) that is *expected to be smaller under the alternative hypothesis.* (In a two-tailed test, we define V simply as the smaller of $\Sigma R+$ or $\Sigma R-$.)

In this case, $V = \Sigma R- = 9.5$.

Because V is defined so that small values always favor the alternative hypothesis, the P value table for V is simpler than the P value table for the W statistic in the Wilcoxon rank sum test. Turning to Chart 13-13, reproduced from Appendix Table F, we initially seek the row with the appropriate sample size entry n. Note that we now use n to represent the number of nonzero ranks, rather than the original number of difference scores. In our example, $n = 9$.

P values are determined by selecting the appropriate column of the chart. At $n = 9$, the column associated with a one-tailed P value $= 0.05$ contains the entry $V = 8$. From this we learn that at $n = 9$ (implying an expected value $V = 22.5$), the chance that V will be ≤ 8 is no more than 0.05 and that for any value $V > 8$, the P value will be > 0.05.

We found $V = 9.5$. Consequently, we cannot reject the null hypothesis without facing more than a 5-percent risk of rejecting a true claim. In this case, the additional data (repeated measurements) acquired for the Wilcoxon signed rank test did not change the conclusion of the Wilcoxon rank sum test at the 5-percent level of significance. However, at 9.5, V is smaller than 10, which is the V score associated with a one-tailed P value of 0.10. If we were willing to take as high as a 10-percent risk of rejecting a true null hypothesis, we would declare that the discrepancy was statistically significant—that Breathe-Ease does reduce the severity of asthmatic response more than the placebo does.

**chart 13-13
Table of the
Signed Rank
Statistic V**

ONE-TAILED P VALUE	0.10	0.05	0.025	0.01	0.005
TWO-TAILED P VALUE	0.20	0.10	0.05	0.02	0.01
n					
5	2	0	—	—	—
6	3	2	0	—	—
7	5	3	2	0	—
8	8	5	3	1	0
9	10	8	5	3	1
10	14	10	8	5	3
11	17	13	10	7	5
12	21	17	13	9	7
13	26	21	17	12	9
14	31	25	21	15	12
15	36	30	25	19	15
16	42	35	29	23	19
17	48	41	34	27	23
18	55	47	40	32	27
19	62	53	46	37	32
20	69	60	52	43	37

Each entry represents the largest value of V with a P value less than or equal to the P value in the column heading.

Adapted from Table II in F. Wilcoxon, S.K. Katti, and Roberta A. Wilcox, *Critical Values and Probability Levels for the Wilcoxon Rank Sum Test and the Wilcoxon Signed Rank Test*, American Cyanamid Company (Lederle Laboratories Division, Pearl River, N.Y.) and The Florida State University (Department of Statistics, Tallahassee, Fla.), August 1963. Used with the permission of American Cyanamid Company and The Florida State University.

**EXERCISES
13-33**

A comparison was to made of the felony arrest records of male and female police officers on a metropolitan force containing hundreds of policemen but only six policewomen. Each of the policewomen was matched with a male officer whose prior experience on the job was comparable. The number of felony arrests made by the individual member's of each pair during a 12-month period follow. (Each officer's years of prior experience on the job is shown in parentheses.)

MATCHED-PAIR NUMBER	NUMBER OF FELONY ARRESTS POLICEWOMAN	POLICEMAN
1	16 (11)	18 (11)
2	13 (6)	17 (6)
3	13 (4)	18 (4)
4	12 (2)	8 (2)
5	9 (1)	7 (1)
6	6 (0)	7 (0)

It was believed that the policewomen were not achieving the felony arrest records of their male counterparts. A Wilcoxon signed rank test is to be employed to analyze this issue.

(a) State the null and alternative hypotheses.

(b) Calculate the d score (Policewoman − Policeman) of the number of felony arrests for each pair and rank the d scores from 1 (lowest) to 6 (highest), ignoring the sign of the d score. Then calculate $\Sigma R+$ and $\Sigma R-$.

(c) What is the expected value of the rank sum of either $\Sigma R+$ or $\Sigma R-$ if the null hypothesis is true?

(d) Would you expect $\Sigma R+$ or $\Sigma R-$ to have the smaller value if the alternative hypothesis is true? Does the discrepancy between the observed and the expected results favor the alternative hypothesis?

(e) Can we reject the chance explanation for the discrepancy?

13-34   (*Continuation of Exercise 13-21, page 411*) The alimony settlements documented in Exercise 13-21 were subsequently paired on the basis of the husbands' income levels; that is, each pair contained a settlement of Judge Artis and a settlement of Judge Bloom, both expressed in dollars per month.

MATCHED-PAIR NUMBER	SETTLEMENT OF JUDGE ARTIS (per month)	SETTLEMENT OF JUDGE BLOOM (per month)
1	$ 10	$ 0
2	115	20
3	100	110
4	220	25
5	300	50
6	650	140

(a) Perform a Wilcoxon signed rank test to determine whether the null hypothesis of identical distributions of alimony settlements can be rejected with less than a 5-percent risk of rejecting a true hypothesis. Recall that the alternative hypothesis was that Judge Artis's settlements are larger.

(b) After comparing your results in (a) with the results of the Wilcoxon rank sum test in Exercise 13-21, can you conclude that the matching procedure helped you detect a difference between the two distributions of alimony settlements?

13-35   Data on the costs per patient day for eight matched pairs of private and community hospitals follow. The matching was based on several variables, including hospital size, the presence or absence of specific types of equipment, and population of the region served.

| MATCHED-PAIR | COST PER PATIENT DAY | |
NUMBER	PRIVATE HOSPITAL	COMMUNITY HOSPITAL
1	$175	$140
2	130	145
3	216	190
4	340	270
5	210	215
6	155	150
7	175	175
8	280	260

Would a Wilcoxon signed rank test permit you to conclude that there *is* an inherent difference in costs per patient day between private and community hospitals?

Caveats on the Design of Two-Group Comparisons

CAVEAT 1:   Check the experimental design.

> Unfortunately, the arithmetic computational part of statistical methods can be applied to any set of numbers no matter how derived. The validity of the method is dependent, however, on how the data are collected. Inappropriate data will yield a numerical solution as readily as appropriate data. The temptation to use these methods regardless of the appropriateness of the data is obviously very great, as one may judge from current literature. In interpreting such literature it therefore becomes necessary to ask first not what the test of significance shows — but what justification there is for calculating it.[6]

This statement appeared in the *New England Journal of Medicine* several years ago, but it applies equally well today.

[6] H.M. Schoolman, "Statistics in Medical Research," *New England Journal of Medicine*, Vol. 280, No. 4, p. 218 (January 23, 1969).

chart 13-14

	LARGE SAMPLES	SMALL SAMPLES
Randomized Design (test subjects are unmatched)	Two-Sample t Test or Wilcoxon Rank Sum Test	Wilcoxon Rank Sum Test
Matched-Pairs Design	Paired t Test or Wilcoxon Signed Rank Test	Wilcoxon Signed Rank Test

An important part of statistical analysis involves the selection of the correct statistical procedure. In this chapter, we have presented four procedures for comparing the effects of two treatments. Chart 13-14 summarizes the guidelines for choosing among these methods.

First, check the experimental design. If the test subjects have been matched in pairs, a randomized design test (two-sample t test or Wilcoxon rank sum test) will produce faulty and imprecise inferences.

Second, check the sample-size requirements for the t tests. Unless sample sizes are large (unless $df \geq 15$ for each sample or for the sample of matched pairs) or unless the population of inference is virtually *normal* in shape, the t tests should be avoided. With large samples, the t tests and the Wilcoxon tests are about equally reliable.

CAVEAT 2:   Check the issue being analyzed.

At times, none of the four statistical procedures discussed in this chapter can be appropriately applied to the issue under analysis. In this caveat, we present two such cases.

Case 1. A financial analyst for a Wall Street brokerage house wished to compare the average returns from two investment strategies: buying call options versus writing call options. However, she knew that the buying strategy must be riskier because the returns varied more (had a larger standard deviation). In other words, before she started to collect data, the analyst knew that the distributions of the potential returns of the two strategies would not have equivalent variability.

Neither the two-sample t test nor the Wilcoxon rank sum test should be used when the standard deviations of the two distributions of potential scores differ greatly.

Both procedures are designed for calculating P values under the assumption that the two populations of inference have identical distributions. If this condition is violated, using either of these tests can result in erroneous P values. Other test procedures, such as the Behrens–Fisher test, can be used when the standard deviations of two distributions differ greatly. These tests are described in more advanced textbooks.[7]

Case 2. Following the adoption of a new system for distributing state grants to local school districts, an analyst wished to evaluate the impact on local school budgets. The key issue was whether and, if so, by how much the new system had reduced the *variability* from school district to school district in money spent per pupil on public education. The null hypothesis was that the variability of local school expenditure per pupil had not been reduced.

Neither the t tests nor the Wilcoxon tests can be used to compare *variability*, such as the standard deviations of two groups. The procedures in this chapter are designed to analyze potential differences in the average of groups of scores, as measured by means or rank sums. Different techniques must be used to test hypotheses comparing variability.

COMING
TO
TERMS

Difference Score *d* In analyzing data from a matched-pairs design, the difference in the responses within each pair. All further analyses are based on these difference scores.

Double-Blind Experiment An experiment in which neither the subject nor the researcher knows which treatment the subject is receiving. A double-blind experiment is used to eliminate the possibility of subconscious bias.

Matched-Pairs Design An experimental design in which test subjects are matched so that, as nearly as possible, they are identical with respect to the characteristics believed to affect the response to treat-

[7] See George W. Snedecor and William G. Cochran, *Statistical Methods*, Sixth Edition (Ames, Iowa: Iowa State University Press, 1967).

ment. One member of each pair is randomly assigned to Treatment A; the other, to Treatment B.

Paired t Test   A procedure used with a matched-pairs design to determine whether or not the mean of the difference scores $\bar{d}$ is significantly different from 0.

Placebo   A preparation known to have no inherent effect on the subject's response.

Pooled Estimate of σ, $\hat{\sigma}_{\text{pool}}$ An estimate of the population standard deviation σ based on an average (or a weighted average) of the sample variances $\hat{\sigma}_A^2$ and $\hat{\sigma}_B^2$.

Precise Hypothesis Test   A hypothesis test that has a high probability of declaring that treatment differences exist when they do; that is, a test with a low probability of a Type II error.

Random Assignment   An extension of random sampling. A chance mechanism—rather than the judgment of the researcher or the preferences of the subjects themselves—is used to place each subject in a group.

Randomized Design   An experimental design in which test subjects are randomly allocated to treatment groups or are randomly selected from the populations of inference.

Response   A synonym for *score* or *observation*.

Standard Error of $(\bar{X}_A - \bar{X}_B)$, $\sigma(\bar{X}_A - \bar{X}_B)$ The standard deviation of the sampling distribution of $\bar{X}_A - \bar{X}_B$, which represents the "average" error made in using $\bar{X}_A - \bar{X}_B$ to estimate $\mu_A - \mu_B$.

Summary Statistics   The statistics used to summarize the observed responses to treatment.

Test Subject   An individual whose response to treatment is to be measured.

Treatment   In group comparisons, the characteristic that defines the groups being compared; examples include sex, educational level, and type of razor blade used.

Two-Sample t Test   A procedure for determining whether or not a significant difference exists between the means $\bar{X}_A$ and $\bar{X}_B$ of two unmatched samples of responses.

Wilcoxon Rank Sum Test   A two-sample test procedure based on the ranks of the scores or responses. Can be used for all sample size combinations, regardless of the shape of the population's distribution. Can also be used when the data are from a ranking rather than a measurement process.

Wilcoxon Signed Rank Test   A test procedure used with a matched-pairs design and based on the ranks of the difference scores. Can be used for any sample size, regardless of the shape of the population's distribution. Can also be used when the data are from a ranking rather than a measurement process.

1. A group of 32 recent college graduates served as test subjects in a four-week, memory training program. The effectiveness of two experimental techniques felt to have promise in expanding the mind's capacity to retain factual information were to be evaluated. Key facts about the operation of the program follow:

   (1) The test subjects were randomly assigned (unmatched) to two groups of equal size. Group MIT was programmed to follow the Memory Induction Technique, and Group MDT was programmed to follow the Memory Deduction Technique.
   (2) A pretest was administered to all 32 subjects to measure each subject's factual-retention capability before the program began.
   (3) At the conclusion of the program, each subject's factual-retention capability was retested.

   The results of the pretest and retest are summarized in the following chart, where 100 = perfect score on the factual-retention test and 0 = absence of demonstrated factual-retention capability.

		MIT	MDT
Pretest	Mean	69.0	72.5
	Standard Deviation	13.3	14.1
Retest	Mean	75.5	76.0
	Standard Deviation	11.8	12.9
Change (Retest − Pretest)	Mean	6.5	3.5
	Standard Deviation	3.8	4.1

   For each of the following issues, state the null and alternative hypotheses, perform an appropriate hypothesis test, and report your conclusion about the issue raised:

   (a) Does the Memory Induction Technique improve factual-retention capability on the average?
   (b) Does the Memory Deduction Technique improve factual-retention capability on the average?
   (c) Does one of the two techniques MIT or MDT show greater promise for improving factual-retention capability on the average?

2. The two-sample t test of a null hypothesis that two treatments are not inherently different was presented in Section 13.3. This comparison also can be performed using a regression procedure. Consider the regression model

$$Y = \beta_0 + \beta_1 X + Noise$$

where Y denotes the response of each test subject, irrespective of the treatment received, and X is a *yes–no* variable where

$X = 1$ if a test subject received Treatment A

$X = 0$ if a test subject received Treatment B

A total of 40 test subjects were randomly assigned (20 to Treatment A and 20 to Treatment B), and their responses were used to estimate the β's of the regression model. The results follow.

DEPENDENT VARIABLE Y

INDEPENDENT VARIABLE	REGRESSION COEFFICIENT	STANDARD ERROR	t SCORE
X	$b_1 = -2.25$	$\hat{\sigma}(b_1) = 3.61$	$t_{b_1} = -0.623$
Intercept	$b_0 = -7.30$		

(a) What response would you predict for a test subject who received Treatment B? (*Hint:* Substitute 0 for X.)

(b) What response would you predict for a test subject who received Treatment A? (*Hint:* Substitute 1 for X.)

(c) Based on your answers to (a) and (b), interpret the value of b_1.

(d) Express the null hypothesis of no inherent difference between Treatment A and Treatment B in terms of β_1.

(e) Test this null hypothesis against the alternative two-tailed hypothesis to determine whether or not you can reject the null hypothesis with less than a 10-percent risk of rejecting a true claim.

(f) We now tell you that the regression results reported here are based on the data in Chart 13-1 (page 384). The two-sample t test performed on these data is summarized on page 395. Comparing the regression results with this summary of the two-sample t test, you see that $\overline{X}_B$ is equivalent in the regression procedure to b_0. State symbolically the regression statistics that are equivalent to:

(1) $\overline{X}_A$

(2) $\overline{X}_A - \overline{X}_B$

(3) $\hat{\sigma}(\overline{X}_A - \overline{X}_B)$

(4) $t_{\overline{x}_A - \overline{x}_B}$

Improved Methods for Tobacco Growers

In an ongoing experiment, tobacco growers are comparing three methods of preparing the tobacco leaf for market. According to the experimental design, one-third of each farmer's crop is prepared by the old hand-tied method, one-third by the baling method, and one-third by the sheeting method in which the leaf is placed in burlap bags. The tobacco growers wish to determine if the new baling and sheeting methods will shorten preparation time without damaging the leaf.

Hand-tied method.

Baling method.

Sheeting method.

Behind the Headline

In Chapter 13, we examined various techniques for comparing two treatments. The experiment described in the headline, however, involves a comparison of three treatments: the hand-tied, sheeting, and baling methods of preparing the tobacco leaf for market. Moreover, studies designed to compare four, five, and six treatments are not uncommon. In Chapter 14, we will introduce a general method for the comparison of three or more groups of measurements—the method called the *analysis of variance*.

the analysis of variance 14

"My process of thought starts upon the supposition that when you have eliminated all which is impossible, then whatever remains, however improbable, must be the truth. It may well be that several explanations remain, in which case one tries test after test until one or the other of them has a convincing amount of support."

Holmes to Watson
in *The Blanched Soldier*

14.1 a three-group comparison

At the same time that experiments are being conducted to improve the profitability of tobacco crops, other experiments are being designed to study the adverse effects of cigarette smoking. In one of these—the pulmonary-function study (PFS)—data were gathered on the lung capacity of individuals who were not afflicted with a known respiratory disease. One of the primary purposes was to determine whether cigarette smoking impairs lung function to the point that a smoker becomes unable to exhale air effectively, and, if so, whether the lung function returns to normal after an individual gives up smoking.

In an initial phase of the PFS, employees from one large company were selected as test subjects. From information provided in the company's employee medical records, three groups of test subjects were formed. Each of these groups included 25 employees:

Group A: Nonsmokers (never have smoked cigarettes)

Group B: Exsmokers

Group C: Smokers

Each subject's lung function was measured in terms of "forced vital capacity" by a pulmonary function test and was recorded as a percent of the normal lung capacity for a person of that sex, age, and height. The results are tabulated in Chart 14-1 in order of increasing lung-function score. A score of 0.648, for example, indicates that the subject's "forced vital capacity" was 64.8 percent of the normal capacity—not as good as a score of 1.137, which indicates that the subject's lung capacity was 13.7 percent above normal.

If we had only two treatments to compare (smokers versus nonsmokers), it would be logical for us to use the two-sample t test. We would test the null hypothesis that there is no inherent difference in lung function between smokers and nonsmokers against the alternative hypothesis that the mean of the distribution of smokers' lung-function scores is lower than the mean of the nonsmoking population. We would have a *randomized* design (separate random samples from each population), and our sample sizes would be large enough to justify the two-sample t test.

However, we have three groups to compare, not just two.

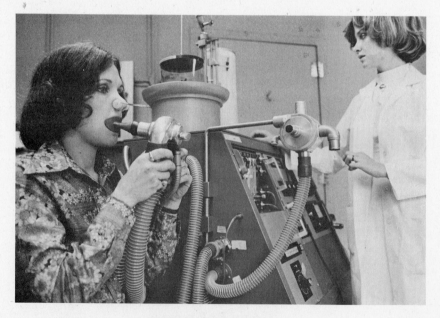

Lung-function test.

chart 14-1

GROUP A (Nonsmokers)	GROUP B (Exsmokers)	GROUP C (Smokers)
0.648	0.759	0.610
0.785	0.770	0.626
0.967	0.807	0.643
1.038	0.814	0.707
1.050	0.848	0.732
1.051	0.867	0.809
1.101	0.925	0.837
1.137	0.955	0.846
1.143	0.984	0.977
1.144	0.985	0.989
1.150	1.043	1.005
1.155	1.068	1.008
1.157	1.070	1.009
1.160	1.073	1.011
1.170	1.106	1.023
1.172	1.131	1.037
1.172	1.139	1.059
1.206	1.157	1.133
1.209	1.194	1.182
1.231	1.204	1.207
1.252	1.211	1.214
1.267	1.226	1.224
1.277	1.251	1.341
1.319	1.286	1.376
1.356	1.323	1.338
$\overline{X}_A = 1.133$	$\overline{X}_B = 1.048$	$\overline{X}_C = 0.998$
$\hat{\sigma}_A^2 = 0.0238$	$\hat{\sigma}_B^2 = 0.0286$	$\hat{\sigma}_C^2 = 0.0513$

So we need to extend the principles of the two-sample t test to cover the case of a three-group comparison. In doing this, we arrive at a procedure that can be used to compare any number of groups—a procedure called ANOVA, the *analysis of variance.*

14.2 the ANOVA procedure

The analysis of variance follows the same logical process that we use for two-group comparisons. We begin by computing summary statistics to represent the results *observed* from our samples. We then determine the nature of the sample results we would *expect* to find if the null hypothesis is true and measure the discrepancy between the observed and the expected results. Finally, we apply laws of chance variation to determine the probability (the P value) that this discrepancy will arise by chance in random sampling. If the P value is very low, we can rule out chance as an explanation for the discrepancy and conclude that there is an inherent difference among the mean effects of the treatments.

1. State the null and alternative hypotheses.

    The null hypothesis asserts that the differences among treatments are of no consequence. A person's lung capacity is not affected by that person's status as a smoker, an exsmoker, or a nonsmoker.

    > The null hypothesis implies that the distributions of lung-function scores for the three populations are identical. The three populations have a common mean μ, a common standard deviation σ or variance σ^2, and a common shape. If the null hypothesis is true, the three samples of lung-function scores are merely three random samples from a common population of inference.

    Chart 14-2 illustrates the implication of the null hypothesis.

    The alternative hypothesis asserts that the lung-function scores of at least one population differ on the average from the lung-function scores of the other populations. In direct parallel to the alternative hypothesis we formulated for the two-sample t test on page 389, the alternative hypothesis in a multigroup comparison assumes that the population distributions still have a common variance σ^2 — or a common standard deviation σ — and a common shape.

    > To support the alternative hypothesis, we require evidence that the mean of the distribution of lung-function scores for at least one population — smokers, exsmokers, or nonsmokers — differs from the mean of the other populations.

2. Summarize our samples of observed responses by summary statistics.

Since we are interested in possible differences in *mean* lung capacity, we summarize our observed results by computing the three sample means. From Chart 14-1:

NONSMOKERS SAMPLE	EXSMOKERS SAMPLE	SMOKERS SAMPLE
$\bar{X}_A = 1.133$	$\bar{X}_B = 1.048$	$\bar{X}_C = 0.998$

Clearly, the means of the samples of lung-function scores differ. The mean of the forced vital capacity scores of the 25 nonsmokers is 13.3 percent above normal, whereas the mean of the forced vital capacity scores of the 25 smokers is 0.2 percent below normal. The exsmokers' scores fall in between, lying 4.8 percent above normal on the average.

**chart 14-2
Common
Population
of Inference**

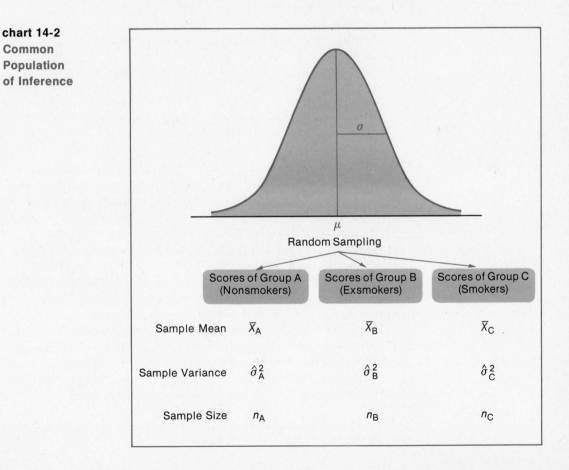

3. Determine the expected values of the summary statistics if the null hypothesis is true.

To determine the expected values of the three sample means, we need to study the implications of the null hypothesis. As we said earlier, if the null hypothesis is true, the three samples of lung-function scores are merely three random samples from a common population of inference. In this circumstance, we can expect each sample mean ($\bar{X}_A$, $\bar{X}_B$, $\bar{X}_C$) to be close to the mean of the common population, which we have labeled μ. But we don't know the value of μ. Our best estimate of μ would be based on the average of all the observations we have — all 75 scores. Such an average is called the *grand mean* and is denoted by $\bar{\bar{X}}$ ("X double bar").

GRAND MEAN The mean of *all* the observations sampled. Under the null hypothesis, $\bar{\bar{X}}$ is an estimate of the mean μ of the common population of inference.

With equal sample sizes,[1] $n_A = n_B = n_C$, the grand mean can be found most simply by

$$\bar{\bar{X}} = \frac{\bar{X}_A + \bar{X}_B + \bar{X}_C}{3} = \frac{1.133 + 1.048 + 0.998}{3} = 1.060 \qquad (14\text{-}1)$$

So, if the null hypothesis is true, we would expect all the individual sample means to be close to the grand mean; that is, we would expect the $\bar{X}$'s to vary only slightly about $\bar{\bar{X}}$.

4. Measure the discrepancy between the observed and the expected results and see if the discrepancy favors the alternative hypothesis.

If the null hypothesis is true, we can expect all the individual sample means to be close to the grand mean. If the null hypothesis is false and the treatments do have an inherent effect on lung function, then the three sample means will tend to vary widely from the grand mean. We can measure the discrepancy between each sample mean and the grand mean by calculating the difference $\bar{X} - \bar{\bar{X}}$. However, we need to combine these differences into a single measure of discrepancy. For

[1]We will examine the modified equations for cases of unequal sample size in Section 14.5.

theoretical reasons, we use a particular statistic — the *between groups variance* — to do this.[2])

(BETWEEN GROUPS VARIANCE   The measure of the discrepancy between the observed and the expected results used in ANOVA. When all samples are of equal size *n,* Equation (14-2) defines the between groups variance as

$$\text{Between groups variance} = \frac{n[\Sigma(\bar{X} - \bar{\bar{X}})^2]}{\text{No. of groups} - 1} \quad (14\text{-}2))$$

(If all individual sample means are equal to the grand mean, then the between groups variance will be equal to 0. If the discrepancy between the observed and the expected results is small, the between groups variance will be small. So low values for the between groups variance are consistent with the null hypothesis.)
(The differences $\bar{X} - \bar{\bar{X}}$ are squared to prevent large negative differences from canceling large positive differences and thereby giving a false impression that the overall discrepancy is small. As a result, no matter what signs the individual differences have, the between groups variance will be a positive quantity, and so we do not need to ask whether the discrepancy favors the alternative hypothesis. Any nonzero result for the between groups variance favors the alternative hypothesis that at least one of the population means differs from the others.)
In the lung-function example, we have three samples of 25 subjects each, and Equation (14-2) reduces to

$$\text{Between groups variance} = \frac{n[\Sigma(\bar{X} - \bar{\bar{X}})^2]}{\text{No. of groups} - 1}$$

$$= \frac{25[\Sigma(\bar{X} - \bar{\bar{X}})^2]}{2}$$

Chart 14-3 provides a worksheet for the computation of the between groups variance using the lung-function data.

[2] As we will see later in the chapter, the between groups variance, when properly standardized, follows a well-known sampling distribution called the *F* distribution.

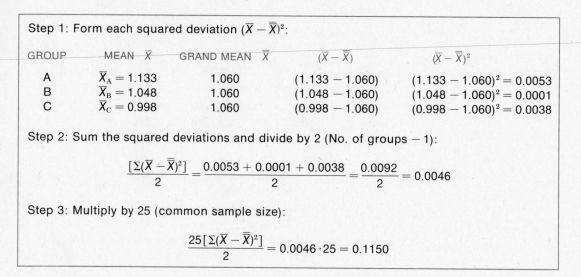

Step 1: Form each squared deviation $(\overline{X} - \overline{\overline{X}})^2$:

GROUP	MEAN $\overline{X}$	GRAND MEAN $\overline{\overline{X}}$	$(\overline{X} - \overline{\overline{X}})$	$(\overline{X} - \overline{\overline{X}})^2$
A	$\overline{X}_A = 1.133$	1.060	$(1.133 - 1.060)$	$(1.133 - 1.060)^2 = 0.0053$
B	$\overline{X}_B = 1.048$	1.060	$(1.048 - 1.060)$	$(1.048 - 1.060)^2 = 0.0001$
C	$\overline{X}_C = 0.998$	1.060	$(0.998 - 1.060)$	$(0.998 - 1.060)^2 = 0.0038$

Step 2: Sum the squared deviations and divide by 2 (No. of groups − 1):

$$\frac{[\Sigma(\overline{X} - \overline{\overline{X}})^2]}{2} = \frac{0.0053 + 0.0001 + 0.0038}{2} = \frac{0.0092}{2} = 0.0046$$

Step 3: Multiply by 25 (common sample size):

$$\frac{25[\Sigma(\overline{X} - \overline{\overline{X}})^2]}{2} = 0.0046 \cdot 25 = 0.1150$$

chart 14-3

The between groups variance is equal to 0.115. (The values for the $\overline{X}$'s are not identical, so the between groups variance is not zero. But chance would make it unlikely that this value would be 0. What we need to know is how large a value for the between groups variance is large enough to reject the null hypothesis (to reject the chance explanation for the discrepancy beyond a reasonable doubt).)

(5. Determine the P value to measure the plausibility of the chance explanation for the discrepancy.)
(The amount by which the between groups variance will differ from zero merely by chance depends on the variance of the scores in the common population of inference σ^2, as well as on sample size n. This relationship follows from the Standard Error Law for $\overline{X}_n$ (see page 255), which tells us that the means of random samples of size n that are selected from the *same* population of inference vary from μ by an "average" of $\sigma/\sqrt{n}$. For a fixed n, the variability of the $\overline{X}$'s about μ will be larger as σ becomes larger; or—another way of saying the same thing—the variability of the $\overline{X}$'s about μ, as estimated by the between groups variance, will be larger as σ^2 becomes larger.)

(To determine a P value, therefore, we must initially estimate the population variance σ^2 and express the between groups variance relative to the size of this estimate.)

Pooled Variance

(An unbiased estimate of the common population variance can be obtained from the statistic called the *pooled variance,* denoted by $\hat{\sigma}^2_{\text{pool}}$.)

(**POOLED VARIANCE   An estimate of the common population variance. When sample sizes are equal, the pooled variance is the average (mean) of the individual sample variances.**)

(Equation (14-3) defines the pooled variance for the case of *three* samples of equal size:

$$\hat{\sigma}^2_{\text{pool}} = \frac{\hat{\sigma}^2_{A} + \hat{\sigma}^2_{B} + \hat{\sigma}^2_{C}}{3} \quad) \qquad (14\text{-}3)$$

From the information on lung-function scores given in Chart 14-1, we extract

$$\hat{\sigma}^2_{A} = 0.0238$$

$$\hat{\sigma}^2_{B} = 0.0286$$

$$\hat{\sigma}^2_{C} = 0.0513$$

So from Equation (14-3)

$$\hat{\sigma}^2_{\text{pool}} = \frac{\hat{\sigma}^2_{A} + \hat{\sigma}^2_{B} + \hat{\sigma}^2_{C}}{3} = \frac{0.0238 + 0.0286 + 0.0513}{3} = 0.0346$$

The *F* Ratio

(If the lung-function scores vary considerably from person to person, $\hat{\sigma}^2_{\text{pool}}$ will be large, implying that the between groups variance can be large just by chance. To adjust for the extent of chance variation, we form the ratio of the between groups variance to the pooled variance. The result is called an F *ratio:*

$$F \text{ ratio} = \frac{\textbf{Between groups variance}}{\textbf{Pooled variance}} \quad) \qquad (14\text{-}4)$$

$\Big($ Low F ratios will be consistent with the null hypothesis that the discrepancy—the between groups variance—is small enough to be attributed to chance. High F ratios favor the alternative hypothesis that the discrepancy is too large to be attributed to chance. $\Big)$

From our data

$$F = \frac{0.1150}{0.0346} = 3.324$$

In this case, the between groups variance is 3.324 times as large as the pooled variance.

the *F* Distribution Law

$\Big($ The distribution of the possible values for the F ratio is governed by the following law of chance variation:

THE F DISTRIBUTION LAW   **If the null hypothesis is true—if the samples of scores are all random samples from a common population with a mean of μ and a variance of σ^2—and if sample sizes are sufficiently large, then the value of the F ratio will be a point from an F distribution with**

Numerator df = No. of groups $-$ 1
Denominator df = df of $\hat{\sigma}^2_{\text{pool}}$ $\Big)$

$\Big($ df of $\hat{\sigma}^2_{\text{pool}}$ is simply the sum of the degrees of freedom for the individual sample variances used in the computation of $\hat{\sigma}^2_{\text{pool}}$. Since the number of degrees of freedom for each sample is 1 less than the sample size for that group, we also can state that

Denominator $df = (n_{\text{A}} - 1) + (n_{\text{B}} - 1) + \cdots$
$= \text{Total sample size} - \text{No. of groups} \Big)$

$\Big($ As with the t distribution, the applicability of the F distribution depends on both the shape of the population of inference and the size of the samples. If the common population has a *normal* shape, then the possible scores for the F ratio will follow an F distribution for any sample size, no matter how small. Moreover, if the common population is not badly skewed, the F distribution can be applied when each group contributes at least 15 degrees of freedom[3] to $\hat{\sigma}^2_{\text{pool}}$.)

Each of the samples in our lung-function study contains 25 scores, making it reasonably safe for us to use the F Distribution Law. From the F distribution, we can determine the probability (P value) that an F ratio ≥ 3.324 will occur by chance.

(Like the t distribution curves, the F distribution curves are distinguished by their number of degrees of freedom. However, every F distribution curve is associated with a *pair* of degrees of freedom—one for the numerator of the F ratio (numerator df) and one for the denominator of the F ratio (denominator df).)

(Chart 14-4 illustrates the F curves for selected pairs of degrees of freedom. If we look at the chart, two other major differences between the t distribution curves and the F distribution curves become obvious. The t distribution curves are symmetric, whereas the F distribution curves are skewed to the

---

[3] The degrees of freedom requirement decreases from 15 as the number of groups increases.

chart 14-4
Selected F Distributions

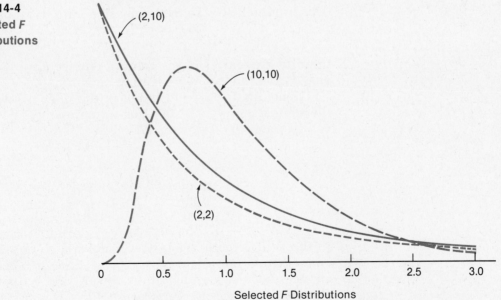

Selected F Distributions

right. A t distribution is centered at 0, whereas all scores in an F distribution are positive.)

In the two-sample t test, we found P values by determining the area in one or both tails of the distribution, depending on whether the alternative hypothesis warranted a one-tailed or a two-tailed test. In contrast, in the analysis of variance procedure, we are always interested in only the right tail area, since F ratios that are close to 0 (in the left tail) are consistent with the null hypothesis.)

Chart 14-5 is one page of the F table; the entire F table appears as Appendix Table G, page 500. The body of the table contains three entries for each pair of degrees of freedom. The first row supplies the values of the F ratio required to place 10 percent of the area under an F curve in the right tail. We use these values to determine whether or not we can reject the null hypothesis with less than a 10-percent risk of rejecting a true claim. The second and third rows, respectively, supply the values of the F ratios required to place 5 percent and 1 percent of the area under an F curve in the right tail.

To use the F table we:

1. Determine numerator df and denominator df:
   Numerator df = No. of groups $- 1 = 3 - 1 = 2$
   Denominator df = Total sample size $-$ No. of groups
   $= 75 - 3 = 72$

2. Identify the entries in the column representing numerator df and the row representing denominator df.
   Denominator $df = 72$ is not in the table, so we use the closest entry $df = 60$. For numerator $df = 2$ and denominator $df = 60$, we read

P VALUE	ENTRY
0.10	2.39
0.05	3.15
0.01	4.98

If the null hypothesis is true, there is a 10-percent chance that the F ratio will be ≥ 2.39, a 5-percent chance that the F ratio will be ≥ 3.15, and a 1-percent chance that F will be ≥ 4.98.

chart 14-5
The F Table

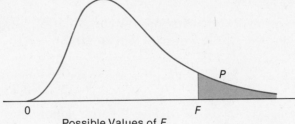

Possible Values of F

Denominator df	P	Numerator df 1	2	3	4	5	6	7	8	9	10
26	0.10	2.91	2.52	2.31	2.17	2.08	2.01	1.96	1.92	1.88	1.86
	0.05	4.22	3.37	2.98	2.74	2.59	2.47	2.39	2.32	2.27	2.22
	0.01	7.72	5.53	4.64	4.14	3.82	3.59	3.42	3.29	3.17	3.09
27	0.10	2.90	2.51	2.30	2.17	2.07	2.00	1.95	1.91	1.87	1.85
	0.05	4.21	3.35	2.96	2.73	2.57	2.46	2.37	2.30	2.25	2.20
	0.01	7.68	5.49	4.60	4.11	3.79	3.56	3.39	3.26	3.14	3.06
28	0.10	2.89	2.50	2.29	2.16	2.06	2.00	1.94	1.90	1.87	1.84
	0.05	4.20	3.34	2.95	2.71	2.56	2.44	2.36	2.29	2.24	2.19
	0.01	7.64	5.45	4.57	4.07	3.76	3.53	3.36	3.23	3.11	3.03
29	0.10	2.89	2.50	2.28	2.15	2.06	1.99	1.93	1.89	1.86	1.83
	0.05	4.18	3.33	2.93	2.70	2.54	2.43	2.35	2.28	2.22	2.18
	0.01	7.60	5.42	4.54	4.04	3.73	3.50	3.33	3.20	3.08	3.00
30	0.10	2.88	2.49	2.28	2.14	2.05	1.98	1.93	1.88	1.85	1.82
	0.05	4.17	3.32	2.92	2.69	2.53	2.42	2.34	2.27	2.21	2.16
	0.01	7.56	5.39	4.51	4.02	3.70	3.47	3.30	3.17	3.06	2.98
40	0.10	2.84	2.44	2.23	2.09	2.00	1.93	1.87	1.83	1.79	1.76
	0.05	4.08	3.23	2.84	2.61	2.45	2.34	2.25	2.18	2.12	2.07
	0.01	7.31	5.18	4.31	3.83	3.51	3.29	3.12	2.99	2.89	2.80
60	0.10	2.79	2.39	2.18	2.04	1.95	1.87	1.82	1.77	1.74	1.71
	0.05	4.00	3.15	2.76	2.52	2.37	2.25	2.17	2.10	2.04	1.99
	0.01	7.08	4.98	4.13	3.65	3.34	3.12	2.95	2.82	2.72	2.63
120	0.10	2.75	2.35	2.13	1.99	1.90	1.82	1.77	1.72	1.68	1.65
	0.05	3.92	3.07	2.68	2.45	2.29	2.18	2.09	2.02	1.96	1.91
	0.01	6.85	4.79	3.95	3.48	3.17	2.96	2.79	2.66	2.56	2.47
∞	0.10	2.71	2.30	2.08	1.94	1.85	1.77	1.72	1.67	1.63	1.60
	0.05	3.84	2.99	2.60	2.37	2.21	2.09	2.01	1.94	1.88	1.83
	0.01	6.63	4.61	3.78	3.32	3.02	2.80	2.64	2.51	2.41	2.32

This table provides the values of F that correspond to a given right tail area P and a specified numerator df and denominator df.

Source: Compiled from *Biometrika Tables for Statisticians*, Volume 1, Third Edition, 1966, Table 18, Pearson and Hartley, eds., and Maxine Merrington and Catherine M. Thompson, "Tables of Percentage Points of the Inverted Beta (F) Distribution," *Biometrika*, 33:73, 1943. Used by permission.

3. Report the P value by comparing the F ratio with the entries in the table:

$$F \text{ ratio} = \frac{\text{Between groups variance}}{\text{Pooled variance}} = 3.324$$

The P value is between 0.01 and 0.05.

If the null hypothesis is true, an F ratio ≥ 3.324 will occur by chance less than 5 percent but more than 1 percent of the time. Consequently, if we reject the null hypothesis, we face less than a 5-percent but more than a 1-percent risk of rejecting a true claim.

Since it is unlikely (there is less than a 5-percent probability) that the variation we observed in the mean lung-function scores of our samples of smokers, exsmokers, and nonsmokers is the result of chance, we can conclude that this study, like many others, implicates smoking as a factor that affects lung capacity. By rejecting the null hypothesis, we accept the alternative hypothesis that the lung-function scores of at least one of the three populations differ on the average from the lung-function scores of the other populations.

to review the
ANOVA steps

1. State the null and alternative hypotheses:
   The null hypothesis is that there is no inherent difference in lung capacity among the populations of smokers, exsmokers, and nonsmokers. The alternative hypothesis is that the lung-function scores of at least one population differ on the average from the lung-function scores of the other populations.

2. Calculate the values of the summary statistics—the sample means:

$$\bar{X}_A = 1.133 \qquad \bar{X}_B = 1.048 \qquad \bar{X}_C = 0.998$$

3. Determine the expected values of the summary statistics if the null hypothesis is true:
   If the null hypothesis is true, then all three sample means should be close to the grand mean

$$\bar{\bar{X}} = \frac{\bar{X}_A + \bar{X}_B + \bar{X}_C}{3} = 1.060$$

4. Measure the discrepancy between the observed and the expected results by computing the between groups variance:

From Equation (14-2)

$$\text{Between groups variance} = \frac{n[\Sigma(\overline{X} - \overline{\overline{X}})^2]}{\text{No. of groups} - 1}$$

$$= \frac{25[\Sigma(\overline{X} - \overline{\overline{X}})^2]}{2}$$

$$= 0.1150$$

5. Determine the P value to measure the plausibility of the chance explanation for this discrepancy:

From the F Distribution Law, we know that the F ratio

$$F = \frac{\text{Between groups variance}}{\text{Pooled variance}}$$

will be a point from an F distribution with numerator df = No. of groups − 1 and denominator df = Total sample size − No. of groups.

(a) Compute the pooled variance $\hat{\sigma}^2_{\text{pool}}$:
From Equation (14-3)

$$\hat{\sigma}^2_{\text{pool}} = \frac{\hat{\sigma}^2_A + \hat{\sigma}^2_B + \hat{\sigma}^2_C}{3} = 0.0346$$

(b) Calculate the F ratio:
From Equation (14-4)

$$F \text{ ratio} = \frac{\text{Between groups variance}}{\text{Pooled variance}} = \frac{0.1150}{0.0346} = 3.324$$

(c) Obtain the appropriate P value:

Numerator df = No. of groups − 1 = 3 − 1 = 2
Denominator df = Total sample size − No. of groups
$$= 75 - 3 = 72$$

From the F table at the entry nearest $df = 2$ and $df = 72$

$$3.15 < 3.324 < 4.98$$

implying a P value between 0.01 and 0.05.

EXERCISES

14-1   The results of four analysis of variance studies are presented here. In each case, determine whether the discrepancy between the observed and the expected results is:

(1) Statistically significant at the 1-percent level.

(2) Statistically significant at the 5-percent level but not at the 1-percent level.

(3) Statistically significant at the 10-percent level but not at the 5-percent level.

(4) Not statistically significant at the 10-percent level.

ANOVA STUDY	F RATIO	NUMERATOR df	DENOMINATOR df
(a)	2.75	2	40
(b)	3.50	3	60
(c)	2.00	3	120
(d)	3.50	4	120

14-2   (*Continuation of Exercise 14-1*) Referring to the preceding ANOVA results, determine in each case:

(a) The number of groups being compared.

(b) The total sample size.

14-3   Three samples of equal size ($n_A = n_B = n_C$) are selected randomly from the same population of inference, which has a mean equal to μ and a variance equal to σ^2.

(a) Denoting the sample means by $\overline{X}_A$, $\overline{X}_B$, and $\overline{X}_C$, respectively, how would you compute an estimate of the population mean μ?

(b) Denoting the sample variances by $\hat{\sigma}^2_A$, $\hat{\sigma}^2_B$, and $\hat{\sigma}^2_C$, respectively, how would you compute an estimate of the population variance σ^2?

*14-4   The results from a comparison of the dropout rates in central city, suburban, and rural high schools are summarized in the following chart. The dropout rate is defined as the number of students who quit a high school during the year divided by the total enrollment in the high school at the start of the year. The data for the study were acquired by a random sampling of 18 high schools from each of the three categories.

Dropout Rates

	CENTRAL CITY HIGH SCHOOLS	SUBURBAN HIGH SCHOOLS	RURAL HIGH SCHOOLS
Mean	$\overline{X}_A = 8.5\%$	$\overline{X}_B = 4.6\%$	$\overline{X}_C = 5.2\%$
Variance	$\hat{\sigma}_A^2 = 7.84\%$	$\hat{\sigma}_B^2 = 5.76\%$	$\hat{\sigma}_C^2 = 6.25\%$
Sample Size (No. of high schools)	$n_A = 18$	$n_B = 18$	$n_C = 18$

An analysis of variance is to be performed to compare dropout rates among the three categories of high schools.

(a) State the null and alternative hypotheses.

(b) Interpret the values of $\overline{X}_A$, $\overline{X}_B$, and $\overline{X}_C$ — the summary statistics for the analysis of variance procedure.

(c) Estimate the expected value of each of the summary statistics if the null hypothesis is true by calculating $\overline{\overline{X}}$.

(d) Measure the discrepancy between the observed and the expected results by calculating the between groups variance.

(e) To measure the plausibility of the chance explanation for the discrepancy:

  (1) Calculate the pooled variance $\hat{\sigma}_{pool}^2$.

  (2) Calculate the F ratio.

  (3) Determine numerator df and denominator df for the F ratio.

  (4) Determine the P value from the F table.

(f) What conclusion can you reach about the issue under examination?

*14-5   Data on the number of automobile accidents (the accident count) that occurred along one segment of a commuter route are presented here. The accident counts have been adjusted to eliminate the effects of time (year-to-year changes). This is why the annual totals are identical.

Accident Counts

YEAR	QUARTER I (Winter)	QUARTER II (Spring)	QUARTER III (Summer)	QUARTER IV (Fall)	ANNUAL TOTAL
1974	17	12	16	18	63
1975	23	8	14	18	63
1976	16	14	12	21	63
1977	20	10	10	23	63
Quarterly Mean	$\overline{X}_I = 19.00$	$\overline{X}_{II} = 11.00$	$\overline{X}_{III} = 13.00$	$\overline{X}_{IV} = 20.00$	
Quarterly Variance	$\hat{\sigma}_I^2 = 10.00$	$\hat{\sigma}_{II}^2 = 6.67$	$\hat{\sigma}_{III}^2 = 6.67$	$\hat{\sigma}_{IV}^2 = 6.00$	

An analysis of variance is to be performed to test for *seasonal* differences in accident counts.

(a) State the null and alternative hypotheses.
(b) If the null hypothesis is true, the seasonal means $\bar{X}_I$, $\bar{X}_{II}$, $\bar{X}_{III}$, and $\bar{X}_{IV}$ have a common expected value μ. Calculate an estimate of μ.
(c) Complete the ANOVA procedure by calculating:
  (1) The between groups variance.
  (2) The pooled variance.
  (3) The *F* ratio.
(d) Can the null hypothesis be rejected with less than a 5-percent risk of rejecting a true claim? With less than a 1-percent risk of rejecting a true claim?
(e) Since the sample sizes were extremely small ($n_I = n_{II} = n_{III} = n_{IV} = 4$), what assumption did you have to make to use the *F* distribution to answer (d)?

*14-6   A random sample of 11 commercial banks from each of six regions in the United States provided analysts with data to compare the regional levels of commercial loan rates during a recent year. The summary statistics follow.

Commercial Loan Rate

	REGION I	REGION II	REGION III	REGION IV	REGION V	REGION VI
No. of Banks	11	11	11	11	11	11
Sample Mean	7.83%	7.34%	7.96%	7.42%	7.58%	7.47%

(a) State the null and alternative hypotheses for an analysis of variance of the regional differences in commercial loan rates.
(b) What is the lowest score for the *F* ratio that would permit you to reject the null hypothesis with no more than a 10-percent risk of rejecting a true claim?
(c) Given $\hat{\sigma}^2_{pool} = 0.50$, can you reject the null hypothesis?

14.3 the ANOVA table

Like regression analysis, the analysis of variance procedure requires considerable effort if calculations are made by hand. Therefore, we rarely perform an ANOVA test without computer assistance. In this section, we will show you how ANOVA information typically is reported on computer printouts as well

GROUP	COUNT	MEAN	STANDARD DEVIATION
GRP. 1 (NONSMOKERS)	25	1.132	0.154
GRP. 2 (EXSMOKERS)	25	1.048	0.169
GRP. 3 (SMOKERS)	25	0.998	0.227
TOTAL	75	1.059	0.192

ANOVA TABLE

---ONEWAY---

SOURCE	DF	SUM OF SQUARES	MEAN SQUARES	F RATIO	F PROB
BETWEEN GROUPS	2	0.233	0.1163	3.362	0.0402
WITHIN GROUPS	72	2.492	0.0346		
TOTAL	74	2.724			

chart 14-6

as in the published research. Chart 14-6 illustrates a conventional format for presenting ANOVA results.

We can interpret the first part of the chart immediately. Each group has been described separately, by the count (number of scores in the group), the mean $\bar{X}$, and the standard deviation $\hat{\sigma}$. The values of $\hat{\sigma}$ are the square roots of the values of $\hat{\sigma}^2$ reported in Chart 14-1 (page 437). The row marked TOTAL completes these same calculations when the smoking status of the individuals is ignored and all 75 individuals are combined into a single sample. This sample mean, then, is the grand mean $\bar{\bar{X}}$. According to this chart, $\bar{\bar{X}} = 1.059$, although we found $\bar{\bar{X}} = 1.060$. As we continue to examine the ANOVA entries, we will find similar discrepancies. They are due to round-off errors in the hand calculations and indicate another advantage of the computer.

The second half of the chart is called the *ANOVA table*. It is labeled ONEWAY because we are considering the influence of only *one factor*—smoking history—on lung capacity. The table contains all the information that we need to complete the ANOVA procedure. Let's look at each column.

SOURCE:
Between Groups refers to the steps leading to the calculation of the between groups variance.

Within Groups refers to the steps leading to the calculation of the pooled variance $\hat{\sigma}^2_{pool}$.

$\Big($ *Total* refers to the total sample of all 75 individuals.

DF (the number of degrees of freedom):
Between Groups: Numerator df

Within Groups: Denominator df

Total: numerator df + denominator $df.\Big)$

$\Big($ SUM OF SQUARES:
Intermediate results sometimes used in computing the between groups variance and the pooled variance.

$$\text{Between groups variance} = \frac{\text{Sum of squares between groups}}{\text{Numerator } df}$$

$$\text{Pooled variance } \hat{\sigma}^2_{\text{pool}} = \frac{\text{Sum of squares within groups}}{\text{Denominator } df}$$

$$\text{Sum of squares, total} = \text{Sum of squares between groups}$$
$$+ \text{Sum of squares within groups} \Big)$$

$\Big($ MEAN SQUARES:
Between Groups: Another name for between groups variance.

Within Groups: Another name for pooled variance $\hat{\sigma}^2_{\text{pool}}$.

Again note the round-off error leading to differences between hand-computed and machine-computed results. $\Big)$

$\Big($ F RATIO:

$$\frac{\text{Mean squares between groups}}{\text{Mean squares within groups}} = \frac{\text{Between groups variance}}{\text{Pooled variance}} = F \Big)$$

$\Big($ F PROB:
The P value associated with the F ratio. The F table supplies only three P values (0.1, 0.05, 0.01), but the printout lists the actual significance level: P value = 0.0402. So rejection of the null hypothesis implies a 4.02 percent risk of rejecting a true claim. $\Big)$

EXERCISES
14-7

(*Continuation of Exercise 14-4, page 450*) Use the results you obtained in Exercise 14-4 to construct an ANOVA table, following the format of the ANOVA table in Chart 14-6 (page 453). You may omit the entries in the Sum of Squares column.

14-8   (*Continuation of Exercise 14-5, page 451*) Prepare an ANOVA table for the analysis of variance in Exercise 14-5, following the format of Chart 14-6. You may omit the entries in the Sum of Squares column.

14-9   Assume that each tobacco grower who participated in the experimental program to compare three methods of tobacco leaf preparation (see the Headline on page 434) was randomly assigned to one of the three methods and that 21 participants used each method. Based on scores denoting *leaf-preparation time*, analysis of variance computations produced these results:

Between Groups Variance = 68
Pooled Variance = 20

Complete the following ANOVA table and state your conclusion about the results.

LEAF-PREPARATION TIME

---ONEWAY---

SOURCE	DF	SUM OF SQUARES	MEAN SQUARES	F RATIO	F PROB
BETWEEN GROUPS	————		————	————	————
WITHIN GROUPS	————	OMIT	————		
TOTAL	————				

14.4   pairwise comparisons

We can conclude from our analysis of variance that smoking history does make a difference in lung-function scores. The null hypothesis of no inherent difference between the lung-function scores of smokers, exsmokers, and nonsmokers can be rejected with less than a 5-percent risk of error. But the test based on the F ratio doesn't tell us *where* the difference lies. We haven't learned if there is a significant difference specifically between smokers and exsmokers, implying that potentially beneficial effects do result from breaking the habit. We don't know if a significant difference exists specifically between exsmokers and nonsmokers, possibly implying that the lung function doesn't fully recover after the habit is broken. And we can't say *how large* a difference there is on the average between the lung-function scores of smokers and nonsmokers.

Questions like these require comparisons between *pairs* of groups. With three groups, up to three pairwise comparisons are possible:

Smokers versus Exsmokers
Smokers versus Nonsmokers
Exsmokers versus Nonsmokers

(Pairwise comparisons can take the form of hypothesis tests (the null hypothesis being that no inherent difference exists between a pair of groups) or estimation (computing a confidence interval for the difference between group means).)

a pairwise
hypothesis test

Having inferred from the ANOVA F test that the means of our groups of smokers, exsmokers, and nonsmokers differ significantly (at the 5-percent level), we can proceed to test for significant differences between pairs of groups.

Let's consider the comparison of smokers (Group C) and exsmokers (Group B).

1. State the null and alternative hypotheses.
   Null Hypothesis: The distribution of the lung-function scores of the populations of smokers and exsmokers are identical, implying, among other things, that $\mu_B = \mu_C$.

   Alternative Hypothesis: $\mu_B > \mu_C$, or the mean of the population of exsmokers is higher than the mean of the population of smokers. (Breaking the habit improves lung capacity.))

   We now have a problem that seems suited to the two-sample t test. Only two groups are being compared, and each group is of adequate sample size ($n_B = 25$ and $n_C = 25$). This is precisely the procedure we use, with one small modification, which will enter in Step 5.

2. Calculate the summary statistics $\overline{X}_B$ and $\overline{X}_C$.
   From the ANOVA printout in Chart 14-6)

$$\overline{X}_B \text{ (Exsmokers)} = 1.048$$
$$\overline{X}_C \text{ (Smokers)} \ = 0.998$$

3. Determine the expected values of $\overline{X}_B$ and $\overline{X}_C$ if the null hypothesis is true.

   If the null hypothesis is true, we would expect $\overline{X}_B$ to be close to $\overline{X}_C$, or equivalently, $\overline{X}_B - \overline{X}_C$ to be close to 0.

4. Measure the discrepancy between the observed and the expected results and see if the discrepancy favors the alternative hypothesis.

   The discrepancy is measured by calculating $\overline{X}_B - \overline{X}_C$:

$$\overline{X}_B - \overline{X}_C = 1.048 - 0.998 = 0.05$$

   The discrepancy does favor the alternative hypothesis that the mean lung-function score for exsmokers is higher than the mean for the population of smokers.

5. Calculate the P value to measure the plausibility of the chance explanation for the discrepancy.

   To determine the P value, we need to calculate the t score:

$$t = \frac{\overline{X}_B - \overline{X}_C}{\hat{\sigma}(\overline{X}_B - \overline{X}_C)}$$

(a) Compute $\hat{\sigma}(\overline{X}_B - \overline{X}_C)$.
    From Equation (13-4)

$$\hat{\sigma}(\overline{X}_B - \overline{X}_C) = \hat{\sigma}_{\text{pool}} \cdot \sqrt{\frac{1}{n_B} + \frac{1}{n_C}} = \sigma_{\text{pool}} \cdot \sqrt{\frac{2}{n}}$$

Following the two-sample t test in Chapter 13—where we started with only two groups—$\hat{\sigma}_{\text{pool}}$ would be computed using Equation (13-3):

$$\hat{\sigma}_{\text{pool}} = \sqrt{\frac{\hat{\sigma}_B^2 + \hat{\sigma}_C^2}{2}}$$

Although we are now performing a two-group comparison (smokers and exsmokers), we began initially with three groups. The standard deviation of the nonsmokers sample, $\hat{\sigma}_A$, also provides an estimate of σ and, accordingly, should be taken into account. So this is our one modification in the two-sample t test:

(Compute $\hat{\sigma}_{\text{pool}}$ based on *all* the groups in the original study — not just on the pair of groups we are comparing.)

From Chart 14-6 (page 453)

$$\hat{\sigma}^2_{\text{pool}} = \text{Within groups mean squares} = 0.0346$$

So

$$\hat{\sigma}_{\text{pool}} = \sqrt{0.0346} = 0.1860$$

and

$$\hat{\sigma}(\overline{X}_B - \overline{X}_C) = \hat{\sigma}_{\text{pool}} \cdot \sqrt{\frac{2}{n}}$$

$$= 0.1860 \cdot \sqrt{\frac{2}{25}} = 0.0526$$

(b) Compute the t score of $(\overline{X}_B - \overline{X}_C)$:

$$t_{\overline{x}_B - \overline{x}_C} = \frac{\overline{X}_B - \overline{X}_C}{\hat{\sigma}(\overline{X}_B - \overline{X}_C)} = \frac{0.05}{0.0526} = 0.951$$

So the value of $\overline{X}_B - \overline{X}_C$ lies less than 1 standard error above 0.

(c) Determine the probability (P value) that a t score ≥ 0.951 can arise by chance.

NOTE 1: The alternative hypothesis $\mu_B > \mu_C$ implies a one-tailed test, so we seek a one-tailed P value.

NOTE 2: The number of degrees of freedom is the number associated with the computation of $\hat{\sigma}_{\text{pool}}$. We used the data from all three groups — all 75 scores — so that

$$df = 75 - 3 = 72$$

Checking the t table at $df = 70$ (the row closest to 72), we find that the one-tailed P value is greater than 0.10. So we cannot reject the null hypothesis without facing more than a 10-percent risk of rejecting a true claim. (The discrepancy $\overline{X}_B - \overline{X}_C$ is not statistically significant even at the 10-percent level of significance.)

> We cannot conclude that the lung-function scores of ex-smokers and smokers are different. Based on the data, we cannot say that breaking the habit produces an improvement in lung capacity.

We have just outlined the procedure for a pairwise test that *follows* the discovery, using the ANOVA F test, that there is a difference in treatments. If the F ratio had not been significant, we would have concluded that there was *no* evidence of treatment differences and this pairwise test would not have been performed. (More about this in Caveat 3 at the end of the chapter.)

a pairwise confidence interval

Let's estimate the size of the difference between μ_A and μ_C, the mean of the population of lung-function scores of non-smokers and smokers, respectively. The confidence interval expression is given by

$$(\overline{X}_A - \overline{X}_C) \pm t \cdot \hat{\sigma}(\overline{X}_A - \overline{X}_C) \qquad (14\text{-}6)$$

Using the ANOVA printout given in Chart 14-6, we obtain

$$\overline{X}_A - \overline{X}_C = 1.132 - 0.998 = 0.134$$

$$\hat{\sigma}(\overline{X}_A - \overline{X}_C) = \hat{\sigma}_{\text{pool}} \cdot \sqrt{\frac{2}{n}}$$

$$= \sqrt{\hat{\sigma}^2_{\text{pool}} \cdot \frac{2}{n}}$$

Again, we use the value of $\hat{\sigma}^2_{\text{pool}}$ reported in Chart 14-6.

$$\hat{\sigma}(\overline{X}_A - \overline{X}_C) = \sqrt{0.0346 \cdot \frac{2}{25}}$$
$$= 0.0526$$

$\Big($ NOTE: This is the same result that is found for $\hat{\sigma}(\overline{X}_B - \overline{X}_C)$, which is not surprising because (1) $n_A = n_B = n_C$, and (2) all three sample variances are used to compute $\hat{\sigma}_{pool}$.$\Big)$

For 95 percent confidence, $t = 1.994$ (using $df = 70$, since 72 is not available). So our 95-percent confidence interval for $\mu_A - \mu_C$ becomes

$$(\overline{X}_A - \overline{X}_C) \pm t \cdot \hat{\sigma}(\overline{X}_A - \overline{X}_C) = 0.134 \pm 1.994 \cdot 0.0526$$

$$= 0.134 \pm 0.105$$

$$= \{0.030 \text{ to } 0.240\}$$
$$\text{percentage points}$$

We can be 95 percent confident that the mean of the lung-function scores of nonsmokers exceeds the mean of the lung-function scores of smokers by 0.030 to 0.240 percentage points.

EXERCISES

14-10   (*Continuation of Exercise 14-4, page 450*)

(a) Does the result of the ANOVA F test permit you to conclude specifically that $\mu_B \neq \mu_C$, where μ_B and μ_C are the population means of the distribution of suburban and rural high school dropout rates, respectively?

(b) Test the null hypothesis that the distributions of suburban and rural high school dropout rates are identical versus the alternative hypothesis that $\mu_B \neq \mu_C$.

14-11   (*Continuation of Exercise 14-4, page 450*) Calculate a 99-percent confidence interval for $\mu_A - \mu_B$, the difference between the means of the distributions of the central city and suburban high school dropout rates.

14-12   Referring to Chart 14-6 (page 453), which summarizes the lung-function comparison of smokers, exsmokers, and nonsmokers, test the null hypothesis that there is no inherent difference in lung function between exsmokers and nonsmokers versus the alternative hypothesis that $\mu_B < \mu_A$. Is the discrepancy $\overline{X}_B - \overline{X}_A$ statistically significant at the 5-percent level?

14-13   Referring to Chart 14-6 on page 453, calculate a 95-percent confidence interval for $\mu_B - \mu_A$ and interpret your result.

14.5   ANOVA: the case of unequal sample sizes

The lung-function data set that we have analyzed in this chapter contains groups of equal size: Each sample of smokers, exsmokers, and nonsmokers numbers 25. (Cases of unequal sample sizes present no new problems if the requisite calculations are performed by the computer, but they do require the use of new equations if the computations are to be done by hand. These equations are introduced in the following example.)

A pulmonary-function study performed at another industrial site generated these summary statistics (as before, the subscripts, A, B, and C denote, respectively, smokers, exsmokers, and nonsmokers):

$$\overline{X}_A = 1.32 \qquad \overline{X}_B = 1.10 \qquad \overline{X}_C = 0.95$$

$$\hat{\sigma}_A^2 = 0.035 \qquad \hat{\sigma}_B^2 = 0.031 \qquad \hat{\sigma}_C^2 = 0.040$$

$$n_A = 20 \qquad n_B = 25 \qquad n_C = 35$$

(To perform the ANOVA F test, we need to calculate:

1. The grand mean $\overline{\overline{X}}$.
2. The between groups variance.
3. The pooled variance $\hat{\sigma}_{pool}^2$.

WORKSHEET 1: THE GRAND MEAN

$$\overline{\overline{X}} = \frac{\Sigma(n \cdot \overline{X})}{\Sigma n} \tag{14-7}$$

$$\Sigma(n \cdot \overline{X}) = (n_A \cdot \overline{X}_A) + (n_B \cdot \overline{X}_B) + (n_C \cdot \overline{X}_C)$$

$$= (20 \cdot 1.32) + (25 \cdot 1.10) + (35 \cdot 0.95)$$

$$= 87.15$$

$$\Sigma n = n_A + n_B + n_C$$

$$= 20 + 25 + 35$$

$$= 80$$

$$\overline{\overline{X}} = \frac{\Sigma(n \cdot \overline{X})}{\Sigma n} = \frac{87.15}{80} = 1.09$$

WORKSHEET 2: THE BETWEEN GROUPS VARIANCE

$$\text{Between groups variance} = \frac{\Sigma[n \cdot (\overline{X} - \overline{\overline{X}})^2]}{\text{No. of groups} - 1} \quad (14\text{-}8)$$

$$\Sigma[n \cdot (\overline{X} - \overline{\overline{X}})] = n_A(\overline{X}_A - \overline{\overline{X}})^2 + n_B(\overline{X}_B - \overline{\overline{X}})^2 \\ + n_C(\overline{X}_C - \overline{\overline{X}})^2$$

$$= 20(1.32 - 1.09)^2 + 25(1.10 - 1.09)^2 \\ + 35(0.95 - 1.09)^2$$

$$= 20(0.0529) + 25(0.0001) \\ + 35(0.0196)$$

$$= 1.7465$$

$$\text{No. of groups} - 1 = 3 - 1 = 2$$

$$\text{Between groups variance} = \frac{\Sigma[n \cdot (\overline{X} - \overline{\overline{X}})^2]}{\text{No. of groups} - 1}$$

$$= \frac{1.7465}{2} = 0.873$$

WORKSHEET 3: THE POOLED VARIANCE

$$\hat{\sigma}^2_{\text{pool}} = \frac{\Sigma[(n - 1) \cdot \hat{\sigma}^2]}{\Sigma n - \text{No. of groups}} \quad (14\text{-}9)$$

$$\Sigma[(n - 1) \cdot \hat{\sigma}^2] = [(n_A - 1) \cdot \hat{\sigma}^2_A] + [(n_B - 1) \cdot \hat{\sigma}^2_B] + \\ [(n_C - 1) \cdot \hat{\sigma}^2_C]$$

$$= (19 \cdot 0.035) + (24 \cdot 0.031) + (34 \cdot 0.040)$$

$$= 2.769$$

$$\Sigma n - \text{No. of groups} = n_A + n_B + n_C - 3$$

$$= 20 + 25 + 35 - 3$$

$$= 77$$

$$\hat{\sigma}^2_{\text{pool}} = \frac{\Sigma[(n - 1) \cdot \hat{\sigma}^2]}{\Sigma n - \text{No. of groups}}$$

$$= \frac{2.769}{77} = 0.036$$

With the results from these three worksheets, we can calculate the F ratio:

$$F \text{ ratio} = \frac{\text{Between groups variance}}{\text{Pooled variance}}$$

$$= \frac{0.873}{0.036} = 24.25$$

Numerator $df =$ No. of groups $- 1 = 2$

Denominator $df = \Sigma n -$ No. of groups $= 77$

From the F table, we determine that $P < 0.01$. This study also supports the alternative hypothesis that smoking affects lung capacity.

EXERCISE
14-14

The 80 students who had just completed a survey course in computer science were placed in one of four groups according to their final grades: A, B, C, or Less than C. Each student was asked to rate the overall capability of the instructor from $10 =$ superb to $1 =$ horrible. The results follow.

GROUP	COUNT	MEAN	VARIANCE
A	16	7.5	2.80
B	28	7.2	3.40
C	24	7.0	3.20
Less than C	12	6.6	3.70

To see whether the data support the hypothesis that students' evaluations of their instructors are influenced by the grades they receive, an analysis of variance is to be performed.
(a) State the null and alternative hypotheses.
(b) Calculate the grand mean $\overline{\overline{X}}$.
(c) Calculate the between groups variance.
(d) Calculate the pooled variance $\hat{\sigma}^2_{\text{pool}}$.
(e) Calculate the F ratio and determine the appropriate P value.
(f) State your conclusion about the issue under examination.

Caveats on Interpreting ANOVA Results

CAVEAT 1:   One-factor versus multifactor ANOVA

In oneway ANOVA, we consider that only one variable— *smoking history*— possibly influences lung-function scores. Potentially, however, other factors, such as work environment, may influence these scores. We speak of oneway ANOVA as a one-factor model, which is analogous to simple regression. Multifactor ANOVA corresponds to multiple regression, in which we account for several variables at once and consider how they jointly affect the variable of concern—in this case, lung capacity.

In ANOVA, as in regression, if factors that influence the variable under study are ignored, the results can be seriously biased. Do not conduct a one-factor analysis when a multifactor analysis is required.

CAVEAT 2:   Sample-size considerations

The F Distribution Law includes a constraint on sample size. As is true of the two-sample t test, an alternative procedure should be used when sample size is small. The alternative procedure is analogous to the Wilcoxon rank sum test and is known as the *Kruskal–Wallis test*.

When the shape of the population and limitations on sample size make the use of the oneway ANOVA questionable, the Kruskal–Wallis procedure can be used.

Many computer programs now offer the Kruskal–Wallis procedure as an alternative to ANOVA, thereby freeing the researcher from having to make rather extensive calculations.

CAVEAT 3:   Constraint on pairwise comparisons

We pointed out earlier that if the ANOVA F test fails to reject the null hypothesis of no difference, we should *not* conduct pairwise tests for the differences between groups. An insignificant F ratio tells us that the data do not demonstrate the existence of an inherent difference between treatments.

Performing a pairwise test under this circumstance amounts to manipulation of the data. If several sample means differ from one another by chance, the smallest and the largest means, for example, may be far enough apart to produce a significant difference in a pairwise comparison. However, if we declared these two groups to be significantly different, we would be ignoring the data from the other groups, which indicated via ANOVA that all discrepancies could be attributed to chance variation. We could legitimately be accused of selectively ignoring the data that did not fit our conclusions.

Believe the results of the ANOVA F Test. Unless an overall significant difference is found between sample means, do not proceed to test for pairwise differences.

EXERCISES

1. (*Continuation of Exercise 14-4, page 450*) Suppose that the variables *dropout rate* and *size of school* are positively correlated. Would this lead you to be skeptical about the results of the analysis of variance performed in Exercise 14-4?

2. (*Continuation of Exercise 14-5, page 451*) If the accident counts had not been adjusted to eliminate the effects of year-to-year changes, would it still have been appropriate to conduct a oneway analysis of variance? Explain your answer.

3. (*Continuation of Exercise 14-6, page 452*)
   (a) Based on the results of the ANOVA F test for regional differences in commercial loan rates, would you have tested for pairwise differences between regions? Why or why not?

(b) Show that if you ignore the results of the F test and perform pairwise comparisons, the difference between the largest and the smallest regional means would be declared statistically significant at the 10-percent level.

COMING
TO
TERMS

Between Groups Variance   The ANOVA measure of the discrepancy between the observed and the expected results if the null hypothesis is true. On computer printouts, this is frequently called the *mean squares between groups*.

Common Population of Inference   Under the null hypothesis, the samples taken from each population can simply be considered samples from a single population. We call this population the *common population of inference*.

F Distributions   A class of distributions that represent the sampling distribution of F ratios.

F Ratio   The ratio Between groups variance/Pooled variance. A measure of the size of the discrepancy between the observed and the expected results relative to the extent of chance variation in the sample scores.

Grand Mean $\overline{\overline{X}}$ An estimate of the mean of the common population of inference under the assumption that the null hypothesis is true. The grand mean is defined as the sum of all the sample scores divided by the total number of scores sampled.

Pooled Variance $\hat{\sigma}^2_{\text{pool}}$ An unbiased estimate of the common population variance σ^2, calculated as a (weighted) average of the individual sample variances. Also called the *mean squares within groups*.

ISSUE
FOR
ANALYSIS

(*Continuation of Exercise 14-5 on page 451*) The ANOVA F test in Exercise 14-5 permitted you to reject the null hypothesis of no inherent difference among seasonal accident counts with less than a 1-percent risk of rejecting a true hypothesis. But this result does not specifically indicate whether, for example, the mean accident count in Winter, $\overline{X}_I$, differs by a statistically significant amount from the mean accident count in Fall, $\overline{X}_{IV}$. In all, six pairwise comparisons can be made:

(1) $\overline{X}_I - \overline{X}_{II}$ (Winter versus Spring)
(2) $\overline{X}_I - \overline{X}_{III}$ (Winter versus Summer)
(3) $\overline{X}_I - \overline{X}_{IV}$ (Winter versus Fall)
(4) $\overline{X}_{II} - \overline{X}_{III}$ (Spring versus Summer)
(5) $\overline{X}_{II} - \overline{X}_{IV}$ (Spring versus Fall)
(6) $\overline{X}_{III} - \overline{X}_{IV}$ (Summer versus Fall)

(a) Calculate each pairwise difference between seasonal means and enter the numeric value of the result (ignoring its sign) in the appropriate location in the following chart:

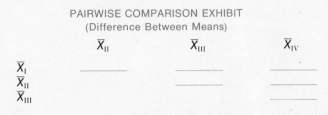

PAIRWISE COMPARISON EXHIBIT
(Difference Between Means)

	$\overline{X}_{II}$	$\overline{X}_{III}$	$\overline{X}_{IV}$
$\overline{X}_{I}$	_____	_____	_____
$\overline{X}_{II}$		_____	_____
$\overline{X}_{III}$			_____

To determine which pairwise differences are statistically significant, you could follow the procedure given in Section 14-4 and determine six P values. Alternatively, suppose you have decided to use a two-tailed P value $= 0.01$ as the cutoff point for declaring that a difference is statistically significant. Then, when sample sizes are equal, a simpler approach is to determine the smallest difference that will give you a P value $= 0.01$ and to compare each of the pairwise differences to this value, called the *least significant difference* LSD.

$$LSD = t \cdot \hat{\sigma}(\overline{X}_i - \overline{X}_j) \qquad (i \text{ and } j \text{ refer to any two seasons})$$

$$= t \cdot \hat{\sigma}_{pool} \cdot \sqrt{\frac{2}{n}}$$

where t is the entry in the t table that is associated with:
(1) two-tailed P value $= 0.01$
(2) $df =$ number of degrees of freedom of $\hat{\sigma}_{pool}$

(b) Calculate the LSD.
(c) Place an asterisk * after each entry in the Pairwise Comparison Exhibit in (a) that exceeds the value of the LSD.
(d) Determine the LSD at the 5-percent level of significance (two-tailed P value $= 0.05$). Is there any entry in the Pairwise Comparison Exhibit that exceeds this LSD but that does not exceed the LSD for the 1-percent level of significance?

Marriage Has Opposite Effects On Buckling Up

WASHINGTON (AP)— Marriage makes young men more likely to buckle their seat belts in cars, but it has an opposite effect on their new wives, a new study indicates.

The percentage of men who say they never use seat belts drops from 50 percent to 40 percent after a trip to the altar. Those figures are reversed for women, according to two Johns Hopkins University researchers.

The study was conducted among 1,000 persons in western Maryland in 1974. Overall, 47 percent said they never used belts, 27 percent

Used by permission of The Associated Press.

used them sometimes, and 26 percent, always.

The researchers found, as past studies have shown, that more women than men refuse to buckle up . . . .

The researchers found that nonuse of seat belts was higher among people who went to church infrequently or who were dissatisfied with their station in life. Persons who continued their education beyond high school used seat belts more often.

The study by Knud J. Helsing and Dr. George W. Comstock appears in the November [1977] issue of the *American Journal of Public Health* . . . .

Behind the Headline

Percentages like the ones in this study appear in the news regularly. But do these and similar results represent real differences? Does marriage really affect the safety patterns of men and women differently? Do more women than men refuse to use safety belts, or are the differences reported here just due to chance variation?

Until now, our group comparisons have been based on observations generated by a measurement process: For example, we *measured* the blood pressure responses to guanadrel and guanethidine. Now we are concerned with observations that result from a classification process. An individual is classified man, woman, married, not married, seat belt user, not seat belt user. In this chapter, we will examine an inferential procedure that is used to compare the frequencies of scores among categories. This procedure is called the *chi-square test.*

the chi-square test 15

"Education never ends, Watson. It is a series of lessons, with the greatest for the last. This is an instructive case."

The Adventure of the Red Circle

15.1 comparing two proportions

Social scientists have devoted extensive study to the behavior of people at work. One of their interests has been the factors that influence occupational performance. Why do some employees achieve superior ratings while others are rated adequate at best?

One study examined the role played by the variable *outside interests* in motivating or hindering job performance. Three hypotheses were considered:

1. *The Stimulation Hypothesis:* Individuals with major outside interests — those who view non-job-related activities as a major contributor to personal fulfillment — tend to be relatively good performers. Outside interests stimulate on-the-job performance.
2. *The Diversion Hypothesis:* Individuals with major outside interests tend to be relatively poor performers. Outside interests divert energies from on-the-job activity.
3. *The Null Hypothesis:* Job performance is unaffected by the extent of outside interests.

The test subjects in this study were a group of 105 Ph.D. scientists, all under 35 years old and employed in research and development for a large company. Based on a questionnaire designed to determine the extent of outside interests, each subject was placed in one of two groups: Group A (outside interests are Major), or Group B (outside interests are Minor).

Based on the amount and quality of work accomplished, each subject was rated as a Superior performer or as a Not Superior performer (no better than an adequate performer). The results are summarized in Chart 15-1.

(Chart 15-1 is called a 2×2 ("2 by 2") *cross-classification table,* since there are two groups of test subjects (Group A and Group B) and two categories of job performance (Superior and Not Superior). Each of the four boxes is called a *cell,* and the entry in each cell is referred to as the *cell frequency* (number of subjects).)

The results in Chart 15-1 appear to support the stimulation hypothesis because the proportion of scientists in Group A who are Superior performers (33 of 50 = 66 percent) is much larger than the proportion of scientists in Group B who are Superior performers (18 of 55 = 32.7 percent).

But, as always, we need to consider the role played by chance in generating our sample results. It may be that in the population of Ph.D. scientists working in research and development, the proportion of Superior performers is the same in the two interest groups and that only the chance selection of this sample of scientists led to the differences observed.

(To test the null hypothesis of equality between proportions, we could extend the laws of chance variation we discussed in Chapter 12 from the case of a proportion in a single group to the case of comparing proportions in two groups. Although this approach is conceptually straightforward, we frequently choose another: the *chi-square test.* The advantage of the

chart 15-1
Observed
Frequencies

PERFORMANCE CATEGORY	OUTSIDE INTEREST GROUP		ROW TOTAL
	Group A (Major)	Group B (Minor)	
Superior	33	18	51
Not Superior	17	37	54
COLUMN TOTAL	50	55	105 GRAND TOTAL

chi-square test is its flexibility. Of the two approaches, the chi-square test can more easily be extended to comparisons of more than two groups as well as to cases in which the scores fall into more than two categories.)

15.2   the chi-square test

(1. State the null and alternative hypotheses.)

(In a 2×2 cross classification, the null hypothesis asserts equality between two population proportions $\pi_A = \pi_B$. In our example, the proportion of interest is the proportion of Superior performers, and $\pi_A = \pi_B$ implies that the proportion of Superior performers is the same for the two interest groups. (The null hypothesis also implies that the proportion of Not Superior performers is the same for the two interest groups.)

The alternative hypothesis is that the population proportions are not equal $\pi_A \neq \pi_B$, implying that the proportion of Superior performers is not the same for the two interest groups./

(2. Summarize our samples of observed responses by summary statistics.

The cell frequencies provide the summary statistics for a chi-square test.) In this case, our summary statistics are the four cell frequencies given in Chart 15-1. In Group A, 33 scientists were Superior performers and 17 were not. In Group B, 18 scientists were Superior performers and 37 were not.

(3. Determine the expected values of the summary statistics if the null hypothesis is true.)

If extent of outside interests makes no inherent difference in job performance (the null hypothesis), then the Superior performers should be divided between Group A and Group B in proportion to group size. Overall, 48.6 percent (51 out of 105) of the scientists are Superior performers. So if the null hypothesis is true, 48.6 percent of the scientists in Group A as well as 48.6 percent of the scientists in Group B should be Superior performers.

To find the expected *frequency* (number of scientists) of Superior performance in Group A, we take 48.6 percent of the number of scientists in Group A. 48.6 percent of 50 is 24.3. In Chart 15-2, we enter 24.3 as the expected frequency in the first cell.

**chart 15-2
Expected
Frequencies**

PERFORMANCE CATEGORY	OUTSIDE INTEREST GROUP		ROW TOTAL
	Group A (Major)	Group B (Minor)	
Superior	24.3	26.7	51
Not Superior	25.7	28.3	54
COLUMN TOTAL	50	55	105
			GRAND TOTAL

The frequencies in the remaining three cells of the table can be determined by following the same reasoning. However, we can take a shortcut. Since we know the row and column totals—the total number of scientists in each group and in each category—we can calculate the remaining cell frequencies simply by subtraction. So, under the null hypothesis, we expect:

(a) $51 - 24.3 = 26.7$ Superior performers in Group B
(b) $50 - 24.3 = 25.7$ Not Superior performers in Group A
(c) $55 - 26.7$ or $54 - 25.7 = 28.3$ Not Superior performers in Group B.

We have determined the expected frequencies in all four cells by calculating the expected frequency in only one cell directly.

4. Measure the discrepancy between the observed and the expected results and see if the discrepancy favors the alternative hypothesis.

We now have computed both the expected frequencies E implied by the null hypothesis (Chart 15-2) and the observed frequencies O in our sample (Chart 15-1). We need a statistic to measure the discrepancy between the observed and the expected results. This is the *chi-square statistic*, denoted by χ^2.

For each cell, we calculate the difference between the observed and the expected frequency, $O - E$. Then we square each difference and divide by the expected frequency, $(O - E)^2/E$. Adding these values for all cells gives us χ^2.

$$\chi^2 = \Sigma \frac{(O - E)^2}{E} \qquad (15\text{-}1)$$

If the observed frequency in each cell were *identical* to the expected frequency in that cell, there would be no discrepancy and χ^2 would equal 0. So values of χ^2 that are close to 0 are consistent with the null hypothesis. On the other hand, if the null hypothesis is false, it is unlikely that the observed and the expected frequencies will match; so χ^2 will be a large number. Because the difference between the observed and the expected frequency of each cell is squared in the computation of χ^2, we will obtain a large *positive* result for χ^2 when the truth is either that $\pi_A > \pi_B$ (the stimulation hypothesis) or that $\pi_B > \pi_A$ (the diversion hypothesis). For this reason, the question of whether the discrepancy favors the alternative hypothesis does not apply in this test procedure.

To make it easier to compute χ^2, it is helpful to place both the observed and the expected cell frequencies in the same table. We have done this in Chart 15-3.

Using Equation 15-1 we can proceed to compute χ^2:

$$\chi^2 = \frac{(33 - 24.3)^2}{24.3} + \frac{(18 - 26.7)^2}{26.7} + \frac{(17 - 25.7)^2}{25.7} + \frac{(37 - 28.3)^2}{28.3}$$

$$= 3.11 + 2.83 + 2.95 + 2.67$$

$$= 11.56$$

5. Determine the P value to measure the plausibility of the chance explanation for the discrepancy.

We need to determine the probability that χ^2 will be as large as 11.56 when the null hypothesis is true. To do this, we use the *Chi-Square Distribution Law.*

THE CHI-SQUARE DISTRIBUTION LAW **Provided our expected frequencies are sufficiently large, the χ^2 score will be a point from a chi-square distribution with $df = (c - 1) \cdot (r - 1)$, where c is the number of columns and r is the number of rows in the cross-classification table.**

chart 15-3
Observed and [Expected] Frequencies

PERFORMANCE CATEGORY	OUTSIDE INTEREST GROUP	
	Group A (Major)	Group B (Minor)
Superior	33 [24.3]	18 [26.7]
Not Superior	17 [25.7]	37 [28.3]

**chart 15-4
Chi-Square
Distribution
Curves**

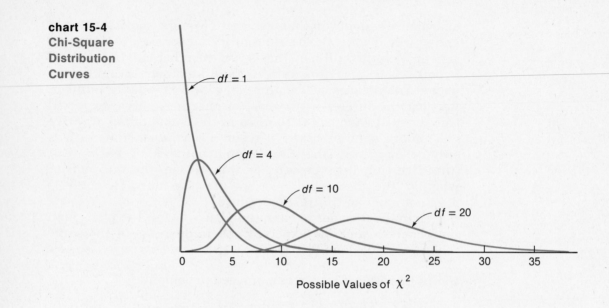

As the law implies, there is a different chi-square distribution for each number of degrees of freedom. Chart 15-4 illustrates the frequency curves of a chi-square distribution for 1, 4, 10, and 20 degrees of freedom (Like the F distribution curves, the chi-square distributions are skewed to the right and contain no scores less than 0.)

(In the case of χ^2 we can think of the degrees of freedom as the number of *free cells*) When we set up Chart 15-2, the expected frequencies in each cell, we found that once we calculated the expected frequency for one cell, we could determine the remaining cell frequencies by subtraction. So in a 2×2 cross classification, there is one free cell, and the value we obtain for χ^2 ($\chi^2 = 11.56$) is a point from a chi-square distribution with $df = 1$.

We obtain the same result from the Chi-Square Distribution Law, which tells us that $df = (c - 1) \cdot (r - 1)$. With two columns, $c - 1 = 2 - 1 = 1$, and with two rows, $r - 1 = 2 - 1 = 1$. So $df = 1 \cdot 1 = 1$.

(The χ^2 scores associated with selected P values are presented in Appendix Table H, which is reproduced here as Chart 15-5. The columns list the P values for the right tail (the blue area in the graph); the row is determined by the number of degrees of freedom.) Referring to the row for $df = 1$, we see that the highest entry for χ^2 is 6.635 in the column P value $= 0.01$.

chart 15-5
Chi-Square
Table

Possible Values of χ^2

DEGREES OF FREEDOM df	RIGHT-TAIL AREA P			
	0.10	0.05	0.02	0.01
1	2.706	3.841	5.412	6.635
2	4.605	5.991	7.824	9.210
3	6.251	7.815	9.837	11.345
4	7.779	9.488	11.668	13.277
5	9.236	11.070	13.388	15.086
6	10.645	12.592	15.033	16.812
7	12.017	14.067	16.622	18.475
8	13.362	15.507	18.168	20.090
9	14.684	16.919	19.679	21.666
10	15.987	18.307	21.161	23,209
11	17.275	19.675	22.618	24.725
12	18.549	21.026	24.054	26.217
13	19.812	22.362	25.472	27.688
14	21.064	23.685	26.873	29.141
15	22.307	24.996	28.259	30.578
16	23.542	26.296	29.633	32.000
17	24.769	27.587	30.995	33.409
18	25.989	28.869	32.346	34.805
19	27.204	30.144	33.687	36.191
20	28.412	31.410	35.020	37.566
21	29.615	32.671	36.343	38.932
22	30.813	33.924	37.659	40.289
23	32.007	35.172	38.968	41.638
24	33.196	36.415	40.270	42.980
25	34.382	37.652	41.566	44.314
26	35.563	38.885	42.856	45.642
27	36.741	40.113	44.140	46.963
28	37.916	41.337	45.419	48.278
29	39.087	42.557	46.693	49.588
30	40.256	43.773	47.962	50.892

This table contains the values of χ^2 that correspond to a specific right-tail area P and specific numbers of degrees of freedom df.
Source: From Table IV of R. A. Fisher and F. Yates, *Statistical Tables for Biological, Agricultural and Medical Research*, Sixth Edition (London: Longman Group Limited, 1974), page 47. Used by permission.

(If the null hypothesis is true, there is a 1-percent chance
of finding $\chi^2 \geq 6.635$. Therefore, $\chi^2 \geq 11.56$ cannot arise
by chance as much as 1 percent of the time, and so the
discrepancy between the observed and the expected results is statistically significant at the 1-percent level.)

(Rejection of the null hypothesis permits us to infer that one
of the alternative hypotheses is true. To decide which alterna-
tive is true, we compare $\hat{\pi}_A$ with $\hat{\pi}_B$. Since $\hat{\pi}_A > \hat{\pi}_B$, we accept
the stimulation hypothesis: We have found that the observed
proportion of Superior performers in Group A, $\hat{\pi}_A$, is significantly *higher* than that expected under the null hypothesis.
This result, of course, implies that the proportion of Superior
performers in Group B is significantly *lower* than that expected
under the null hypothesis.)

(Let's review the steps we took in the chi-square test for a
2×2 cross classification:

1. State the null and alternative hypotheses.
   Null Hypothesis: $\pi_A = \pi_B$. The proportion of Superior performers is the same
   in Group A and Group B.
   Alternative Hypothesis: $\pi_A \neq \pi_B$. The proportion of Superior performers is not the
   same in the two groups. •

2. Summarize the observed responses by using the cell fre-
   quencies as the summary statistics.
   The observed cell frequencies are presented in
   Chart 15-1.

3. Determine the expected values of the cell frequencies if
   the null hypothesis is true.
   The expected cell frequencies are determined by
   calculating:
   (a) The overall proportion of Superior performers:

$$\text{Proportion Superior} = \frac{51}{105} = 48.6 \text{ percent}$$

   (b) The expected frequency of Superior performers in
   Group A, based on the overall proportion:

$$50 \cdot 48.6\% = 24.3$$

(c) The other three cell frequencies by subtracting from the appropriate row and column totals:

$51 - 24.3 = 26.7$ Superior performers in Group B
$50 - 24.3 = 25.7$ Not Superior performers in Group A
$55 - 26.7 = 28.3$ Not Superior performers in Group B

4. Measure the discrepancy between the observed and the expected results by calculating the χ^2 statistic:

$$\chi^2 = \Sigma \frac{(O - E)^2}{E} = \frac{(33 - 24.3)^2}{24.3} + \frac{(18 - 26.7)^2}{26.7} + \frac{(17 - 25.7)^2}{25.7} + \frac{(37 - 28.3)^2}{28.3}$$

$$= 11.56$$

5. Determine the P value to measure the plausibility of the chance explanation for the discrepancy:

$$df = (c - 1) \cdot (r - 1) = 1 \cdot 1 = 1$$

$$P \text{ value} < 0.01$$

required
frequencies

The Chi-Square Distribution Law contains the proviso that the χ^2 score will be a point from a chi-square distribution *if the expected frequencies are sufficiently large*. What is sufficiently large?

Experience has shown that the chi-square distribution will yield a good approximation to the true probabilities if the *expected frequency in each cell is at least 5.* In our illustrative example, the smallest expected frequency is 24, so that we can quite safely obtain P values from the chi-square distribution.

Alternative procedures are available if the expected frequencies do not meet the criterion of the Chi-Square Distribution Law. Generically, these procedures are labeled *randomization tests*. In the case of a 2×2 cross classification, we can use the version called Fisher's Exact Test. Such procedures are covered in more advanced textbooks.

EXERCISES

15-1   If the null hypothesis $\pi_A = \pi_B$ is true and we measure the discrepancy between the observed and the expected sample results by χ^2, how small can χ^2 be and still permit us to declare that the discrepancy is statistically significant:

(a) at the 10-percent level?
(b) at the 5-percent level?
(c) at the 1-percent level?

15-2   (a) Based on the 2 × 2 cross-classification table below, calculate the value of the χ^2 statistic.
(b) Complete the following statement:
The result is statistically significant at the _____ percent level but not at the _____-percent level.

Observed and [Expected] Frequencies

CATEGORY	GROUP	
	A	B
This	7[10]	13[10]
That	13[10]	7[10]

15-3   The row and column totals for a 2 × 2 cross-classification table are:

CATEGORY	GROUP		
	A	B	ROW TOTAL
Success	_____	_____	60
Failure	_____	_____	120
COLUMN TOTAL	96	84	180
			GRAND TOTAL

(a) Determine the expected frequency of successes in Group A under the null hypothesis that the proportion of successes is equal for the two groups ($\pi_A = \pi_B$).
(b) Enter the expected frequencies for the remaining cells.

*15-4   Sixty individuals (20 who were overweight, 40 who were of normal weight) participated in an experiment designed to determine if overweight people are more responsive to food stimuli (the sight, smell, or discussion of food) than people of normal weight are. After viewing a film featuring culinary delights from many countries, each subject was asked whether the film had stimulated the appetite. The responses, categorized as strong or weak, follow.

APPETITE RESPONSE	WEIGHT GROUP	
	A (Overweight)	B (Normal Weight)
Strong	12	12
Weak	8	28

A chi-square test is to be performed.

(a) What proportion of the overweight subjects had strong responses? Denote this proportion by $\hat{\pi}_A$. What proportion of the subjects of normal weight had strong responses? Denote this proportion by $\hat{\pi}_B$.

(b) State the null and alternative hypotheses.

(c) Determine the expected cell frequencies if the null hypothesis is true.

(d) Measure the discrepancy between the observed and the expected cell frequencies by calculating χ^2.

(e) How plausible is the chance explanation for the discrepancy?

(f) What conclusion about the issue under examination can you draw if you are willing to accept a 5-percent risk of rejecting a true null hypothesis?

15-5   An experimental drug's effectiveness in preventing epileptic attacks was tested. The results follow.

Observed Results

CATEGORY	TREATMENT GROUP	
	Drug	Placebo
Attack	11	7
No Attack	29	43

Do the results permit you to conclude that the experimental drug differs in effectiveness from the placebo?

15-6   Do right-to-work laws—laws that prohibit employees from being required to join a union as a condition of employment—affect the availability of manufacturing jobs? A recent survey of the 50 states (20 with right-to-work laws, 30 without) revealed the following changing patterns of job availability:

	STATES WITH RIGHT-TO-WORK LAWS	STATES WITHOUT RIGHT-TO-WORK LAWS
Job Market Decline	0	19
Job Market Growth	20	11

From these data, would you conclude that right-to-work laws affect job availability?

15.3   the chi-square test: multiple groups

In another phase of the study, each scientist was placed in one of the following six personality groups based on responses to a psychological test:

> Realistic (aggressive, materialistic)
> Investigative (curious, introverted)
> Artistic (independent, self-expressive)
> Social (responsive, humanistic)
> Enterprising (dominant, insightful)
> Conventional (defensive, subordinate)

The question of interest: Do the various personality types differ in job performance? The sample data are presented in Chart 15-6.

Chart 15-6 is called a 2×6 cross-classification table, because it contains two categories of performance and six personality groups (two rows and six columns). Although this table is composed of 12 cells instead of four, the chi-square test procedure can still be applied.

(1. State the null and alternative hypotheses.)

The null hypothesis is that the proportion of Superior performers is the same for all personality groups—that personality does not make a difference in job performance. This is a direct extension of the two-group case where we considered the proportion Superior to be the same for both interest groups. The alternative hypothesis is that the proportion of Superior performers is not the same for all the personality groups.

**chart 15-6
Observed and
[Expected]
Results**

(2. Summarize our samples of observed responses by summary statistics.)

PERFORMANCE CATEGORY	PERSONALITY GROUP						ROW TOTAL
	Realistic	Investi-gative	Artistic	Social	Enter-prising	Conven-tional	
Superior	8[9.7]	9[7.8]	11[9.7]	7[9.2]	10[7.3]	6[7.3]	51
Not Superior	12[10.3]	7[8.2]	9[10.3]	12[9.8]	5[7.7]	9[7.7]	54
COLUMN TOTAL	20	16	20	19	15	15	105 GRAND TOTAL

Again, the cell frequencies serve to summarize the 105 observations.

(3. Determine the expected values of the summary statistics if the null hypothesis is true.)
We use the same reasoning to calculate the expected cell frequencies here that we used in the two-group case. The overall percentage of Superior performers is 48.6 percent, as we calculated earlier, and 20 of the Superior scientists are Realistic. So the number of performers in that group who are expected to be Superior is $20 \cdot 0.486 = 9.7$. Proceeding across the cells of the first row, we enter the expected frequency of 7.8 for Investigative ($48.6\% \cdot 16$). By the time we have calculated the expected frequencies for the first five cells in the row, we find that we can determine the remaining frequencies by subtraction from the totals.

(4. Measure the discrepancy between the observed and the expected results.)
(We define the χ^2 statistic as our measure of discrepancy, just as we did in our 2×2 cross classification earlier:

$$\chi^2 = \Sigma \frac{(O - E)^2}{E})$$

$$\chi^2 = \frac{(8 - 9.7)^2}{9.7} + \frac{(9 - 7.8)^2}{7.8} + \frac{(11 - 9.7)^2}{9.7} +$$

$$\frac{(7 - 9.2)^2}{9.2} + \frac{(10 - 7.3)^2}{7.3} + \frac{(6 - 7.3)^2}{7.3}$$

$$= \frac{(12 - 10.3)^2}{10.3} + \frac{(7 - 8.2)^2}{8.2} + \frac{(9 - 10.3)^2}{10.3} +$$

$$\frac{(12 - 9.8)^2}{9.8} + \frac{(5 - 7.7)^2}{7.7} + \frac{(9 - 7.7)^2}{7.7}$$

and we obtain the result $\chi^2 = 4.69$.

(5. Determine the P value to measure the plausibility of the chance explanation for the discrepancy.)
(Before we apply the Chi-Square Distribution Law, we must check to see that each *expected* frequency is at least 5)(it is; the smallest is 7.3), and we must compute $df = (c - 1) \cdot (r - 1) =$

$5 \cdot 1 = 5$. (Note that we had five free cells when we computed the expected frequencies.)

So the result $\chi^2 = 4.69$ can be considered a point from a chi-square distribution with $df = 5$. Referring to Chart 15-5, we see that a value for $\chi^2 \geq 9.236$ will arise by chance 10 percent of the time. We found $\chi^2 = 4.69$. The table tells us that this result will arise more than 10 percent of the time (P value > 0.10). If we reject the null hypothesis, we face in excess of a 10-percent risk of rejecting the truth. Accordingly, we conclude that our evidence does not refute the hypothesis that the proportion Superior is equal among all personality groups. Personality does not appear to make a difference in job performance.

EXERCISES

15-7   For each of the following cases, determine:

(a) The number of degrees of freedom associated with the χ^2 statistic.

(b) The minimum value of χ^2 required to declare the discrepancy between the observed and the expected results statistically significant at the 5-percent level.

Case 1: 2×5 cross classification.

Case 2: 2×7 cross classification.

Case 3: 2×9 cross classification.

15-8   A 2×3 cross-classification table of observed and expected cell frequencies follows.

CATEGORY	GROUP		
	A	B	C
Normal	10[10]	15[20]	20[15]
Abnormal	10[10]	25[20]	10[15]

(a) How many free cells are there? Explain your answer.

(b) Calculate χ^2.

(c) Calculate the number of degrees of freedom based on the Chi-Square Distribution Law. Is your result consistent with your answer to (a)?

(d) Can you conclude that there is a difference among the groups in the proportion of scores classified as normal?

15-9   In an experiment designed to compare the effectiveness of different brands of sleeping pills, a group of 120 individuals who complained of chronic insomnia were randomly assigned to four nightly treatments: two tablets of Brands A, B, or C (three nonprescription sleeping pills) or of a placebo. After four successive nights of use, each subject was asked whether there had been any improvement in ability to fall asleep. The responses were classified as Yes (improvement) or No.

IMPROVEMENT	SLEEPING PILL TREATMENTS			
	Brand A	Brand B	Brand C	Placebo
Yes	8	9	7	6
No	22	21	23	24

(a) State the null and alternative hypotheses.

(b) Determine the expected cell frequencies if the null hypothesis is true.

(c) Calculate χ^2 to measure the discrepancy between the observed and the expected cell frequencies.

(d) Do the results of the experiment permit you to reject the null hypothesis with less than a 10-percent risk of rejecting a true claim?

*15-10   Students majoring in business administration at a midwestern university must take concentrated courses in one of five functional areas: accounting, finance, management of people, management of operations, or marketing. A questionnaire was sent to recent graduates of the College of Business Administration asking whether the graduate's initial employment primarily involved activities within the *same* functional area that he or she concentrated on in college or activities within a *different* functional area. The responses are tabulated in the following cross-classification table.

PRIMARY FUNCTIONAL AREA OF INITIAL EMPLOYMENT	FUNCTIONAL AREA CONCENTRATION IN COLLEGE				
	Accounting	Finance	Management of People	Management of Operations	Marketing
Same as College Concentration	30	16	10	9	10
Different from College Concentration	5	4	2	3	11

You wish to compare the five college concentration groups to determine whether or not they differ in the proportion of graduates whose initial employment area is the same as their college concentration area.

(a) State the null and alternative hypotheses.

(b) Perform a chi-square test and report your conclusion.

(c) Were the expected cell frequencies sufficient to justify using the chi-square distribution?

15.4   extensions of the chi-square test

In both previous examples, the job performance of a scientist was rated Superior or Not Superior, but many organizations give more detailed performance ratings. For example, we could create five job-performance categories:

$$
\left.\begin{array}{l}\text{Distinguished}\\\text{Commendable}\end{array}\right\}\text{Superior}
$$

$$
\left.\begin{array}{l}\text{Competent}\\\text{Marginal}\\\text{Probationary}\end{array}\right\}\text{Not Superior}
$$

With this more detailed categorization, the study of the effect of *outside interests* on job performance would lead to the creation of a 5 × 2 cross-classification table containing five categories, or rows, and two groups, or columns. The personality study would then evolve into a 5 × 6 cross-classification table (five categories and six personality groups).

In principle, the same chi-square test can accommodate any number of groups and any number of categories. However, as more groups and categories are added, new problems may arise.

1. The expected frequencies may be spread too thinly among the cells.

Recall that the Chi-Square Distribution Law requires the expected frequency of each cell to be at least 5. As the number of groups and/or categories increases, the number of cells increases. Given the number of test subjects available for study, as the number of cells increases, the expected frequency in each cell becomes smaller.

Suppose that we observed the following frequencies for our sample of 105 Ph.D. scientists:

Distinguished	26
Commendable	25
Competent	25
Marginal	22
Probationary	7

As before, we'll assume that 50 subjects were classified as having Major outside interests and that 55 were classified as

chart 15-7

PERFORMANCE CATEGORY	OUTSIDE INTEREST GROUP		ROW TOTAL
	Group A (Major)	Group B (Minor)	
Distinguished	12.38	13.62	26
Commendable	11.90	13.10	25
Competent	11.90	13.10	25
Marginal	10.48	11.52	22
Probationary	3.33	3.67	7
COLUMN TOTAL	50	55	105 GRAND TOTAL

having Minor outside interests. The *expected* frequencies under the null hypothesis appear in Chart 15-7.

Two of these cells contain expected frequencies of less than 5. Many statisticians would argue that the chi-square test can still be applied in this case. Only two of the ten cells fail to meet the requirement of an expected frequency of at least 5, and the frequencies of these two cells are close to this value. In this case, however, we can make certain of the validity of the chi-square distribution by combining the Marginal and the Probationary categories into the single category Less than Competent.

(When not all expected cell frequencies are ≥5, try to consolidate the categories, thereby reducing the number of cells.)

2. Another difficulty arises if the categories are ordered.

A rating of Distinguished is higher than a rating of Commendable. Commendable, in turn, is a higher rating than Competent, and so on. (Categories that can be ranked in this fashion are called *ordered categories*.)

(**ORDERED CATEGORIES** Categories that have an inherent ranking from lowest to highest based on some criterion.)

If we perform a chi-square test on ordered categories, we must interpret the results with restraint. Suppose that the χ^2 score for the 5×2 cross classification in Chart 15-7 is statistically significant at the 1-percent level. This result implies that the distribution of scientists among the performance categories is *different* for Group A than for Group B. But the result does not necessarily imply that one group of scientists tends to perform *better on the average* than the other group.

The statistically significant score for χ^2 can result when the scientists with Major outside interests are extreme performers—either Distinguished or Probationary—and the scientists with Minor outside interests are primarily Competent (the middle category). On the average, the performance ranks of the two groups can still be identical.

(The chi-square test should not be used when the issue of interest is whether the subjects in one group rank more highly on the average than do the subjects in the other group. For this purpose, we should apply tests that explicitly utilize the ordering of categories; that is, we should apply tests based on ranks, such as the Wilcoxon rank sum test.)

EXERCISES

15-11   A new drug was being studied to see if it helped prevent heart attacks. Researchers also wanted to know if the drug influenced the type of death in patients who eventually died of heart problems. They had obtained the following information on heart-related deaths.

TYPE OF DEATH	TREATMENT GROUP	
	Active Drug	Placebo
"Sudden Death" Heart Attack	87	39
Myocardial Infarction	36	27
Other Heart Complications	9	6

(a) Calculate the χ^2 statistic for these data.

(b) Based on the result you obtained in (a), would you conclude that the distribution of heart-related deaths among the three categories differed depending on treatment?

15-12 (*Continuation of Exercise 15-10, page 483*)

(a) Consolidate the two management groups into a single functional area labeled Management, and prepare a revised cross-classification table.

(b) Perform a chi-square test on the revised data.

(c) Compare the P value and the χ^2 value from the test you performed in (b) with the P value and the χ^2 value for the test you performed in Exercise 15-10. The χ^2 values are nearly the same. Explain the difference in the associated P values.

15-13 (*Continuation of Exercise 15-4, page 478*) Suppose that the appetite responses had been categorized as

Very Strong	10
Fairly Strong	15
Fairly Weak	20
Very Weak	15

The figures represent the total number of individuals with each response.

(a) Construct a 4×2 cross-classification table of the expected frequencies.

(b) Are the expected cell frequencies sufficient to justify using the chi-square test?

(c) Regardless of your answer to (b), would it be appropriate to conduct a chi-square test in this case?

Caveat on the Calculation of χ^2

It is quite common to see cross-classification tables in which the cell entries are not absolute frequencies but percentages or proportions (see Chart 15-8).

When group sizes differ as they do in this study of outside interests and job performance (50 subjects in Group A; 55 in Group B), tables that contain percentages as cell entries allow us to make visual comparisons more readily. We don't have to

**chart 15-8
Observed
Percentages**

PERFORMANCE CATEGORY	OUTSIDE INTEREST GROUP	
	Group A (Major)	Group B (Minor)
Superior	66%	33%
Not Superior	34%	67%

compute the percentages from the frequencies; they are already listed. However, these tables present a potential pitfall if we wish to perform a chi-square test.

> The chi-square test must be based on the *frequency* in each cell—not on the *percentage* in each cell. We cannot calculate χ^2 from Chart 15-8 without knowing actual group sizes.

If we ignore this rule and use the percentages in Chart 15-8, we obtain

$$\chi^2 = 21.78$$

which differs greatly from the value of $\chi^2 = 11.56$ we calculated in Section 15.2 using the frequency in each cell.

> When a cross-classification table contains percentages, we must convert the percentages to frequencies before we can compute χ^2.

EXERCISE

(a) From the information given in the headline article on page 468, complete the following table.

SEAT BELT USER	RELATIVE FREQUENCY	
	Married Man	Unmarried Man
Yes	_____	_____
No	_____	_____

(b) From the information in the article, can you determine if the difference in the proportions of seat belt users among married and unmarried men in this sample could be due to chance?

(c) Suppose that 400 of the 1,000 people were married men and that 100 of the 1,000 were unmarried men. Redo the table you completed in (a) so that it displays the observed and the expected results that are required to complete the chi-square test.

COMING
TO
TERMS

Cell   Another name for a box in a cross-classification table.

Cell Frequency   The number of observations in a cell.

Chi-Square Distribution   A class of distributions representing the sampling distributions of the χ^2 statistic.

Cross-Classification Table   A table used to summarize categorical data. Each entry in the table represents the frequency of observations fitting a certain classification.

Expected Cell Frequency E The number of observations expected to be found in a particular cell if the null hypothesis is true.

Number of Free Cells   In a cross-classification table, the number of cells whose expected frequencies must be calculated directly before the remainder of the cell frequencies can be determined by subtraction from the appropriate row and column totals. Equals the number of degrees of freedom for χ^2.

Observed Cell Frequency O The number of observations that are classified as belonging in a particular cell. The word "observed" is used to point out that this is the *actual* number rather than the *expected* number of observations.

Ordered Categories   Categories that have an inherent ranking from lowest to highest based on some criterion.

ISSUE
FOR
ANALYSIS

The information from Chart 1-2 (pages 9–10) on cause of termination for Supreme Court justices is divided into two groups here: Pre-Civil War Appointment (by 1864) and Post-Civil War Appointment.

CAUSE OF TERMINATION	PRE-CIVIL WAR	POST-CIVIL WAR	TOTAL
Death	25	23	48
Retirement	2	22	24
Resignation	12	5	17
Disablement	0	3	3

The expected cell frequency for the Disablement category is less than 5, so the table must be consolidated to justify using the chi-square test. Consider two strategies:

(1) Consolidate the categories into Death and Other.
(2) Combine the two categories with the lowest frequencies into a category labeled Other.

(a) Which of these strategies do you feel would be better? Why?
(b) Use a chi-square test to compare the causes of termination of Pre-Civil War and Post-Civil War justices based on both strategies (1) and (2).
(c) Is the outcome in (b) consistent with your comments in (a)? If not, discuss why you think the results differed depending on the strategy used.

appendix tables

Table A
Squares and
Square Roots

X	X^2	$\sqrt{X}$	$\sqrt{10X}$	X	X^2	$\sqrt{X}$	$\sqrt{10X}$
1.0	1.00	1.000	3.162	5.5	30.25	2.345	7.416
1.1	1.21	1.049	3.317	5.6	31.26	2.366	7.483
1.2	1.44	1.095	3.464	5.7	32.49	2.387	7.550
1.3	1.69	1.140	3.606	5.8	33.64	2.408	7.616
1.4	1.96	1.183	3.742	5.9	34.81	2.429	7.681
1.5	2.25	1.225	3.873	6.0	36.00	2.449	7.746
1.6	2.56	1.265	4.000	6.1	37.21	2.470	7.810
1.7	2.89	1.304	4.123	6.2	38.44	2.490	7.874
1.8	3.24	1.342	4.243	6.3	39.69	2.510	7.937
1.9	3.61	1.378	4.359	6.4	40.96	2.530	8.000
2.0	4.00	1.414	4.472	6.5	42.25	2.550	8.062
2.1	4.41	1.449	4.583	6.6	43.56	2.569	8.124
2.2	4.84	1.483	4.690	6.7	44.89	2.588	8.185
2.3	5.29	1.517	4.796	6.8	46.24	2.608	8.246
2.4	5.76	1.549	4.899	6.9	47.61	2.627	8.307
2.5	6.25	1.581	5.000	7.0	49.00	2.646	8.367
2.6	6.76	1.612	5.099	7.1	50.41	2.665	8.426
2.7	7.29	1.643	5.196	7.2	51.84	2.683	8.485
2.8	7.84	1.673	5.292	7.3	53.29	2.702	8.544
2.9	8.41	1.703	5.385	7.4	54.76	2.720	8.602
3.0	9.00	1.732	5.477	7.5	56.25	2.739	8.660
3.1	9.61	1.761	5.568	7.6	57.76	2.757	8.718
3.2	10.24	1.789	5.657	7.7	59.29	2.775	8.775
3.3	10.89	1.817	5.745	7.8	60.84	2.793	8.832
3.4	11.56	1.844	5.831	7.9	62.41	2.811	8.888
3.5	12.25	1.871	5.916	8.0	64.00	2.828	8.944
3.6	12.96	1.897	6.000	8.1	65.61	2.846	9.000
3.7	13.69	1.924	6.083	8.2	67.24	2.864	9.055
3.8	14.44	1.949	6.164	8.3	68.89	2.881	9.110
3.9	15.21	1.975	6.245	8.4	70.56	2.898	9.165
4.0	16.00	2.000	6.325	8.5	72.25	2.915	9.220
4.1	16.81	2.025	6.403	8.6	73.96	2.933	9.274
4.2	17.64	2.049	6.481	8.7	75.69	2.950	9.327
4.3	18.49	2.074	6.557	8.8	77.44	2.966	9.381
4.4	19.36	2.098	6.633	8.9	79.21	2.983	9.434

(Continued)

Source: Bernard W. Lindgren. *Basic Ideas of Statistics* (New York: Macmillan, 1975), Table 10, page 340. Copyright © The Macmillan Company. Used by permission.

Table A
(Continued)

X	X^2	$\sqrt{X}$	$\sqrt{10X}$	X	X^2	$\sqrt{X}$	$\sqrt{10X}$
4.5	20.25	2.121	6.708	9.0	81.00	3.000	9.487
4.6	21.16	2.145	6.782	9.1	82.81	3.017	9.539
4.7	22.09	2.168	6.856	9.2	84.64	3.033	9.592
4.8	23.04	2.191	6.928	9.3	86.49	3.050	9.644
4.9	24.01	2.214	7.000	9.4	88.36	3.066	9.695
5.0	25.00	2.236	7.071	9.5	90.25	3.082	9.747
5.1	26.01	2.258	7.141	9.6	92.16	3.098	9.798
5.2	27.04	2.280	7.211	9.7	94.09	3.114	9.849
5.3	28.09	2.302	7.280	9.8	96.04	3.130	9.899
5.4	29.16	2.324	7.348	9.9	98.01	3.146	9.950

Examples for $X = 1.5$

To find $\sqrt{100X}$, multiply $\sqrt{X}$ by 10; $\sqrt{150} = 10 \cdot \sqrt{1.5} = 12.25$

To find $\sqrt{1,000X}$, multiply $\sqrt{10X}$ by 10; $\sqrt{1,500} = 10\sqrt{10 \cdot 1.5} = 38.73$

To find $\sqrt{X/10}$, divide $\sqrt{10X}$ by 10; $\sqrt{0.15} = 10\sqrt{1.5/10} = 0.3873$

Table B
The Normal
Table

Z_U	SECOND DECIMAL PLACE OF Z_U									
	0	1	2	3	4	5	6	7	8	9
0.0	.0000	.0040	.0080	.0120	.0160	.0199	.0239	.0279	.0319	.0359
0.1	.0398	.0438	.0478	.0517	.0557	.0596	.0636	.0675	.0714	.0753
0.2	.0793	.0832	.0871	.0910	.0948	.0987	.1028	.1064	.1103	.1141
0.3	.1179	.1217	.1255	.1293	.1331	.1368	.1406	.1443	.1480	.1517
0.4	.1554	.1591	.1628	.1664	.1700	.1736	.1772	.1808	.1844	.1879
0.5	.1915	.1950	.1985	.2019	.2054	.2088	.2123	.2157	.2190	.2224
0.6	.2257	.2291	.2324	.2357	.2389	.2422	.2454	.2486	.2517	.2549
0.7	.2580	.2611	.2642	.2673	.2704	.2734	.2764	.2794	.2823	.2852
0.8	.2881	.2910	.2939	.2967	.2995	.3023	.3051	.3078	.3106	.3133
0.9	.3159	.3186	.3212	.3238	.3264	.3289	.3315	.3340	.3365	.3389
1.0	.3413	.3438	.3461	.3485	.3508	.3531	.3554	.3577	.3599	.3621
1.1	.3643	.3665	.3686	.3708	.3729	.3749	.3770	.3790	.3810	.3830
1.2	.3849	.3869	.3888	.3907	.3925	.3944	.3962	.3980	.3997	.4015
1.3	.4032	.4049	.4066	.4082	.4009	.4115	.4131	.4147	.4162	.4177
1.4	.4192	.4207	.4222	.4236	.4251	.4265	.4279	.4292	.4306	.4319
1.5	.4332	.4345	.4357	.4370	.4382	.4394	.4406	.4418	.4429	.4441
1.6	.4452	.4463	.4474	.4484	.4495	.4505	.4515	.4525	.4535	.4545
1.7	.4554	.4564	.4573	.4582	.4591	.4599	.4608	.4616	.4625	.4633
1.8	.4641	.4649	.4656	.4664	.4671	.4678	.4686	.4693	.4699	.4706
1.9	.4713	.4719	.4726	.4732	.4738	.4744	.4750	.4756	.4761	.4767
2.0	.4772	.4778	.4783	.4788	.4793	.4798	.4803	.4808	.4812	.4817
2.1	.4821	.4826	.4830	.4834	.4838	.4842	.4846	.4850	.4854	.4857
2.2	.4861	.4864	.4868	.4871	.4875	.4878	.4881	.4884	.4887	.4890
2.3	.4893	.4896	.4898	.4901	.4904	.4906	.4909	.4911	.4913	.4916
2.4	.4918	.4920	.4922	.4925	.4927	.4929	.4931	.4932	.4934	.4936
2.5	.4938	.4940	.4941	.4943	.4945	.4946	.4948	.4949	.4951	.4952
2.6	.4953	.4955	.4956	.4957	.4959	.4960	.4961	.4962	.4963	.4964
2.7	.4965	.4966	.4967	.4968	.4969	.4970	.4971	.4972	.4973	.4974
2.8	.4974	.4975	.4976	.4977	.4977	.4978	.4979	.4979	.4980	.4981
2.9	.4981	.4982	.4982	.4983	.4984	.4984	.4985	.4985	.4986	.4986
3.0	.4987									
3.5	.4998									
4.0	.49997									
4.5	.499997									
5.0	.4999997									

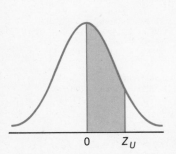

The entries in this table are the proportion of the observations from a *normal* distribution that have Z scores between 0 and Z_U. This proportion is represented by the colored area under the curve in the figure.

Reprinted with permission from *CRC Standard Mathematical Tables*, Fifteenth Edition (West Palm Beach, Fla.: Copyright The Chemical Rubber Co., CRC Press, Inc.).

**Table C
Random Digits
Table**

12651	61646	11769	75109	86996	97669	25757	32535	07122	76763
81769	74436	02630	72310	45049	18029	07469	42341	98173	79260
36737	98863	77240	76251	00654	64688	09343	70278	67331	98729
82861	54371	76610	94934	72748	44124	05610	53750	95938	01485
21325	15732	24127	37431	09723	63529	73977	95218	96074	42138
74146	47887	62463	23045	41490	07954	22597	60012	98866	90959
90759	64410	54179	66075	61051	75385	51378	08360	95946	95547
55683	98078	02238	91540	21219	17720	87817	41705	95785	12563
79686	17969	76061	83748	55920	83612	41540	86492	06447	60568
70333	00201	86201	69716	78185	62154	77930	67663	29529	75116
14042	53536	07779	04157	41172	36473	42123	43929	50533	33437
59911	08256	06596	48416	69770	68797	56080	14223	59199	30162
62368	62623	62742	14891	39247	52242	98832	69533	91174	57979
57529	97751	54976	48957	74599	08759	78494	52785	68526	64618
15469	90574	78033	66885	13936	42117	71831	22961	94225	31816
18625	23674	53850	32827	81647	80820	00420	63555	74489	80141
74626	68394	88562	70745	23701	45630	65891	58220	35442	60414
11119	16519	27384	90199	79210	76965	99546	30323	31664	22845
41101	17336	48951	53674	17880	45260	08575	49321	36191	17095
32123	91576	84221	78902	82010	30847	62329	63898	23268	74283
26091	68409	69704	82267	14751	13151	93115	01437	56945	89661
67680	79790	48462	59278	44185	29616	76531	19589	83139	28454
15184	19260	14073	07026	25264	08388	27182	22557	61501	67481
58010	45039	57181	10238	36874	28546	37444	80824	63981	39942
56425	53996	86245	32623	78858	08143	60377	42925	42815	11159
82630	84066	13592	60642	17904	99718	63432	88642	37858	25431
14927	40909	23900	48761	44860	92467	31742	87142	03607	32059
23740	22505	07489	85986	74420	21744	97711	36648	35620	97949
32990	97446	03711	63824	07953	85965	87089	11687	92414	67257
05310	24058	91946	78437	34365	82469	12430	84754	19354	72745
21839	39937	27534	88913	49055	19218	47712	67677	51889	70926
08833	42549	93981	94051	28382	83725	72643	64233	97252	17133
58336	11139	47479	00931	91560	95372	97642	33856	54825	55680
62032	91144	75478	47431	52726	30289	42411	91886	51818	78292
45171	30557	53116	04118	58301	24375	65609	85810	18620	49198
91611	62656	60128	35609	63698	78356	50682	22505	01692	36291
55472	63819	86314	49174	93582	73604	78614	78849	23096	72825
18573	09729	74091	53994	10970	86557	65661	41854	26037	53296
60866	02955	90288	82136	83644	94455	06560	78029	98768	71296
45043	55608	82767	60890	74646	79485	13619	98868	40857	19415
17831	09737	79473	75945	28394	79334	70577	38048	03607	06932
40137	03981	07585	18128	11178	32601	27994	05641	22600	86064
77776	31343	14576	97706	16039	47517	43300	59080	80392	63189
69605	44104	40103	95635	05635	81673	68657	09559	23510	95875
19916	52934	26499	09821	87331	80993	61299	36979	73599	35055
02606	58552	07678	56619	65325	30705	99582	53390	46357	13244
65183	73160	87131	35530	47946	09854	18080	02321	05809	04898
10740	98914	44916	11322	89717	88189	30143	52687	19420	60061
98642	89822	71691	51573	83666	61642	46683	33761	47542	23551
60139	25601	93663	25547	02654	94829	48672	28736	84994	13071

Source: The Rand Corporation, *A Million Random Digits with 100,000 Normal Deviates* (New York: The Free Press, 1955). Reproduced with permission of The Rand Corporation.

Table D
The *t* Table

One-tailed *P* value   Two-tailed *P* value

DEGREES OF FREEDOM	ONE-TAILED P VALUE					
	0.25	0.10	0.05	0.025	0.01	0.005
	TWO-TAILED P VALUE					
	0.50	0.20	0.10	0.05	0.02	0.01
1	1.000	3.078	6.314	12.706	31.821	63.657
2	0.816	1.886	2.920	4.303	6.965	9.925
3	.765	1.638	2.353	3.182	4.541	5.841
4	.741	1.533	2.132	2.776	3.747	4.604
5	.727	1.476	2.015	2.571	3.365	4.032
6	.718	1.440	1.943	2.447	3.143	3.707
7	.711	1.415	1.895	2.365	2.998	3.499
8	.706	1.397	1.860	2.306	2.896	3.355
9	.703	1.383	1.833	2.262	2.821	3.250
10	.700	1.372	1.812	2.228	2.764	3.169
11	.697	1.363	1.796	2.201	2.718	3.106
12	.695	1.356	1.782	2.179	2.681	3.055
13	.694	1.350	1.771	2.160	2.650	3.012
14	.692	1.345	1.761	2.145	2.626	2.977
15	.691	1.341	1.753	2.131	2.602	2.947
16	.690	1.337	1.746	2.120	2.583	2.921
17	.689	1.333	1.740	2.110	2.567	2.898
18	.688	1.330	1.734	2.101	2.552	2.878
19	.688	1.328	1.729	2.093	2.539	2.861
20	.687	1.325	1.725	2.086	2.528	2.845
21	.686	1.323	1.721	2.080	2.518	2.831
22	.686	1.321	1.717	2.074	2.508	2.819
23	.685	1.319	1.714	2.069	2.500	2.807
24	.685	1.318	1.711	2.064	2.492	2.797
25	.684	1.316	1.708	2.060	2.485	2.787
26	.684	1.315	1.706	2.056	2.479	2.779
27	.684	1.314	1.703	2.052	2.473	2.771
28	.683	1.313	1.701	2.048	2.467	2.763
29	.683	1.311	1.699	2.045	2.462	2.756
30	.683	1.310	1.697	2.042	2.457	2.750
35	.682	1.306	1.690	2.030	2.438	2.724
40	.681	1.303	1.684	2.021	2.423	2.704
45	.680	1.301	1.680	2.014	2.412	2.690
50	.680	1.299	1.676	2.008	2.403	2.678
55	.679	1.297	1.673	2.004	2.396	2.669
60	.679	1.296	1.671	2.000	2.390	2.660
70	.678	1.294	1.667	1.994	2.381	2.648
80	.678	1.293	1.665	1.989	2.374	2.638
90	.678	1.291	1.662	1.986	2.368	2.631
100	.677	1.290	1.661	1.982	2.364	2.625
120	.677	1.289	1.658	1.980	2.358	2.617
∞	.674	1.282	1.645	1.960	2.326	2.576

Source: Adapted from *Scientific Tables*, published by Ciba-Geigy Limited, 1962, pages 32–34.

Table E
Table of the
Rank Sum W

Sample Sizes:	P Value	(3, 3)		(3, 4)		(3, 5)		(3, 6)	
Two-Tailed	One-Tailed	L	U	L	U	L	U	L	U
0.20	0.10	7	14	7	17	8	19	9	21
0.10	0.05	6	15	6	18	7	20	8	22
0.05	0.025	*	*	*	*	6	21	7	23
0.02	0.01	*	*	*	*	*	*	*	*

Sample Sizes:	P Value	(4, 4)		(4, 5)		(4, 6)		(4, 7)	
Two-Tailed	One-Tailed	L	U	L	U	L	U	L	U
0.20	0.10	13	23	14	26	15	29	16	32
0.10	0.05	11	25	12	28	13	31	14	34
0.05	0.025	10	26	11	29	12	32	13	35
0.02	0.01	*	*	10	30	11	33	11	37

Sample Sizes:	P Value	(5, 5)		(5, 6)		(5, 7)		(5, 8)	
Two-Tailed	One-Tailed	L	U	L	U	L	U	L	U
0.20	0.10	20	35	22	38	23	42	25	45
0.10	0.05	19	36	20	40	21	44	23	47
0.05	0.025	17	38	18	42	20	45	21	49
0.02	0.01	16	39	17	43	18	47	19	51

Sample Sizes:	P Value	(6, 6)		(6, 7)		(6, 8)		(6, 9)	
Two-Tailed	One-Tailed	L	U	L	U	L	U	L	U
0.20	0.10	30	48	32	52	34	56	36	60
0.10	0.05	28	50	29	55	31	59	33	63
0.05	0.025	26	52	27	57	29	61	31	65
0.02	0.01	24	54	25	59	27	63	28	68

* Scores with P values this small do not exist for this sample size.

Adapted from W.J. Dixon and F.J. Massey, Jr., *Introduction to Statistical Analysis*, Second Edition, page 445. Copyright © McGraw-Hill Book Company, 1957. Used with permission of McGraw-Hill Book Company.

Table F
Table of the
Signed Rank
Statistic *V*

ONE-TAILED *P* VALUE	0.10	0.05	0.025	0.01	0.005
TWO-TAILED *P* VALUE	0.20	0.10	0.05	0.02	0.01
n					
5	2	0	—	—	—
6	3	2	0	—	—
7	5	3	2	0	—
8	8	5	3	1	0
9	10	8	5	3	1
10	14	10	8	5	3
11	17	13	10	7	5
12	21	17	13	9	7
13	26	21	17	12	9
14	31	25	21	15	12
15	36	30	25	19	15
16	42	35	29	23	19
17	48	41	34	27	23
18	55	47	40	32	27
19	62	53	46	37	32
20	69	60	52	43	37

Each entry represents the largest value of *V* with a *P* value less than or equal to the *P* value in the column heading.
Adapted from Table II in F. Wilcoxon, S.K. Katti, and Roberta A. Wilcox, *Critical Values and Probability Levels for the Wilcoxon Rank Sum Test and the Wilcoxon Signed Rank Test,* American Cyanamid Company (Lederle Laboratories Division, Pearl River, N.Y.) and The Florida State University (Department of Statistics, Tallahassee, Fla.), August 1963. Used with the permission of American Cyanamid Company and The Florida State University.

Table G
The *F* Table

Possible Values of *F*

Denominator df	P	1	2	3	4	5	6	7	8	9	10
6	0.10	3.78	3.46	3.29	3.18	3.11	3.05	3.01	2.98	2.96	2.94
	0.05	5.99	5.14	4.76	4.53	4.39	4.28	4.21	4.15	4.10	4.06
	0.01	13.70	10.90	9.78	9.15	8.75	8.47	8.26	8.10	7.98	7.87
7	0.10	3.59	3.26	3.07	2.96	2.88	2.83	2.78	2.75	2.72	2.70
	0.05	5.59	4.74	4.35	4.12	3.97	3.87	3.79	3.73	3.68	3.64
	0.01	12.20	9.55	8.45	7.85	7.46	7.19	6.99	6.84	6.72	6.62
8	0.10	3.46	3.11	2.92	2.81	2.73	2.67	2.62	2.59	2.56	2.54
	0.05	5.32	4.46	4.07	3.84	3.69	3.58	3.50	3.44	3.39	3.35
	0.01	11.30	8.65	7.59	7.01	6.63	6.37	6.18	6.03	5.91	5.81
9	0.10	3.36	3.01	2.81	2.69	2.61	2.55	2.51	2.47	2.44	2.42
	0.05	5.12	4.26	3.86	3.63	3.48	3.37	3.29	3.23	3.18	3.14
	0.01	10.60	8.02	6.99	6.42	6.06	5.80	5.61	5.47	5.35	5.26
10	0.10	3.28	2.92	2.73	2.61	2.52	2.46	2.41	2.38	2.35	2.32
	0.05	4.96	4.10	3.71	3.48	3.33	3.22	3.14	3.07	3.02	2.98
	0.01	10.00	7.56	6.55	5.99	5.64	5.39	5.20	5.06	4.94	4.85
11	0.10	3.23	2.86	2.66	2.54	2.45	2.39	2.34	2.30	2.27	2.25
	0.05	4.84	3.98	3.59	3.36	3.20	3.09	3.01	2.95	2.90	2.85
	0.01	9.65	7.21	6.22	5.67	5.32	5.07	4.89	4.74	4.63	4.54
12	0.10	3.18	2.81	2.61	2.48	2.39	2.33	2.28	2.24	2.21	2.19
	0.05	4.75	3.89	3.49	3.26	3.11	3.00	2.91	2.85	2.80	2.75
	0.01	9.33	6.93	5.95	5.41	5.06	4.82	4.64	4.50	4.39	4.30
13	0.10	3.14	2.76	2.56	2.43	2.35	2.28	2.23	2.20	2.16	2.14
	0.05	4.67	3.81	3.41	3.18	3.03	2.92	2.83	2.77	2.71	2.67
	0.01	9.07	6.70	5.74	5.21	4.86	4.62	4.44	4.30	4.19	4.10
14	0.10	3.10	2.73	2.52	2.39	2.31	2.24	2.19	2.15	2.12	2.10
	0.05	4.60	3.74	3.34	3.11	2.96	2.85	2.76	2.70	2.65	2.60
	0.01	8.86	6.51	5.56	5.04	4.70	4.46	4.28	4.14	4.03	3.94

(Continued)

This table provides the values of *F* that correspond to a given right tail area *P* and a specified numerator *df* and denominator *df*.

Source: Compiled from *Biometrika Tables for Statisticians,* Volume 1, 1966, Third Edition, Table 18, Pearson and Hartley, eds., and Maxine Merrington and Catherine M. Thompson, "Tables of Percentage Points of the Inverted Beta (*F*) Distribution," *Biometrika,* 33:73, 1943. Used by permission.

Numerator *df*

Denominator df	P	1	2	3	4	5	6	7	8	9	10
15	0.10	3.07	2.70	2.49	2.36	2.27	2.21	2.16	2.12	2.09	2.06
	0.05	4.54	3.68	3.29	3.06	2.90	2.79	2.71	2.64	2.59	2.54
	0.01	8.68	6.36	5.42	4.89	4.56	4.32	4.14	4.00	3.89	3.80
16	0.10	3.05	2.67	2.46	2.33	2.24	2.18	2.13	2.09	2.06	2.03
	0.05	4.49	3.63	3.24	3.01	2.85	2.74	2.66	2.59	2.54	2.49
	0.01	8.53	6.23	5.29	4.77	4.44	4.20	4.03	3.89	3.78	3.69
17	0.10	3.03	2.64	2.44	2.31	2.22	2.15	2.10	2.06	2.03	2.00
	0.05	4.45	3.59	3.20	2.96	2.81	2.70	2.61	2.55	2.49	2.45
	0.01	8.40	6.11	5.19	4.67	4.34	4.10	3.93	3.79	3.68	3.59
18	0.10	3.01	2.62	2.42	2.29	2.20	2.13	2.08	2.04	2.00	1.98
	0.05	4.41	3.55	3.16	2.93	2.77	2.66	2.58	2.51	2.46	2.41
	0.01	8.29	6.01	5.09	4.58	4.25	4.01	3.84	3.71	3.60	3.51
19	0.10	2.99	2.61	2.40	2.27	2.18	2.11	2.06	2.02	1.98	1.96
	0.05	4.38	3.52	3.13	2.90	2.74	2.63	2.54	2.48	2.42	2.38
	0.01	8.19	5.93	5.01	4.50	4.17	3.94	3.77	3.63	3.52	3.43
20	0.10	2.97	2.59	2.38	2.25	2.16	2.09	2.04	2.00	1.96	1.94
	0.05	4.35	3.49	3.10	2.87	2.71	2.60	2.51	2.45	2.39	2.35
	0.01	8.10	5.85	4.94	4.43	4.10	3.87	3.70	3.56	3.46	3.37
21	0.10	2.96	2.57	2.36	2.23	2.14	2.08	2.02	1.98	1.95	1.92
	0.05	4.32	3.47	3.07	2.84	2.68	2.57	2.49	2.42	2.37	2.32
	0.01	8.02	5.78	4.87	4.37	4.04	3.81	3.64	3.51	3.40	3.31
22	0.10	2.95	2.56	2.35	2.22	2.13	2.06	2.01	1.97	1.93	1.90
	0.05	4.30	3.44	3.05	2.82	2.66	2.55	2.46	2.40	2.34	2.30
	0.01	7.95	5.72	4.82	4.31	3.99	3.76	3.59	3.45	3.35	3.26
23	0.10	2.94	2.55	2.34	2.21	2.11	2.05	1.99	1.95	1.92	1.89
	0.05	4.28	3.42	3.03	2.80	2.64	2.53	2.44	2.37	2.32	2.27
	0.01	7.88	5.66	4.76	4.26	3.94	3.71	3.54	3.41	3.30	3.21
24	0.10	2.93	2.54	2.33	2.19	2.10	2.04	1.98	1.94	1.91	1.88
	0.05	4.26	3.40	3.01	2.78	2.62	2.51	2.42	2.36	2.30	2.25
	0.01	7.82	5.61	4.72	4.22	3.90	3.67	3.50	3.36	3.26	3.17
25	0.10	2.92	2.53	2.32	2.18	2.09	2.02	1.97	1.93	1.89	1.87
	0.05	4.24	3.39	2.99	2.76	2.60	2.49	2.40	2.34	2.28	2.24
	0.01	7.77	5.57	4.68	4.18	3.86	3.63	3.46	3.32	3.22	3.13
26	0.10	2.91	2.52	2.31	2.17	2.08	2.01	1.96	1.92	1.88	1.86
	0.05	4.22	3.37	2.98	2.74	2.59	2.47	2.39	2.32	2.27	2.22
	0.01	7.72	5.53	4.64	4.14	3.82	3.59	3.42	3.29	3.17	3.09
27	0.10	2.90	2.51	2.30	2.17	2.07	2.00	1.95	1.91	1.87	1.85
	0.05	4.21	3.35	2.96	2.73	2.57	2.46	2.37	2.30	2.25	2.20
	0.01	7.68	5.49	4.60	4.11	3.79	3.56	3.39	3.26	3.14	3.06

(Continued)

Table G
(Continued)

Denominator df	P	Numerator df 1	2	3	4	5	6	7	8	9	10
28	0.10	2.89	2.50	2.29	2.16	2.06	2.00	1.94	1.90	1.87	1.84
	0.05	4.20	3.34	2.95	2.71	2.56	2.44	2.36	2.29	2.24	2.19
	0.01	7.64	5.45	4.57	4.07	3.76	3.53	3.36	3.23	3.11	3.03
29	0.10	2.89	2.50	2.28	2.15	2.06	1.99	1.93	1.89	1.86	1.83
	0.05	4.18	3.33	2.93	2.70	2.54	2.43	2.35	2.28	2.22	2.18
	0.01	7.60	5.42	4.54	4.04	3.73	3.50	3.33	3.20	3.08	3.00
30	0.10	2.88	2.49	2.28	2.14	2.05	1.98	1.93	1.88	1.85	1.82
	0.05	4.17	3.32	2.92	2.69	2.53	2.42	2.34	2.27	2.21	2.16
	0.01	7.56	5.39	4.51	4.02	3.70	3.47	3.30	3.17	3.06	2.98
40	0.10	2.84	2.44	2.23	2.09	2.00	1.93	1.87	1.83	1.79	1.76
	0.05	4.08	3.23	2.84	2.61	2.45	2.34	2.25	2.18	2.12	2.07
	0.01	7.31	5.18	4.31	3.83	3.51	3.29	3.12	2.99	2.89	2.80
60	0.10	2.79	2.39	2.18	2.04	1.95	1.87	1.82	1.77	1.74	1.71
	0.05	4.00	3.15	2.76	2.52	2.37	2.25	2.17	2.10	2.04	1.99
	0.01	7.08	4.98	4.13	3.65	3.34	3.12	2.95	2.82	2.72	2.63
120	0.10	2.75	2.35	2.13	1.99	1.90	1.82	1.77	1.72	1.68	1.65
	0.05	3.92	3.07	2.68	2.45	2.29	2.18	2.09	2.02	1.96	1.91
	0.01	6.85	4.79	3.95	3.48	3.17	2.96	2.79	2.66	2.56	2.47
∞	0.10	2.71	2.30	2.08	1.94	1.85	1.77	1.72	1.67	1.63	1.60
	0.05	3.84	2.99	2.60	2.37	2.21	2.09	2.01	1.94	1.88	1.83
	0.01	6.63	4.61	3.78	3.32	3.02	2.80	2.64	2.51	2.41	2.32

Table H
The Chi-Square
Table

Possible Values of χ^2

DEGREES OF FREEDOM df	RIGHT-TAIL AREA P			
	0.10	0.05	0.02	0.01
1	2.706	3.841	5.412	6.635
2	4.605	5.991	7.824	9.210
3	6.251	7.815	9.837	11.345
4	7.779	9.488	11.668	13.277
5	9.236	11.070	13.388	15.086
6	10.645	12.592	15.033	16.812
7	12.017	14.067	16.622	18.475
8	13.362	15.507	18.168	20.090
9	14.684	16.919	19.679	21.666
10	15.987	18.307	21.161	23,209
11	17.275	19.675	22.618	24.725
12	18.549	21.026	24.054	26.217
13	19.812	22.362	25.472	27.688
14	21.064	23.685	26.873	29.141
15	22.307	24.996	28.259	30.578
16	23.542	26.296	29.633	32.000
17	24.769	27.587	30.995	33.409
18	25.989	28.869	32.346	34.805
19	27.204	30.144	33.687	36.191
20	28.412	31.410	35.020	37.566
21	29.615	32.671	36.343	38.932
22	30.813	33.924	37.659	40.289
23	32.007	35.172	38.968	41.638
24	33.196	36.415	40.270	42.980
25	34.382	37.652	41.566	44.314
26	35.563	38.885	42.856	45.642
27	36.741	40.113	44.140	46.963
28	37.916	41.337	45.419	48.278
29	39.087	42.557	46.693	49.588
30	40.256	43.773	47.962	50.892

This table contains the values of χ^2 that correspond to a specific right-tail area P and specific numbers of degrees of freedom df.
Source: From Table IV of R. A. Fisher and F. Yates, *Statistical Tables for Biological, Agricultural and Medical Research*, Sixth Edition (London: Longman Group Limited, 1974), page 47. Used by permission.

solutions
to
selected
exercises

chapter 1

1-1 (a) Two variables: (1) dollar amount of a state government's grant to institutions of higher education during 1965, and (2) dollar amount of a state government's grant to institutions of higher education during 1975.

(b) 51 observations per variable (one for each state).

1-2 Variable 1: Measurements
Variable 2: Counts
Variable 3: Index numbers
Variable 4: Ranks
Variable 5: Counts
Variable 6: Measurements
Varibale 7: Index numbers

1-3 (a) 1965: $95; 1970: $116; 1975: $161

1-6 (a) 0.29 (b) 0.53 (c) 0.66 (d) 0.87 (e) 0.13

1-7 (a) and (b):

SIZE (NO. OF EMPLOYEES)	FREQUENCY (NO. OF FIRMS)	RELATIVE FREQUENCY (PERCENT OF FIRMS)	CUMULATIVE FREQUENCY (CUMULATIVE NO. OF FIRMS)	CUMULATIVE RELATIVE FREQUENCY (CUMULATIVE PERCENT OF FIRMS)
Less than 10	4	8.0	4	8.0
10 through 49	10	20.0	14	28.0
50 through 99	15	30.0	29	58.0
100 through 999	12	24.0	41	82.0
1,000 or more	9	18.0	50	100.0
	50	100.0		

(c) 24 percent (d) 29 firms (e) 82 percent

1-10

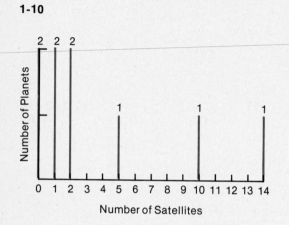

1-14

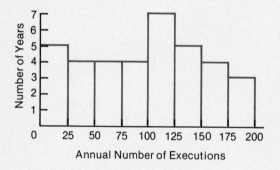

chapter 2

2-1   Midrange = 36 years

2-2   (a) Mode = 5 years
(b) Modal interval = 5 to 10 years
(c) Midpoint of modal interval = 7.5 years

2-3   Median = 78 percent

2-5   (a) Midrange = 7 satellites
(b) There is no mode.
(c) Median = 2 satellites
(d) Mean = 3.89 satellites

2-7   Average = 12 yards per pass (the mean)

2-9   (a) 3,000 push-ups
(b) No (Median = 24 push-ups)
(c) 115 push-ups

2-11   (a) Median = 6 carts
(b) No. You don't know the exact number of carts in Line 3.

2-13   Mean = 3/20, or 0.15, or 15 percent

2-14   (a) $\Sigma X = 15$ pounds (b) $n = 5$ pounds
(c) $\overline{X} = 15/5 = 3$ pounds

2-15   The first fisherman

2-16 $\overline{X} = 2.5$ grade points

2-18 $\overline{X} = 16.67$ years

2-21 $\overline{X} = 75.92$ inches, or approximately 76 inches

2-24   (a) 4 feet from the left end of the beam, since $\overline{X} = 4$
(b) $\Sigma(X - \text{Median}) = +5$;
$\Sigma(X - \text{Midrange}) = -5$

2-25 $\Sigma(X - 4)^2$

2-26   (a) 6 job offers (the mode)
(b) 4 job offers (the midrange)
(c) 4.6 job offers (the mean)

2-27   (a) Skewed to the left
(b) Long tail extending to the left
(c) Mean < Median
(d) Mode > Mean

2-28   (a) ①= Mode; ②= Median; ③= Mean
(b) The distribution is skewed to the right.

2-29   Skewed to the left (Mean $= 2.5$;
Median $=$ Mode $= 3.0$)

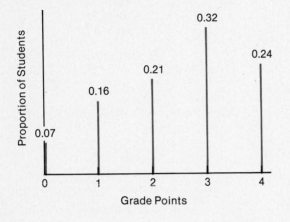

chapter 3

3-1   (a) Range $= 29$ (78 to 107)
(b) 25th percentile $= 87$
(c) 75th percentile $= 100.5$
(d) Interquartile range $= 13.5$ (87 to 100.5)

3-2   (a) Range $= 35$ (1 to 36 years)
(b) 25th percentile $= 6.5$ years
(c) 75th percentile $= 22.5$ years
(d) Interquartile range $= 16$ (6.5 to 22.5 years)

3-3   Range $= \$72.50$ (\$6.75 to \$79.25)

3-4   (a) Interquartile range $= 20$ (8 to 28) mopeds (b) Interdecile range $= 38$ (3 to 41) mopeds
(c) Midquartile $= 18$ mopeds; Median $= 19$ mopeds
(d) If the distribution were symmetric

3-6   (a) $s = 1.414$ (b) $s = 2.828$

3-9   Curve B

3-10 (a) $s = 50.0880$ hours, or approximately 50.09 hours
(b) $s = 50.0880$ hours, or approximately 50.09 hours

3-11 $s = 2$

3-13 $s = 1.212$ grade points, or approximately 1.2 grade points

3-14 $s = 3.85$ inches

3-15 $s/\overline{X} = 0.40$, or 40 percent

3-16 140.34 (percent)

3-19 $s = 25$ units

3-20 True

3-21 (a) $Z_{12} = -1.0$ (b) $Z_{32} = 1.5$
(c) $Z_{20} = 0.0$

3-23 $\overline{X} = 9$

3-24 $Z_A = 1.25$; $Z_B = 0.417$; $Z_C = -0.417$; Z_D -1.25; $Z_F = -2.083$

3-25 (a) Fortas: $Z = -1.175$; Burton: $Z = -0.247$, Black: $Z = 1.918$
(b) Johnson (Justice 83)
(c) 44.5 years

3-26 (a) At least 3/4   (b) At most 1/4
(c) At least 8/9   (d) At most 1/9

3-27 (a) Top 25 percent
(b) Bottom 25 percent
(c) Top 11 percent
(d) Bottom 11 percent
(e) Nothing   (f) 50th percentile

3-29 (a) 0 (and) 34.8 (years)
(b) 98.9 (percent)

3-31 (a) 1950: $Z_{gold} = 1.44$; $Z_{silver} = 1.91$
Relatively more silver.
(b) 1975: $Z_{gold} = -1.56$; $Z_{silver} = -1.91$
Relatively less silver.

3-34 (a) The turkey dinner
(b) The price distributions for the three types of dinners are of similar shape.

chapter 4

4-1

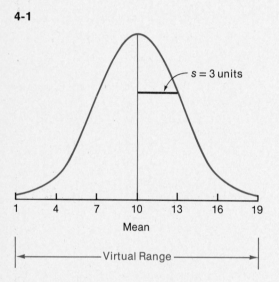

$s = 3$ units

1   4   7   10   13   16   19

Mean

Virtual Range

4-3 (a) $\overline{X} = 55$ (b) $s = 11\frac{2}{3}$

4-5 (a) 68.26 percent (b) 95.44 percent (c) 99.74 percent (d) 95.00 percent (e) 89.90 percent (f) 98.98 percent

4-6 (b) 0.3108, or approximately 31 percent
(c) 0.7698, or approximately 77 percent
(d) 0.1151, or approximately 12 percent

4-8 Steel: 0.8262, or approximately 83 percent
Cloth: 0.5468, or approximately 55 percent

Rubberized cloth: 0.6372, or approximately 64 percent

4-9 (a) 0.50, or 50 percent
(b) 0.1587, or approximately 16 percent
(c) 0.0228, or approximately 2 percent
(d) 0.1587, or approximately 16 percent
(e) 0.0228, or approximately 2 percent

4-10 (a) 0.2119, or approximately 21 percent
(b) 0.0026, or approximately 0.3 percent
(c) 0.0228, or approximately 2 percent
(d) 0.0013, or approximately 0.1 percent

4-12 (a) 0.6826, or approximately 68 percent
(b) 0.1271, or approximately 13 percent
(c) 0.6664, or approximately 67 percent
(d) 0.3423, or approximately 34 percent
(e) 0.0301, or approximately 3 percent

4-13 (a) 27.31 to 36.69 years, or approximately 27 to 37 years
(b) Narrower; $\overline{X} \pm 1.0s$ must be wider than $\overline{X} \pm 0.67s$. Former includes the middle 68 percent of the scores; latter includes the middle 50 percent of the scores.

4-14 (a) 20.66 to 22.34 inches (b) 23.10 inches

4-17 Foundation A: $Z_{10,000} = 1.67$;
Foundation B: $Z_{25,000} = 1.60$
The gift from Foundation A is slightly more generous.

chapter 5

5-1 (a)

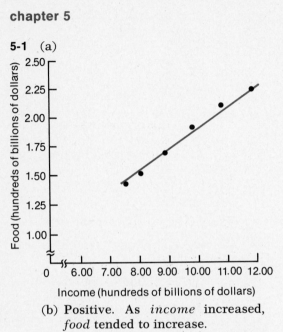

Income (hundreds of billions of dollars)

(b) Positive. As *income* increased, *food* tended to increase.

(c) Yes. There is very little scatter of the data points about the line through the center of the scatter.

5-4

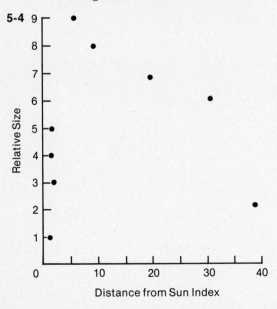

Distance from Sun Index

The random scatter indicates that there is no relationship between the two variables.

5-5 (a) Positively related

(b) It is an exact linear relationship.

(c) 1.8 A one-degree increase in temperature on the Celsius scale corresponds to a 1.8-degree increase in temperature on the Fahrenheit scale.

(d) When the temperature is 0° Celsius, it is 32° Fahrenheit.

chapter 6

6-1 (a) (*c*) (b) (*a*) (c) (*b*) (d) (*c*)

6-2 $r(X, Y) = 0.997$

6-3 $r(X, Y) = -0.863$

6-4 $r(X, Y) = -0.954$

6-7 $r(X, Y) = 0.018$

6-8

(a) X	Y	(b) X	Y
1	2	1	10
2	4	2	8
4	8	4	4
5	10	5	2

(c) X	Y	or	Y
1	4		8
2	10		2
4	2		10
5	8		4

6-10 (a) Yes (b) Need more information
(c) Need more information

6-11 (a) A variable is perfectly correlated with itself.

(b) *dist* and *por;* positive relationship ($r = 0.989$). *size* and *sat:* positive relationship ($r = 0.842$).

6-14 (a) *lat* (b) *elev* and *lat*

chapter 7

7-1 (a) Average number of hours per credit that a student devotes to studies (positive); absentee rate of a student (negative)

(b) Wattage (negative); frequency that light is turned on and off (negative)

(c) Average income level in area of practice (positive); size of practice (positive); length of experience (positive)

(d) Playing time (positive); height (positive)

(e) Latitude of city (positive); elevation of city (positive)

7-2 (a) Food expenditures. At what rate does family food expenditure tend to increase in response to increases in family income? OR How much more does a family tend to spend on food for each additional dollar of family income?

7-3 In $Y = a + bX$, Y represents *observed* values. In $\hat{Y} = a + bX$, $\hat{Y}$ denotes *values of Y predicted* by the equation of the line. $\hat{Y} = Y$ for all values of X if the two variables are perfectly correlated; that is, if $r(X, Y) = \pm 1.0$.

7-4 (a) $b = 0.03$ For each additional book page, we would predict an increase in the book's price of 3 cents; $a = 2.80$. The hypothetical book without pages would be predicted to cost $2.80.

7-5 The distance e is the prediction error $(Y - \hat{Y})$. So the least squares regression line keeps the sum of the squared prediction errors Σe^2, or $\Sigma(Y - \hat{Y})^2$, to a minimum.

7-7 (a) The direction of relationship (positive or negative)

(b) The closeness of the relationship to a line

(c) The rate of response (slope of the regression line)

7-8 (a) $b = 0.200$ (b) $a = -0.086$

(c) $food = -0.086 + 0.200\ income$

(e) 1971: 1.40; 1972: 1.52; 1973: 1.72; 1974: 1.88; 1975: 2.08; 1976: 2.28

(f) 1971: 0.7 percent; 1972: 1.3 percent; 1973: 2.4 percent; 1974: 1.1 percent; 1975: 0.9 percent; 1976: 1.3 percent. In each year, $1Y - \hat{Y}1 \leq 5$ percent of Y.

7-10 (a) $\widehat{mpg\ combined} = 47.339 - 0.00745\ weight$

(b) 0.745 mpg, or approximately 7/10 of a mile per gallon

7-15 (a) Total: $\Sigma(Y - \overline{Y})^2 = 360$

(b) Unexplained: $\Sigma(Y - \hat{Y})^2 = 18.3602$

(c) Explained: $\Sigma(Y - \overline{Y})^2 - \Sigma(Y - \hat{Y})^2 = 341.6398$ (d) $R^2 = 0.949$, or 94.9 percent

7-16 (a) 10   (b) 5   (c) -3 (d) 0

7-18 $R^2 = 57.7$ percent

7-19 (a) Each *buck* ($1,000) is predicted to buy 50 votes. votes.

(e) 200 votes (f) 650 votes

chapter 8

8-1 STATEMENT 1: (a) Refreshment expenditures of the spectators at the Civic Center in 1977. (b) μ (the population mean) (c) Sample mean; $\overline{X} = \$1.62$

STATEMENT 2: (a) Gestation periods for the mares. (b) σ (the population standard deviation) (c) Sample standard deviation; $s = 4$ days

STATEMENT 3: (a) Tuition increase at each public college and university (b) μ (the population mean) (c) Sample mean; $\overline{X} = 38$ percent

8-2 $\mu = 1.015$ cars per family

8-3 $\sigma = 0.731$ cars per family

8-5 Yes. There is a 1/10,000 chance of selecting each four-digit ID.

8-6 126, 516, 164, 611

8-7 (a) 126, 367, 213, 140, 154, 186

8-8 (a) 91, 57, 68, 42, 21, 78, 90, 28, 20, 10
(b) $\overline{X} = \$19,240; \mu = \$18,800$ Difference $(X - \mu) = \$440$

8-13 (a) 0.46, or 46 percent (b) 0.069, or 6.9 percent (c) 0.837, or 83.7 percent

8-14 (a) 0.2116, or 21.16 percent
(b) 0.0324, or 3.24 percent
(c) 0.0185, or 1.85 percent
(d) 0.0185, or 1.85 percent
(e) 0.0370, or 3.70 percent

8-15 25 samples

8-16 (a) 0.0973, or 9.73 percent
(b) 125 samples

chapter 9

9-1 (a) (1) $\mu = 1.0$ newspaper per household
(2) $\sigma = 0.82$ newspaper per household.

(b)

POSSIBLE SAMPLE NO.	NEWSPAPER COUNTS	SAMPLE MEAN	SAMPLE RANGE
1	(0,0)	0.0	0
2	(0,1)	0.5	1
3	(0,2)	1.0	2
4	(1,0)	0.5	1
5	(1,1)	1.0	0
6	(1,2)	1.5	1
7	(2,0)	1.0	2
8	(2,1)	1.5	1
9	(2,2)	2.0	0

(c)

$\overline{X}_2$ POSSIBLE VALUES OF THE SAMPLE MEAN	$p(\overline{X}_2)$ RELATIVE FREQUENCY WITH WHICH EACH VALUE OCCURS
0.0	1/9
0.5	2/9
1.0	3/9
1.5	2/9
2.0	1/9
Any of the above	9/9

sr_2 POSSIBLE VALUES OF THE SAMPLE RANGE	$p(sr_2)$ RELATIVE FREQUENCY WITH WHICH EACH VALUE OCCURS
0	3/9
1	4/9
2	2/9
Any of the above	9/9

(d) (1) 1/9 (2) 1/9 (3) 2/9 (4) 6/9
    (5) 3/9
(e) (1) 3/9 (2) 4/9 (3) 2/9 (4) 7/9
    (5) 2/9

9-4   (a) $\mu(\overline{X}_2) = 1.0$ (b) $\sigma(\overline{X}_2) = 0.58$
(c)

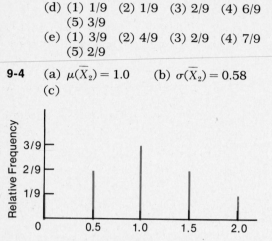

Possible Values of $\overline{X}_2$

9-7   (a) Yes. $\mu(\overline{X}_2) = 1.0 = \mu$ (b) Yes
(c) No. $\mu(sr_2) = 8/9$, or 0.89

9-8   (a) σ and n
(b) (1) $\sigma(\overline{X}_4) = 10$ units
    (2) $\sigma(\overline{X}_{16}) = 5$ units
    (3) $\sigma(X_{64}) = 2.5$ units
(c) $n = 400$

9-12 (a) False (b) False (c) False (d) False

9-13 (a) (1) 38.3 percent (2) 78.9 percent
    (3) 98.8 percent
(b) (1) 38 percent confident that μ
    lies between 49 and 51.
    (2) 79 percent confident that μ
    lies between 47.5 and 52.5.
    (3) 99 percent confident that μ
    lies between 45 and 55.
(c) There is an inverse relationship
    between precision and confi-
    dence level.

9-14 90-percent confidence interval:
{6.18 to 7.82} hamsters
95-percent confidence interval:
{6.02 to 7.98} hamsters
99-percent confidence interval:
{5.71 to 8.29} hamsters

9-16 (a) 96.7 minutes
(b) {91.7 to 101.7} minutes
(c) 98.8 percent

9-18 (a) 25 (b) 15 (c) 60

chapter 10

10-1   (b)

POSSIBLE VALUES OF s	POSSIBLE VALUES OF $\hat{\sigma}$	RELATIVE FREQUENCY $\rho(s)$ OR $\rho(\hat{\sigma})$
0	0	3/9
0.5	0.707	4/9
1.0	1.414	2/9

(c) $\mu(s) = 0.444$ newspaper
    $\mu(\hat{\sigma}) = 0.628$ newspaper

10-2 $\hat{\sigma}(\overline{X}_{25}) = 2$ units

10-3 $\hat{\sigma} = 2$ units $\hat{\sigma}(\overline{X}_{16}) = 0.5$ unit

10-4   (a) $\sigma(\overline{X}_{25}) = 1.3$ units
    $\sigma(X_{25}) = 1.0$ unit
(b) $Z_{\overline{X}_{25}} = -1.54$ units
    $t_{\overline{X}_{25}} = -2.0$ units

10-5 $t_{\overline{X}_4} = 3.375$

10-7   (a) 2.831   (b) 2.947
(c) 2.093   (d) 2.045
(e) 1.684   (f) 1.676

10-8   {6.14 to 7.86} hamsters

10-10 $183.16 to $209.64

10-12 (a) $\mu = 0$ (b) $\mu > 0$ (c) one-tailed
(d) The P value: the probability
    that chance will give us a sam-
    ple whose mean is as far above
    zero as our observed mean in-
    crease in miles per gallon.

10-13 (a) $\mu = 5.7$ years (b) $\mu < 5.7$ years
(c) one-tailed
(d) The P value: the probability that chance will give us a sample whose mean is as far below 5.7 years as our observed mean number of years for promotion to lieutenant colonel.

10-16 (a) $\hat{\sigma}(\bar{X}_{16}) = 3$ units (b) $t_{\bar{X}_{16}} = 1.67$ units
(c) The one-tailed P value is less than 0.10 but greater than 0.05, so the null hypothesis can be rejected with less than a 10-percent risk (but more than a 5-percent risk) of rejecting a true claim.
(d) The two-tailed P value is less than 0.20 but greater than 0.10.
(e) The two-tailed P value is twice as large as the one-tailed P value, since *both* tails of the t distribution support the alternative hypothesis.

10-18 (a) One-tailed P value < 0.005

10-19 (a) No. One-tailed P value > 0.10

chapter 11

11-1 (a) No. The model acknowledges that observed scores on Y will vary from those predicted by the model, $Y = \beta_0 + \beta_1 X_1 + \beta_2 X_2$, because of the influence on Y of excluded variables (*noise*).
(b) No. A unit increase in X_1, holding X_2 constant, will change Y by β_1 units *on the average*.
(c) Yes

11-2 (a) $gpa = \beta_0 + \beta_1 IQ + \beta_2 diversity + \beta_3 ambition + noise$

11-6 (a) $df = 18$ (b) $df = 22$ (c) $df = 55$

11-9 (b) $\{-2.87$ to $-18.91\}$ dollars per resident for each unit increase in *gpa*
(c) $\{-1.3$ to $11.3\}$ cents per resident for each 1-point increase in % *male*
(d) $\{1.3$ to $21.7\}$ cents per resident for each 1-point increase in % *in-state*

11-10 (a) $\{16.32$ to $23.68\}$ cents for each additional dollar of income

11-12 (b) $b_1 = 25$ cents for each additional mile
(d) $\{18.316$ to $31.684\}$ cents for each additional mile

11-13 (b) $t_{b_1} = (b_1 - \beta_1)/\hat{\sigma}(b_1) = 3.00$
(c) Yes. Two-tailed P value < 0.01
(e) Null hypothesis: $\beta_2 = 0$
Alternative hypothesis: $\beta_2 > 0$
(f) $0.01 <$ One-tailed P value < 0.025

11-14 (a) $t_{b_1} = 25.000$
(b) $df = 4$
One-tailed P value < 0.005
(c) $\beta_1 = 0.22$
(d) $t_{b_1} = -2.500$
(e) $\beta_1 \neq 0.22$ Two-tailed test
(f) Yes. Two-tailed P value < 0.10

chapter 12

12-1 (a) $\hat{\pi}_{29} = \dfrac{9}{29} = 0.310$, or 31.0 percent
(b) $\hat{\pi}_{29} = \dfrac{8}{29} = 0.276$, or 27.6 percent
(c) $\hat{\pi}_{29} = \dfrac{21}{29} = 0.724$, or 72.4 percent
(d) $\hat{\pi}_9 = \dfrac{8}{9} = 0.889$, or 88.9 percent

12-2 Between 145 and 580 rapes occurred during 1974.

12-3 (a) $N = 22,000$ employees
(b) $n = 100$ employees
(c) π is the proportion of all company employees who exceeded
the sick-day allotment last year.
(d) $\hat{\pi}_n$ is the proportion of the random sample of 100 employees
who exceeded the sick-day allotment last year.

12-5 (b)

$\hat{\pi}$	$\mu(\hat{\pi}_{100})$	$\sigma(\hat{\pi}_{100})$
0.5	0.5	0.050
0.4	0.4	0.049
0.3	0.3	0.046
0.2	0.2	0.040
0.1	0.1	0.030

12-6 (a) (1) $\sigma(\hat{\pi}_{25}) = 0.10$
(2) $\sigma(\hat{\pi}_{100}) = 0.05$
(3) $\sigma(\hat{\pi}_{10,000}) = 0.005$
(b) $n = 2,500$

12-10 (a) Null hypothesis: $\pi \leq 0.50$
Alternative hypothesis: $\pi > 0.50$
(b) 15.87 percent (c) 5.48 percent
(d) Any proportion ≥ 0.51165, or approximately 51.2 percent.

12-12 (a) Statement (2) (b) Statement (3)
(c) Statement (4) (d) Statement (1)

12-15 (a) $\hat{\pi}_{64} = 0.375$, or 37.5 percent
$\hat{\pi}_{36} = 0.833$, or 83.3 percent
(b) {25.25 to 49.75} percent
(c) {67.0 to 99.6} percent

12-18 (a) $\hat{\pi}_{64} = 0.3125$, or 31.25 percent
(b) {21.75 to 40.75} percent
(c) {46.08 to 66.42} percent

12-20 (a) $n = 1,068$ (b) $n = 9,604$

12-21 (a) (1) $\dfrac{n}{N} = 0.50$

(2) $\sqrt{1 - \dfrac{n}{N}} = 0.707$

(3) $\hat{\sigma}(\hat{\pi}_{100}) = 0.035$

12-23 {350,200 to 394,400} senior citizens

chapter 13

13-1 (a) Death penalty versus absence
of death penalty.
(b) The homicide rate: number of
homicides per 100,000 people

13-3 (a) Private status versus community status of hospitals
(b) hospitals
(c) The cost per patient day
(d) The status of a hospital (private
or community) must be taken
as an established fact.

13-6 (a) Null hypothesis: There is no inherent difference in the distribution of homicide rates between states that do and states
that do not impose the death
penalty.
(b) Alternative hypothesis: Homicide rates in states that do not
impose the death penalty are
higher on the average than
those in states that do impose
the death penalty.
(c) A one-tailed test

13-8 (a) Null hypothesis: There is no inherent difference in costs per
patient day between private
and community hospitals.
(b) $\mu_P > \mu_C$ On the average, cost
per patient day is higher in private hospitals than in community hospitals. (c) A one-tailed
test

13-11 $\hat{\sigma}(\overline{X}_A - \overline{X}_B) = \sqrt{8}$ = approximately
2.8 units

13-12 (a) $\hat{\sigma}_{\text{pool}} = 5.25$ units
(c) $\hat{\sigma}(\overline{X}_A - \overline{X}_B) = 1.307$ units

13-13 (e) $\hat{\sigma}(\overline{X}_A - \overline{X}_B) = 0.424$ minutes
(f) $t_{\overline{X}_A - \overline{X}_B} = -3.538$
(i) Two-tailed P value < 0.01

13-15 (d) $\hat{\sigma}_{\text{pool}} = 2.588$ shaves per blade;
$\hat{\sigma}(\overline{X}_A - \overline{X}_B) = 0.802$ shaves per blade;
$t_{\overline{X}_A - \overline{X}_B} = 1.122$

(d) $0.05 <$ One-tailed P value < 0.10

13-17 $\{-0.79$ to $-2.21\}$ minutes

13-19 $\{27.72$ to $42.28\}$ gallons

13-20 The probability is:
(a) No more than 1 percent
(b) Between 2.5 and 5 percent
(c) No more than 2.5 percent
(d) No more than 5 percent

13-21 (a) $W = 49$ (b) $\mu(W) = 39$ (c) Yes
(d) $0.05 <$ One-tailed P value < 0.10

13-22 (b) $W = 20$
(c) Two-tailed P value is no more than 0.10
(d) $\mu(W) = 30$

13-27 Size (number of beds), nature of location (population, average income level, . . .), and type of specialty care offered (dialysis unit, . . .)

13-30 (b) $\overline{d} = 0.75$ shaves per blade

(e) (1) $\hat{\sigma}_d = 1.390$ shaves per blade
(2) $\hat{\sigma}(\overline{d}) = 0.348$ shave per blade
(3) $t_d = 2.155$
$0.010 <$ One-tailed P value < 0.025

13-33 (b) $\Sigma R+ = 7$ $\Sigma R- = 14$ (c) 10.5
(d) $\Sigma R+$ is expected to be smaller Yes (e) No.   One-tailed P value > 0.10

chapter 14

14-1 (a) Statement (3) (b) Statement (2)
(c) Statement (4) (d) Statement (1)

14-2

	NUMBER OF GROUPS	SAMPLE SIZE
(a)	3	43
(b)	4	64
(c)	4	124
(d)	5	125

14-3 (a) $\overline{\overline{X}} = \dfrac{(\overline{X}_A + \overline{X}_B + \overline{X}_C)}{3}$

(b) $\hat{\sigma}^2_{\text{pool}} = \dfrac{(\hat{\sigma}^2_A + \hat{\sigma}^2_B + \hat{\sigma}^2_C)}{3}$

14-4 (c) $\overline{\overline{X}} = 6.1$ percent
(d) Between groups variance $= 79.380$
(e) (1) $\hat{\sigma}^2_{\text{pool}} = 6.617$
(2) F ratio $= 11.996$
(3) numerator $df = 2$; denominator $df = 51$
(4) P value < 0.01

14-6 (b) F ratio $= 1.95$

14-7

SOURCE	df	SUM OF SQUARES	MEAN SQUARES	F RATIO	F PROB
BETWEEN GROUPS	2	Omit	79.380	11.996	<0.01
WITHIN GROUPS	51	Omit	6.617		
TOTAL	53	Omit			

14-10 (a) No. The result of the ANOVA F test permits you to conclude just that the dropout rates of at least one of the three populations differs on the average from those of the other two populations.

(b) $t_{\bar{X}_B - \bar{X}_C} = 0.6997$; $0.20 < P$ value < 0.50

14-11 {1.604 to 6.196} percentage points

14-14 (b) $\bar{\bar{X}} = 7.11$ (c) Between groups variance $= 2.024$

(d) $\hat{\sigma}^2_{pool} = 3.264$ (e) F ratio $= 0.620$; P value > 0.10

chapter 15

15-1 (a) $\chi^2 = 2.706$ (b) $\chi^2 = 3.841$
(c) $\chi^2 = 6.635$

15-2 (a) $\chi^2 = 3.6$

15-3

CATEGORY	GROUP	
	A	B
Success	32	28
Failure	64	56

15-4 (a) $\hat{\pi}_A = 12/20 = 0.60$
$\hat{\pi}_B = 12/40 = 0.30$

(b) Null hypothesis: $\hat{\pi}_A = \hat{\pi}_B$
Alternative hypothesis: $\hat{\pi}_A \neq \hat{\pi}_B$

(d) $\chi^2 = 5.0$ (e) The P value is greater than 0.02 but less than 0.05.

15-7 Case 1: $df = 4$; $\chi^2 = 9.488$
Case 2: $df = 6$; $\chi^2 = 12.592$
Case 3: $df = 8$; $\chi^2 = 15.507$

15-8 (a) 2 free cells (b) $\chi^2 = 5.834$
(c) $df = 2$
(d) The P value is greater than 0.05 but less than 0.10.

15-11 (a) $\chi^2 = 2.769$

Illustration Credits

index

within groups mean squares, 454

Y

yes–no variable, 36, 352
   See also proportion

Z

Z score, 82–83, 94
   normal distribution and, 101–12
   relative standing and, 82–88
   t score and, 290

A 8
B 9
C 0
D 1
E 2
F 3
G 4
H 5
I 6
J 7